Regulation and the Evolution of the Global Telecommunications Industry

To my mother and the memory of my father who provided encouragement and guidance in ways they never imagined. Also to my colleague and best friend, Gary Lumsdem, who started his successful new business by taking advantage of lucrative opportunities created by the new regulations of telecommunications industry.
Anastassios Gentzoglanis

To my good colleagues at the Center for Communication, Media and Information Technologies (CMI) at Aalborg University in Copenhagen with whom I have cooperated for many years.
Anders Henten

Regulation and the Evolution of the Global Telecommunications Industry

Edited by

Anastassios Gentzoglanis

University of Sherbrooke, Canada

and

Anders Henten

Aalborg University, Denmark

Edward Elgar

Cheltenham, UK • Northampton, MA, USA

Published by
Edward Elgar Publishing Limited
The Lypiatts
15 Lansdown Road
Cheltenham
Glos GL50 2JA
UK

Edward Elgar Publishing, Inc.
William Pratt House
9 Dewey Court
Northampton
Massachusetts 01060
USA

A catalogue record for this book
is available from the British Library

Library of Congress Control Number: 2009940969

Mixed Sources
Product group from well-managed
forests and other controlled sources
www.fsc.org Cert no. SA-COC-1565
© 1996 Forest Stewardship Council
FSC

ISBN 978 1 84844 588 8

Printed and bound by MPG Books Group, UK

Contents

Contributors

Elias Aravantinos has been Project Manager at the Columbia Institute for Tele-Information on the Ultrabroadband project since 2004. He is also the Managing Director of ExelixisNet which specializes in high-technology strategies and new media products. His clients are carriers, governments, academic institutes and media companies in North America and Europe. From 2006 till 2008, Mr Aravantinos worked for Queens College and the Metropolitan College of New York, teaching several business and technology courses at graduate and undergraduate level. He is completing his PhD thesis in Technology Management at Stevens Institute of Technology, USA with a specialization in Telecommunications. His research interests are new technology projections, strategies and business development, with a focus on 4G strategies. He is a frequent speaker at industry and academic events, with numerous presentations and publications. Mr Aravantinos holds a BSc degree in Electrical Engineering and Computer Technology from the University of Patras, Greece and an MBA in Information Technology from Oklahoma City University, USA.

Bruno Basalisco is a PhD candidate in Economics at Imperial College Business School, London. He conducts research in the field of industrial economics, focusing on innovation, regulation, platforms and business models in the information and communications and network industries in general. His work has been presented at several international conferences and workshops in the field of industrial organization, technology strategy and telecommunications policy, and he has served as referee for leading information and communication technology (ICT) economics and strategy journals. Bruno holds an MSc in Science and Technology Policy from SPRU, University of Sussex and an MA in Economics from Sant'Anna School of Advanced Studies and University of Pisa. He has also briefly worked for NERA Economic Consulting and for Ofcom.

Erik Bohlin is currently Head and Professor in Technology Assessment in the Division of Technology and Society, Department of Technology Management and Economics at Chalmers University of Technology. He has published in a number of areas relating to the information society – policy, strategy, and management. He is Chair of the International

Telecommunications Society and Chief Editor of *Telecommunications Policy*. He served as Special Advisor to IT and Media Commissioner Viviane Reding of the European Commission in 2008–2009. He obtained his graduate degree in Business Administration and Economics at the Stockholm School of Economics (1987) and his PhD at Chalmers University of Technology (1995).

Marcel Boyer holds a PhD in Economics from Carnegie-Mellon University. He has taught economics at York University (1971–73), UQÀM (1973–74) and at the University of Montreal (1974–2008). He held the Bell Canada Chair in Industrial Economics in the Department of Economics at the University of Montreal (2003–08) and the Jarislowsky–SSHRC–NSERC (Social Sciences and Humanities Research Council; Natural Sciences and Engineering Research Council of Canada) Chair in Technology and International Competition at l'École Polytechnique de Montréal (1993–2000). He is presently Emeritus Professor of Economics at the Université de Montréal, Fellow of the C.D. Howe Institute, Centre interuniversitaire de recherche en analyse des organisations (CIRANO) and Centre interuniversitaire de recherche en économie quantitative (CIREQ), Academic Affiliate of The Analysis Group, Senior Economist at the Montreal Economic Institute, member of the board of the Agency for Public–Private Partnerships of Québec, member of the Industry Canada Advisory Committee on Business Strategies and innovation, member of the Governance Committee of the Sustainable Development and Socially Responsible Investment Chairs (École polytechnique de Paris and Université de Toulouse). Marcel Boyer has received numerous prizes for excellence in research. He is author or co-author of over 250 scientific articles and papers and public and private reports. Marcel Boyer has acted as an expert economist on behalf of several national and international corporations and government organizations, and has testified as an expert witness before various organizations and tribunals.

Peter Curwen is Visiting Professor of Telecommunications Strategy at the Strathclyde Business School, Glasgow and also researches and publishes on a private basis. He was previously Professor of Business Economics at Sheffield Hallam University. His primary research interests concerns the manner in which a rapidly changing environment affects the structure of the mobile telecommunications industry and its strategic implications for major companies in that sector. He has published three books on telecommunications, including *Telecommunications Strategy: Cases, Theory and Applications* (Routledge, 2004) with Jason Whalley.

Dieter Elixmann is head of the Corporate Strategies research group at Wissenschaftliches Institut fur Kommunikationsdienste (WIK), Bad

Honnef, Germany. He has been with WIK since 1985. He is also head of the Telecommunications Markets research group. His recent research focuses on econometric estimation of production structures of Deutsche Telekom, demand analysis in the telco sector, internationalization and globalization in the telco sector, and multimedia business strategies.

Morten Falch is Associate Professor at the Center for Communication, Media and Information Technologies (CMI) located at Copenhagen Institute of Technology, Aalborg University, Denmark. He holds a Bachelors degree in Mathematics, a Masters degree in Economics and a PhD. Since 1988 his research has been specialized in socio-economic issues related to information and communication technologies. This includes economic analysis of applications and telecommunication networks and services (for example cost analysis of telecom networks), e-government, regulation of the telecom sector (in particular regulation of interconnection), information and communication technology (ICT) and industry policy, the role of competition in innovation of new services and spectrum management. He has participated in many European Union (EU) funded research projects in the telematics area. He has also conducted a large number of consultancies for national and international organizations such as the International Telecommunication Union (ITU), the United Nations Conference Trade and Development (UNCTAD), the World Bank and the national telecom agencies in Denmark, Norway and Sweden.

Claudio Feijóo is Professor at Universidad Politécnica de Madrid, Spain where he researches on the future socio-economic impact of emerging information society technologies. He recently returned from a highly rewarding two years as a visiting researcher at the Institute for Prospective Technological Studies of the Joint Research Centre of the European Commission. Since 1993 he has been involved in telecommunications and information society development from the academic, the practical implementation and the public administration perspectives.

Hidenori Fuke has been Professor of Info-Communications Industry at the Faculty of Global Media Studies, Komazawa University in Tokyo, Japan since 2007. He has done extensive research on the info-communications industry and regulatory policy in the industry as a member of the board of directors at InfoCom Research and Professor at Kansai University in Osaka. He received his BA in Economics from Tokyo University, and M Litt. from Glasgow University. He also received a PhD in International Public Policy from Osaka University. His major areas of specialization include the economic analysis of the telecommunications industry and regulatory policy. He has written many papers and books in this area in both Japanese and English.

Anastassios Gentzoglanis holds a PhD degree from McGill University, Canada and is a full Professor of Economics and Finance at the University of Sherbrooke. He teaches and conducts research in the area of regulatory economics. He is frequently invited to teach and work on research projects related to the economics of regulation and new technologies in Europe, Asia, Central and Latin America and Africa. He has also received many funded research grants from both domestic and international organizations, published in a number of high-quality scholarly journals and participated in many international conferences. In June 2008, Dr Gentzoglanis organized and chaired the 17th Biennial conference of the International Telecommunications Society (ITS) in Montreal. He is also director of the Centre for the Study of Regulatory Economics and Finance (CEREF) and he organizes courses for executives in the area of regulation of public utilities for the Francophonie.

Toshiya Jitsuzumi is a Professor of Industrial Policy at the Faculty of Economics, Kyushu University, Japan. He has earned an LLB from the University of Tokyo, an MBA from New York University, and a DSc from Waseda University, Japan. Prior to starting his academic career in 2000, Professor Jitsuzumi had served for 15 years at the Ministry of Post and Telecommunications (now Ministry of Internal Affairs and Communication), Japan. The research topics of his interest include public economics and communication and Internet economics. During 2007–08, he was a visiting scholar at the Columbia Institute for Tele-Information (CITI), Columbia University, as an Abe Fellow sponsored by the Social Science Research Council/Japan Foundation Center for Global Partnership. His research work has appeared in various journals, including *Telecommunications Policy*, *Foresight* and *Socio-Economic Planning Sciences*. He is a member of several academic societies, including the International Telecommunications Society and the Regional Science Association International.

José-Luis Gómez-Barroso is Associate Professor at the Universidad Nacional de Educación a Distancia (UNED), Spain. He holds a PhD in Economics, as well as a Masters in Telecommunication Engineering and a Masters in Law. His teaching and research interests lie in the evolution of the information society and electronic communications under the triple perspective of technology, economy and regulation, Dr Gómez-Barroso has contributed significantly in the field of telecommunications by authoring many papers in academic journals and international conferences on the subject.

Anders Henten is Associate Professor at the Center for Communication, Media and Information technologies (CMI) at Aalborg University, Denmark. He is a graduate in Communications and International Development Studies from Roskilde University in Denmark and holds a PhD from the Technical University of Denmark. His main areas of research are service innovation and internationalization, regulation of communications, standardization, and socio-economic implications of information and communication technologies, including e-business and business modeling. Anders Henten has worked professionally in the areas of communications economy and policy for more than 20 years. He has participated in numerous research projects financed by the European Community, the Nordic Council of Ministers, Danish research councils and ministries, and in consultancies, financed by the World Bank, the United Nations Conference on Trade and Development (UNCTAD), the International Telecommunication Union (ITU), Danish ministries, and so on. He has published nationally and internationally – more than 200 academic publications in international journals, books, anthologies, conference proceedings, and so on.

Bronwyn Howell is General Manager of the New Zealand Institute for the Study of Competition and Regulation, and a faculty member of Victoria Management School, Victoria University of Wellington, New Zealand. She teaches and researches in the areas of institutional economics, the information economy and regulation, with specific interests in the development, implementation and performance of institutions and policies governing the operation of information technology and telecommunications markets.

Kenneth Jull practices at Baker & McKenzie LLP, Toronto office, in the area of risk management strategies to promote regulatory and corporate compliance. Mr Jull is the co-author with Justice Todd Archibald and Professor Kent Roach of *Regulatory and Corporate Liability: From Due Diligence to Risk Management*. He is an Adjunct Professor at the University of Toronto, Faculty of Law where he teaches 'Financial Crimes'. Mr Jull has an appointment to the Faculty of Graduate Studies at Osgoode Hall Law School, where he is the Director of the part-time LLM specializing in civil litigation and dispute resolution.

Arata Kamino has been an Executive Researcher of the global information and communication technology (ICT) industry study group at InfoCom Research, Inc. in Tokyo, Japan since 1995. He has been responsible for comprehensive analysis of ICT industry using method of comparative study between major overseas countries and Japan. His main interests cover policies, regulations, market structure and corporate strategies in the field of domestic and global ICT. He received his BA in Economics from Tokyo University. He has written many papers for books, academic journals and magazines.

Scott Marcus is a Senior Consultant for WIK-Consult GmbH, Germany, a research institute in economics and regulatory policy for network industries. Previously, he served as Senior Advisor for Internet Technology for the US Federal Communication Commission (FCC). Prior to that, he was the Chief Technology Officer (CTO) of Genuity, Inc. In 2004, Scott was attached to the European Commission (DG INFSO – Directorate General for information Society and Media) as a Transatlantic Fellow of the German Marshall Fund of the United States. Scott is also a newly appointed Fellow of GLOCOM (the Center for Global Communications, a research institute of the International University of Japan), and a Visiting Fellow of the University of Southern California's Center for Communication Law and Policy. He is co-editor for public policy and regulation for *IEEE Communications Magazine*. He served on the board of the American Registry of Internet Numbers (ARIN) from 2000 to 2002, and on the Meetings and Conference Board of the Institute of Electrical and Electronics Engineers (IEEE) Communications Society from 2001 to 2005, and as Chair of the IEEE Committee on Network Operations and Management (CNOM). He is a Senior Member of the IEEE. He is the author of numerous papers and of a book on data network design: *Designing Wide Area Networks and Internetworks: A Practical Guide* (Addison Wesley, 1999).

Sergio Ramos gained his PhD degree in Telecommunications Engineering from the Universidad Politécnica de Madrid, and he holds an MBA from the Stockholm School of Economics. He worked for the European Commission as Resident Twinning Adviser of a European Commission (EC) Twinning Project for the Public Utilities Commission of Latvia, to design and monitor the transposition process of the European Union (EU) framework into Latvian legislation. He is currently Deputy Director at the Spanish Association of Telecommunication Network Operators (REDTEL).

Andy Reid has the role of Chief Network Services Strategist at British Telecom (BT). Prior to this position he has worked in technical as well as product marketing roles in BT. His professional interests and expertise cover telecommunications technologies, network architecture, service pricing, and links to regulatory and competition law microeconomic analysis. He made major contributions to the development of the telecommunications transmission technology including SONET/SDH (synchronous optical networking/synchronous digital hierarchy) and more recently on Ethernet standards, and has co-authored two major textbooks in this area as well as numerous papers for professional standards bodies and technical forums.

Paul Richards is a Regulatory Economist and has worked for British Telecom (BT) since 1997. Prior to this position he worked in a variety of private and public sector roles in the areas of economic modeling and regulatory analysis.

Olaf Rieck is an Assistant Professor at Nanyang Business School, Singapore. His research focuses on telecommunications economics and policy, as well as on corporate strategy for telecommunications service providers. Dr Rieck studied Industrial Engineering and Management (Universität Karlsruhe/Germany, University of Auckland/New Zealand) and Economics (University of British Columbia/Canada) and holds a PhD in Business Administration (University of British Columbia). Previous to his academic career he worked for Mercedes-Benz AG on strategic information technology (IT) projects in the areas of telecommunications and business process re-engineering. Dr Rieck teaches an introduction to IT, telecommunications industry management, and mobile commerce to undergraduate students and MBAs. He has also held executive teaching assignments with Korea Telecom and LG.

Stephen Schmidt is Chief Regulatory Legal Counsel at TELUS Communications Company. His responsibilities include regulatory, legal and policy submissions to the Canadian Radio-television and Telecommunications Commission (CRTC), the Federal Court of Appeal, the Supreme Court of Canada, the Competition Bureau and other government bodies. Prior to joining TELUS, he worked at AT&T Canada and ACC TelEnterprises, with a focus on regulatory and legal matters respectively. Mr Schmidt holds a BA (with Distinction) from the University of Toronto and an LLB from the University of Manitoba. He is on the board of directors of the International Telecommunications Society and lives in Ottawa, Canada.

Martyn Taylor is a Partner in the competition and market regulation team of Gilbert & Tobin, based in Sydney. He specializes in telecommunications law, antitrust law and international economic law. Martyn has been advising on telecoms law and regulation since 1995, principally within the Asia-Pacific region. His clients have included some of the world's largest technology and telecoms multinationals. Martyn has multiple postgraduate qualifications in law, corporate finance and economics, including a

PhD. He has published extensively, including a book titled *International Competition Law* (Cambridge University Press, 2006) and as a contributor to *Merger Control Worldwide* by Dabbah and Lasok (eds), Cambridge University Press, 2008. He is currently the technology editor for the *Australian Business Law Review*.

William E. Taylor is a Senior Vice-President at NERA Economic Consulting, where he heads the Communications Practice and the Boston office. He received a BA in Economics from Harvard, an MA in Statistics and a PhD in Economics from Berkeley. He taught economics, statistics, and econometrics at Cornell, Louvain and MIT and published research in economics, econometrics and telecommunications policy at Bell Laboratories and Bellcore. At NERA, he has worked and published extensively in telecommunications economics on such issues as access charges, costs, regulatory reform, productivity, competition policy, mergers, vertical integration, interconnection pricing and antitrust issues. He has testified on telecommunications economics before numerous domestic and foreign regulatory authorities, courts and legislative bodies.

Karsten Vandrup, Director at Litepoint Europe, is currently responsible for Litepoint's sites in Israel and Denmark. Prior to this position, Vandrup was Vice-President of Engineering at Litepoint's Headquarters in Silicon Valley, California. Before joining Litepoint, Vandrup held several positions at Nokia including Senior Research Manager in the Nokia System Research, Manager of Strategic Planning in the Nokia Research and Technology Access (Espoo, Finland), Global R&D Cooperation Manager in the Nokia Research and Education Policy Department at the Nokia Head Office (Copenhagen). As a part of the latter, Vandrup was a member of the Career-Space Consortium. He started his career in Nokia as DSP Design Engineer, later Engineering Manager, and holds an *akademiingeniør* degree in Telecommunication and Electronics from the Technical University of Denmark, supplemented with studies at INSEAD, the University of California, Los Angeles (UCLA) and the Swedish School of Economics and Business Administration in Helsinki.

Jason Whalley is Reader in the Department of Management Science, University of Strathclyde, Glasgow. His research interests include the internationalization of mobile operators, as well as the development of telecommunications policy in mountainous developing countries. In addition, his research also examines the diffusion of broadband and the regulatory measures taken to encourage its uptake.

Patrick Xavier is Director of Info-Comm Strategies and Adjunct Professor of Communications Economics, Centre for Communications Economics and Electronic Markets, Curtin University of Technology, Australia. He has served as consultant to a wide range of national and international agencies, including the Organisation for Economic Co-operation and Development (OECD), World Bank, Asia-Pacific Economic Cooperation (APEC) and International Telecommunication of Union (ITU).

Dimitri Ypsilanti is head of the Information, Communications and Consumer Policy Division in the Directorate of Science, Technology and Industry at the OECD. He has specialized in telecommunication policy and regulation.

Foreword

TELUS is very proud to have been the corporate host, along with the Université de Sherbrooke as the academic host, of the 17th Biennial Conference of the International Telecommunications Society (ITS): The Changing Structure of the Telecommunications Industry and the New Role of Regulation, which took place in June 2008, in Montreal, Canada. TELUS invited the ITS conference to Canada because TELUS believes in fostering high-quality analysis of important public policy issues affecting telecommunications both in Canada and around the world.

More than 300 delegates from more than 40 countries came together for four days in Montreal to share new ideas and approaches for dealing with pressing issues in telecommunications, including technological change, new sources of competition, new theoretical developments, and the ways in which government policy can best respond to these issues.

This book flows directly from the ITS conference and effectively captures the character of the conference in microcosm. The 15 chapters selected by Drs Gentzoglannis and Henten reflect the rich, international character and professional diversity of ITS and its members. The present volume contains contributions from the academic, private sector and government communities, providing a crucial window into key contemporary debates in telecommunications. I am confident that this volume will inform and shape future debate, policy-making, business planning, research and strategic thinking.

Stephen Schmidt
Chief Regulatory Legal Counsel
TELUS Communications
Ottawa, Canada

Preface

The International Telecommunications Society (ITS) convened its 17th Biennial Conference in Montréal, 24–27 June 2008. The conference was organized by the Université de Sherbrooke, on the initiative of Professor Anastassios Gentzoglanis, Organizing Committee Chair. There was also a main sponsor and corporate host – TELUS – represented by Dr Stephen Schmidt, Organizing Committee Co-chair, and supported by Organizing Committee Co-chair Stanford L. Levin, Southern Illinois University at Edwardsville, and Program Chair Richard Schultz, McGill University.

Looking back, it was in 1997 that a major ITS conference took place in North America. The industry and research have certainly moved a long way, facing new technologies and regulatory challenges, but the critical methods and policy analyses remain. The title of the conference aptly reflected both the change and the continuity: 'The Changing Structure of the Telecommunications Industry and the New Role of Regulation'.

The conference was a significant success. The theme attracted a large amount of submissions. In fact, it drew one of the largest pool of submissions to an ITS conference ever, and the was one of the largest ITS conferences ever from all over the world. Conference evaluations were very positive after the event. The conference also brought to the fore new research and policy agendas.

This book is based upon selected and edited conference papers. The editors have been careful to select a wide range of chapters in order to capture the convergence issue from a number of perspectives, and to achieve a historical state of the art. The volume succeeds in providing a multifaceted and rich view of both changing industry structure and for new roles of regulation. The editors and the contributors are to be congratulated for a timely book.

Erik Bohlin
ITS Chair
Professor, Chalmers University of Technology, Sweden

Evolving technologies, competition and the new role of regulation: introduction and synopsis of the book

Anastassios Gentzoglanis and Anders Henten

Regulation of the telecommunications industry has traditionally focused on the supply side of the industry, chiefly the retail segment of the market. Since liberalization of the industry has begun, regulation has gradually shifted to the wholesale segment of the market. The regulatory agencies have intervened to regulate access and facilitate entry and, hopefully, investment in infrastructure. This asymmetric regulation has had mixed results. As an answer to that some regulatory agencies have abandoned wholesale regulation (the case of the USA) and some others have shifted from light-handed regulation to heavy-handed regulation with mandatory unbundling of the local loop (the case of Australia). Other regulatory agencies have moved to a more gradual type of wholesale deregulation on the grounds that competition in this segment of the market has not yet fully developed, but as it grows the need for regulation is reduced (the case of Europe and Canada).

Paradoxically, as competition increases in the retail segment of the market, consumers are increasingly 'impaired' in their capacity to make decisions in their best interest. The array of services and suppliers that competition makes available increases consumer choice and with it the difficulties for them to make rational decisions with respect to services, quality and prices, and so on. If consumers are able to make rational decisions, competition among new service suppliers will increase. But if consumers are unable to make rational decisions because of too much choice or poor quality of information or misinformation, competition is dampened and the competitive process is jeopardized. In that context, regulation must emphasize the demand aspects of the industry and consumer protection should become a priority for the regulatory agencies.

Further, next generation mobile networks (NGMNs) depend heavily on the existence of ubiquitous broadband (BB) connectivity, applications and content. Broadband deployment is uneven within and among

1

countries. Without an adequate deployment of BB technologies the economic growth of the countries is jeopardized. Apparently, regulation is one important factor determining the pace of deployment of BB technologies. The regulatory model chosen may impact positively or negatively on investment in BB infrastructure. Additionally, ubiquitous BB and mobile applications create new needs for spectrum availability and spectrum management becomes an important function of the regulatory agencies, particularly at this time of rapid evolution of the mobile technologies. The NGMNs require a new distinctive regulatory and policy framework which will deal explicitly with the issues and opportunities of the next phase of wireless technologies.

Regulation evolves and its evolution is the result and the impetus of change of the telecommunications industry structure and performance. As the industry becomes more mature and incumbents and new entrants get more familiar with the rules of the game, they become able to develop strategies which increase the value of the firm. In a competitive context telecommunications firms will invest only if their investment achieves an average return which is greater than the weighted average cost of capital (WACC). Projects with positive net present values (adjusted for specific risks) are value accretive and therefore it is worth undertaking them. They will bring more wealth to stakeholders through dividends and capital gains. In a regulatory context the investment decisions, particularly for projects of high risk (sunk investments in BB and NGMNs, for instance) may not occur or may be 'unreasonably' delayed. Thus, under specific regulatory frameworks, deployment of BB technologies may not be optimal, NGMNs may not roll out adequately or optimally, consumers may be 'impaired' or harmed; and these are not necessarily the results the regulatory authorities have sought to achieve in the first place. Regulation thus has a role in an evolving global telecommunications industry.

This was precisely the main theme of the International Telecommunications Society (ITS) 17th Biennial Conference which was held in Montreal from 24 to 28 June 2008. This book has been prepared to highlight the main arguments and the richness of ideas about the new issues which arise from the evolving structure of the global telecommunications industry and the role of regulation. The 15 papers from the conference have been peer reviewed and they meet the stringent criteria of the scientific papers and those of Edward Elgar Publishing. We are grateful to the publisher and particularly to Alan Sturner for his meticulous work in the preparation of this volume. We also thank our colleagues and authors of the 15 chapters of this volume for having contributed high-quality chapters. As organizer of the 17th Biennial ITS Conference, Dr Gentzoglanis thanks TELUS, and particularly

Stephen Schmidt, for contributing financially to the publication of this book and its contribution, financial and in kind, in being a corporate host for the very successful ITS Conference. Gratitude is also extended to the Université de Sherbrooke, Office of Research and the Faculty of Business Administration for their financial contribution and support. Dr Gentzoglanis thanks the SSHRC (the Social Sciences and Humanities Research Council of Canada) for its generous contribution to the organization of the ITS Conference and the publication of this book. He also thanks Emmanuella Gentzoglanis and Andrianiaina Rajaobelina for their help and comments and suggestions. This book could not exist without our many devoted manuscript peer reviewers. Sincere thanks to all the reviewers who dedicated their time and shared their knowledge with the authors and the editors of this book.

The first chapter by Anastassios Gentzoglanis and Elias Aravantinos provides an extensive review of the literature on the role of regulation in the deployment of broadband technologies. The chapter starts out with a general introduction to the role of regulation in telecommunications markets and, furthermore, as an input discusses the relationship between competition and investment. It is emphasized that this depends on the degree of competition and the development of the market. An inverted U-shaped curve is suggested as describing the implications of competition on investment. More specifically, the chapter examines whether and how broadband development can be promoted by means of regulation. Two different types of broadband competition are discussed, service-based competition (SBC) and facilities-based competition (FBC). FBC is generally seen to be the most desirable situation or goal as FBC provides a dynamic efficiency while SBC is seen to lead to static efficiency.

The chapter gives focus to the ongoing discussion as to whether the static efficiency associated with SBC can lead to the dynamic efficiency related to FBC. This is the proposition advanced by the theory of the ladder of investment (LoI). According to this theory, new operators to the market will not generally begin with investing in infrastructure but will be more inclined to enter the market using a service-based model. Eventually, such new operators will start investing in infrastructure when they acquire a better understanding of the market conditions. Regulation can facilitate this process by setting the right entry conditions, for example, by starting with a relatively low access fee which is later on increased as the new entrants settle in the market.

The chapter examines the theoretical aspects of the ladder of investment theory and also refers to papers presenting empirically based analyses of the tenability of the theory. Moreover, the chapter introduces the issue of intermodal or interplatform competition. Even though this issue is not absent per se from the LoI theory, nonetheless, when debating LoI, there is a tendency to confine it within a single type of access network. The chapter by Gentzoglanis and Aravantinos contributes to the debate by synthesizing the main arguments pertaining to the LoI theory. In addition, the chapter underlines the fact that BB development in different countries is not the same and this is attributed to the differences in the countries' regulatory policies and the type of their initial technological infrastructure. It should be expected therefore that the implications of a LoI policy will be different for countries with basically just one type of infrastructure (PSTN – public switched telephone network) and countries with competing infrastructures (for example, PSTN and cable).

On the basis of the theoretical propositions and the empirical analyses referred to, the chapter concludes that the actual LoI policies implemented so far have not been successful, but one cannot dismiss the possibility that a LoI policy can work. This presupposes, however, that the timing is right and the policy is fine-tuned precisely with respect to the types of new entrants. Regulators should then ask: Are they first-movers or second-movers? But this sets very high requirements on regulation.

In the second chapter, William Taylor discusses the nature of intermodal competition and its implications for regulation of wholesale services in the context of the US, but given the ubiquity of intermodal technologies, the conclusions of his analysis can find a wider application. Taylor notes that demand for wholesale services is a derived demand, in the sense that wholesale services can be considered 'essential' in the provision of retail services in the downstream market. Competition among retail service providers may curtail retail prices but there is a likelihood of anticompetitive conduct when the incumbent local exchange carrier (ILEC) is not regulated in the form of mandatory unbundling and provision of essential wholesale services at regulated prices.

When competition is fierce among dependent wireline competitive local exchange carriers (CLECs), their returns in the retail market cannot be higher than normal and therefore the ILEC monopolist of wholesale wireline services cannot behave anticompetitively and charge high prices for its essential wholesale services. Since it cannot exercise its market power at the wholesale level, its profits will also be normal. This may have some undesirable effects on investment in infrastructure. The mandatory unbundling of ILEC facilities at regulated rates would reduce the incentive of retail wireline competitors to invest in their own network infrastructures and compete on an end-to-end basis. At the same time, the requirement that ILEC facilities be shared with competitors reduces the ILEC's incentives to introduce and roll out that infrastructure, particularly for services associated with investment that will, eventually, be sunk.

Further, Taylor casts doubts about the necessity of price regulation at multiple stages of production (wholesale and retail). Using examples from the US, he demonstrates that regulation at the wholesale and retail levels becomes a source of unintended consequences in competitive markets. Thus, where intermodal competition is present the ILEC's capacity to exercise its market power is nil and therefore customers cannot be harmed. Accordingly, *ex ante* economic regulation cannot lead to an increase in social welfare and there are not any efficiency arguments that can justify the presence of economic regulation at the wholesale level. In the absence of an *ex ante* wholesale regulation, customers can be protected from competition authorities which will respond *ex post* to specific complaints of anticompetitive behavior on behalf of ILEC. Thus, where intermodal competition exists, *ex post* regulation is a better vehicle to protect customers than ex ante regulation.

In the third chapter Martyn Taylor delves into the question of how to strike a balance between competition, investment in infrastructure and regulation. Using the conventional financial framework for project appraisal, Taylor argues that the returns that a firm expects to get from investment in infrastructure are constrained by market conditions (consumers' willingness to pay), the presence of substitutable services (wireless broadband, for instance) and regulatory policies. These factors increase the perceived risk of the investment in next generation networks (NGNs) and the latter will not materialize unless a firm has reasonable certainty that it will earn a return that exceeds its cost of capital over the project's life.

Taylor argues that well-intended regulation can have unintended adverse effects, chiefly by providing disincentives to investment due to an increase in perceived risks particularly when regulation is asymmetric. Network unbundling and cost-based access regulation are two cases in point. While they contribute to allocative (static) efficiency by increasing the level of competition to which incumbent network owners are subject, they may do so at the expense of long-term investment incentives. The total element long-run incremental cost (TELRIC) approach has been used in the US and its variant – total service long-run incremental cost (TSLRIC) – in Australia as a mechanism to regulate prices based on the incremental costs faced by an efficient cost-minimizing firm with an optimally configured network that uses the best available current technology. Under the TELRIC approach, any investment in network infrastructure would simply earn enough to cover the project's weighted average cost of capital (WACC), bringing thereby the net present value (NPV) of the project near zero. The firm would not have an incentive to invest in infrastructure and the desirable long-term effects of dynamic competition are sacrificed. This was the case with Australia which has applied excessive regulation using the TSLRIC approach. The application of TSLRIC pricing by the Australian Competition and Consumer Commission (ACCC) has deterred investment in infrastructure and investment flows have been distorted away from regulated (and potentially regulated) services towards unregulated services and infrastructure. The negative effects of excessive regulation are illustrated by the example of failed negotiations between Telstra (Australia's incumbent telecommunications carrier) and the Australian government when the former was seeking exceptions from the current regulation in order to implement fiber to the node (FTTN) upgrade investments. Telstra's withdrawal forced the government to intervene and proceed with the investment in partnerships with the private sector. Taylor suggests that policies such as access holidays and public–private partnerships (PPPs) may be some interesting techniques to provide incentives to firms to invest in NGN infrastructure.

Patrick Xavier and Dimitri Ypsilanti's chapter (Chapter 4) examines the demand side of the telecommunications industry. Since liberalization, the number of new entrants in telecommunications markets has increased considerably and service competition has grown significantly. In this context, Xavier and Ypsilanti believe that the consumers' rational decision-making mechanism is seriously impaired amidst the plethora of services and packages offered by the telecommunications firms. This contrasts with the neoclassical view of consumer behavior according to which consumers are able to make rational decisions and choose, all the time, the goods and services which maximize their utility. In reality, consumers' behavior departs significantly from the traditional rational behavior assumed by the conventional neoclassical economic theory. By using the behavioral approach to consumer behavior and statistical examples, the authors demonstrate that consumers choose telecommunications services for a number of reasons which may not be characterized as a 'rational' behavior in the sense of classical theory. For instance, consumers are discouraged from switching to a different service provider because of the perceived or real high switching costs (lengthy and cumbersome switching procedures; early exit charges; confusing products and non-transparent pricing; technical incompatibility of equipment; long-term deals that lock consumers into lengthy relationships with their providers).

The barriers to switching service providers can also have an impact on the supply side of the industry. New entrants could be deterred from entering the market, fearing that they will be unable to persuade customers to switch from their existing provider. 'This could diminish contestability and the effectiveness of competition and limit the benefits that consumers would otherwise derive from it', say Xavier and Ypsilanti. The statistics from various countries (the UK, Portugal, Australia and the US) and from various segments of the market (fixed line, mobile and Internet) illustrate that competition has made the telecommunications industry quite complex (on the demand side) and consumers facing this complexity prefer to adopt consumption strategies which are not necessarily in their best interests. Indeed, consumers will prefer to stay with what they know (no switching) instead of choosing a cheaper alternative.

Xavier and Ypsilanti's analysis is consistent with the argument of behavioral economics according to which an 'endowment factor' serves to influence decisions in favor of the present provider. In that context, regulation has an important role. Regulators have to take a number of measures to assist consumers to make 'rational' decisions in their best interests. Educating consumers; increasing awareness about new services and options; requiring that all major operators provide comparable and complete information about the services, quality and prices; targeting information to most vulnerable consumers; could be some of the new responsibilities of the regulatory agencies in an era of increased competition in the telecommunications industry.

In the fifth chapter Marcel Boyer develops a methodological framework which is used to characterize properly the level of competition in the telecommunications industry, particularly in the residential local access market. He argues that failing to recognize and properly evaluate the nature of competition in the telecommunications industry may lead to an inefficient use of regulation. In competitive telecommunications markets, regulators play a new role and must undertake three new functions, acting as: generators of information for the consumers; managers of the rules of competition among telecommunication players; and promoters of efficient investment programs.

Boyer argues that the telecommunications industry has changed dramatically from the mid-1990s onwards and the technological changes have made it appear much more like an emerging industry than a mature industry. Therefore, the traditional measures used to determine the level of completion in the industry are less relevant. New measures must be used to determine the level of competition in the industry. For instance, the use of market shares as an index of competition may make sense in mature industries where there is a relative stability of market conditions, but applying it to the telecommunications industry characterized by a rapid pace of changes may be misleading. Further, even though the pricing schemes used by competitors are differentiated so that switching among service providers becomes difficult on the part of consumers, the lack of switching does not make the telecommunications markets less competitive. On the contrary, the very existence of price differentiation indicates that the telecommunications industry is indeed competitive. Without it, prices would have been quite low, making the provision of new services a losing opportunity inviting exit from the industry and an ultimate reduction in the number of firms in the industry (less competition).

Although competition may be limited in terms of prices, telecommunication service providers may compete in a number of other areas, such as coverage, type of transmission (digital or analogue); interplatform provision of services (DSL – digital subscriber line; cable modem), security, and so on. Therefore, traditional measures of measuring competition such as relevant market, relevant (substitutable) services available to consumers, relevant set of actual or even potential competitors, and so on, are less relevant during this changing phase of the telecommunications industry. In that context, Boyer proposes a different regulatory mechanism. He argues that, to achieve a proper balance between static (short-term) and dynamic (long-term) goals, regulators must rely on competitive processes, in a sense that instead of micro-managing prices and quantities, they must make sure that these prices and quantities emerge from a competitive environment. Thus regulators have an essential role: they have to act as efficient generators of social efficiency by safeguarding the competitive process in the telecommunications industry. This is achieved by making sure that inter-access to essential facilities is available at non-discriminatory conditions and prices so that only new, more efficient entrants enter the industry. In that way, consumers benefit from the entry of efficient competitors. Thus, viewing the emerging structure of the telecommunications industry as a process, regulators must act as: (1) trusted generators of information for consumers; (2) managers of fair conditions for access to the local loop; and (3) promoters of investment programs which should contain pricing rules designed to include all network access costs and guarantees safeguarding the integrity and reliability of the entire network.

The sixth chapter, contributed by Kenneth Jull and Stephen Schmidt, intends to offer an alternative framework that can be used as a basis for new approaches to the regulation of the telecommunications industry. The authors make a distinction between *ex ante* and *ex post* regulation and they use examples from other industries to illustrate that *ex post* regulation can be an option for the telecommunications industry. Traditionally, the regulatory systems in the telecommunications industry have been *ex ante* systems, but the latter are increasingly criticized as inefficient since they are perceived as being a blunt 'one-size-fits-all' mechanism. In Canada, there are suggestions (Telecommunications Policy Review Panel) to replace this mechanism by a new regulatory framework which would set out broad principles to prohibit anticompetitive conduct instead of detailed *ex ante* rules.

Jull and Schmidt propose three principles that ought to govern the balance between *ex ante* and *ex post* systems in the application of the regulatory policies. According to them, regulators focus ought to be on: (1) the prevention of harm and attainment of specific social objectives; (2) strategies for managing risks; and (3) the adoption of flexible mechanisms (use of multiple models) to reflect the different needs of the stakeholders.

Each system, being *ex ante* or *ex post*, might have two subsets within it, being rules-based and principles-based. A rules-based system (whether it is *ex ante* or *ex post*) is better suited to industries which are stable, technologically simple and where the economic and financial stakes are relatively low. The telecommunications industry is undergoing rapid technological and market structure changes, so this industry is neither simple nor stable and the economic stakes are very high. Because rules are inflexible, they can be overtaken by changing circumstances in fields such as telecommunications. In such a context, a principles-based approach may be more appropriate. As a matter of fact, the choice between principles-based and rules-based systems is a function of social and economic priorities and the level of maturity of the industry. Rules-based systems are more suitable for cases of *social regulation* where serious harm may be prevented whereas principles-based systems are generally more suitable for cases of *economic regulation*. Jull and Schmidt propose the same type of regulation for the Canadian telecommunications industry as the one proposed by Boyer in Chapter 5 but each author draws his conclusions using a different analytical framework.

The seventh chapter by Olaf Rieck analyzes, from the industry's perspective, the strategic activities of the telecommunication carriers to integrate vertically the various segments of the value chain. Few studies examine quantitatively the new strategic directions that telecommunication firms take in the rapid changing environment, and Rieck's study falls into this category. He divides the telecommunications industry into five layers and then uses Fransman's (2007) value chain simplified layer model to assess empirically the impact of various strategic initiatives on the valuation of telecom operators. Rieck is interested in the evolution of the structure of the telecommunications industry and in the strategies of industry players to extend their market power to more layers. This is quite interesting when one views the recent changes in the telecommunications industry structure and its trend to convergence.

Nowadays, the telecommunications industry has many firms which entered from outside the traditional telecommunications services industry, and many are quite new or did not even exist back in 1998. For instance, Google emerged as world leader for online searches and has established itself as a significant player in the telecoms value chain. Google is also involved in the roll-out of broadband infrastructure and the provision of content. Apple launched the iPhone which threatens various traditional players in the value chain. By offering hand phone devices, Apple has effectively become a new player in the hand phone equipment market (Layer 1). By tying Apple's handsets to iTunes (Layer 4) and by striking content deals with content providers (Layer 5), Apple has extended its market power in almost all layers of the value chain. Nokia and Sony Ericsson have followed Apple's lead by launching their own content platforms. Traditional telecommunications carriers, while under threat from all sides to be reduced to 'bit-pipes', have tried to counter the threats by engaging in activities in vertically related markets. This includes initiatives like joining the open handset alliance, the development of mobile portals, or striking deals with content providers such as to strengthen their position in the content integration layer. Do these strategies help to increase the telecommunications firms' reach and market power, or will they reduce their role in the future? Rieck answers these questions by examining the reactions of financial markets to the strategies adopted by the telecommunications companies, particularly Mergers and Acquisitions (M&A), after rival firms have announced their decisions to extend their reach in the vertical value chain of the industry.

Chapter 8, authored by Bronwyn Howell, asks the question whether an industry-specific regulatory regime (a telecommunications regulator) is more able to pursue an economic efficiency (static and dynamic) objective than a competition authority (non industry-specific) without falling into the trap of regulatory capture. To answer this question, Howell examines New Zealand's telecommunications sector in the 1990s and 2000s. She notes that the initial goal of New Zealand's government was to preserve the telecommunications industry's long-run incentives to invest in new networks and technologies using a 'light-handed' regulatory regime. Unfortunately, the objectives of regulators often do not coincide with those of politicians. The latter are subject to more pressures from vested interest groups and more inclined to satisfy their demands by adopting new and/or modifying existing legislation to pursue different sectoral objectives. Given that regulators are the agents of political principals, they may lose their power when politicians decide to change objectives and move from efficiency objetives to distributional ones, as was the case with New Zealand's government in the 2000s. Indeed, prior to the competition law review in 2000, New Zealand had adopted 'light-handed' regulation for its telecommunications sector. Despite this regulation, the telecommunications industry was far from unregulated. Under a contractual arrangement, the incumbent monopolist, Telecom Corporation of New Zealand Limited, had rural–urban universal service obligations, free local calling and a price cap on residential services. This contractual arrangement, known as the 'Kiwi Share', was a type of regulation capable of achieving economic efficiency.

The objectives of regulation have been changed with the change of government in 2000 and the adoption of the Telecommunications Act in 2001. The Act established an apparently independent Telecommunications Commission (within the Commerce Commission) and TSLRIC pricing for 'designated services'. This creation of an apparent independent regulatory body free from risk of capture by vested interests was seen as an 'enlightened' form of industry-specific regulation. Nonetheless, its independence was tested when the Commission undertook a revision of local loop unbundling (LLU). The Commission, applying dynamic efficiency principles, decided not to proceed with unbundling. By contrast, the Commission used a different approach when it had to examine the mobile termination market where serious concerns have been raised concerning the exercise of monopoly power. In a surprising decision, the Commission asserted that the sector's objective was to pursue competition rather than efficiency, and therefore short-run objectives were prioritized. Although making efficiency an explicit regulatory objective is rationally justified from an economic perspective, viewed from a political perspective the efficiency objective is unsustainable in the long run. The New Zealand case clearly illustrates that *ex ante* regulation can be inferior compared to *ex post* regulation – competition law – particularly when the risks of regulatory capture cannot be avoided.

Scott Marcus and Dieter Elixmann contribute the ninth chapter. They argue that the migration of current networks to next generation networks (NGNs) and the issues arising from their access to the fixed network bring new challenges to regulators and policy makers alike. For instance, the use of local loop unbundling (LLU) as a solution to the problem of incumbent's market power is particularly challenged by the migration to FTTC/VDSL (fiber to the curb/cabinet/very high speed digital subscriber line) or to FTTH/FTTB (fiber to the home or fiber to the building) networks. The migration to NGNs is very different from country to country and these differences are due to the existence of various regulatory regimes and the market evolution in each country. Germany and the Netherlands, for instance, have a regulatory framework which incentivizes incumbents to be the 'first-movers' to replace the traditional fixed access network with VDSL-capable networks. In France, the NGNs' deployment has been undertaken by the incumbent as well as its competitors. In Japan, deployment is realized not by the incumbents but by other independent companies. In the US, a change in regulation in 2003 put an end to mandatory broadband unbundling and allowed broadband services to be offered over cable. Such a policy gave incentives to incumbent operators AT&T and Verizon to make substantial investments in fiber. The authors compare the performance of each country trying to identify whether competition and/or regulation are the most important factors for change. After having analyzed in detail the level of deployment of the fiber-based NGNs and the regulatory regimes determining the access conditions in Germany, the Netherlands, France, Japan and the United States, they conclude that regulation is an important input in the deployment process of NGNs but there is no unique model of regulation which fits well to every country. Rather, a number of factors – demographic, geographic and historical (availability of alternative last mile infrastructure) – determine the formulation of the regulatory policy. Once the latter is well conceived and put into play, it is the force of competition which determines the pace of NGNs deployment. Thus, country-specific regulation and competition are the *sine qua non* conditions for a wider deployment of NGNs.

The tenth chapter is by Arata Kamino and Hidenori Fuke, and is the first of three chapters to discuss functional and structural separation and its effects on the roll-out of BB technologies and platforms and NGN applications. Their analysis is a case study of the Japanese telecommunications industry demonstrating that the deployment of BB technologies and the implementation of the world's fastest and cheapest DSL technologies in Japan are attributed to the particularities of the Japanese regulatory regime and market conditions. Indeed, Japan was one of the first industrialized countries to implement the most rigid open network policy for the promotion of service-based competition. For instance, unbundling obligations have been imposed on both copper and fiber loops and the competition that this regulation has entailed resulted in very low LLU fees particularly for shared lines. Given the rapid increase in competition in LLU, no bitstream access competition has been developed in Japan. The European Union (EU) and the US have implemented a similar type of copper LLU but they have not experienced the same degree of DSL as in Japan. Although there is no simple answer to this conundrum, the authors advance the arguments according to which the difference in performance may be attributed to the way this regulation has been applied in different countries and continents. For instance, in Japan, the LLU fees were fixed at an extremely low level favoring service-based competition by new entrants. In addition, the Japanese entrepreneurs may be driven more by a kind of 'animal spirit' which contributes to intensify competition even when profits are not as high as they could be. Indeed, Japanese DSL competitors continued to provide very cheap alternative DSL services despite long-standing fiscal losses during the early 2000s. Competition is effective and works when it is fair and transparent. But it is not entirely clear whether competition is more effective when conduct regulation is imposed on the incumbent or when structural separation is applied. Given that Japan adopted in 1999 the same level of conduct regulation as the EU, but that competition and the deployment of BB technologies have developed faster in Japan than the EU, the authors argue that Japan's structural separation was an effective means to foster competition in the market. Nonetheless, the introduction of FTTx (fiber to the x) and NGNs raises additional issues in the discussion of vertical separation in the telecommunications industry, such as the 'hold-up' and 'coordination' problems, and one should analyze the perspectives of competition which will emerge between traditional carriers and content providers on the basis of new business models before the implementation of a structural separation.

Peter Curwen and Jason Whalley continue the theme of the implementation of functional separation in Chapter 11, this time using the UK experience. In the EU, the implementation of functional separation within fixed telecommunications markets is increasingly seen as a way to resolve the tensions that exist between incumbent operators and those other service providers that require access to incumbents' networks to deliver their own services. In 2005, functional separation was implemented by the British incumbent, British Telecom (BT), after pressure by Ofcom, the regulatory authority, thereby making the UK the de facto European leader in functional separation. Under this agreement, BT created a new company – Openreach – to run BT's local access network. The creation of Openreach was possible after BT has agreed on a series of undertakings. Curwen and Whalley's chapter focuses on the implementation of these undertakings and highlights the difficulties encountered to make the functional separation effective. The authors demonstrate that functional separation is not a simple task. On the one hand, regulators encountered enormous difficulties in making the undertakings operational and, on the other hand, Openreach had difficulties in implementing them.

One of the major tasks in implementing functional separation is to selectively separate those parts of the network that are difficult for other operators to replicate but which they need to access in order to provide their own services. To interpreter such a separation as simply a division of the incumbent's wholesale and retail businesses from one another – either in the form of accounting, corporate or type of service (local from long distance; mobile from fixed; local from broadband, and so on.) – is completely misleading. The way functional separation is defined – in a broad or narrow sense – has important implications for the actual form of separation. Regardless of the extent to which functional separation is implemented, regulators should provide incentives so that the separated network could act in the interests of all its customers, internal and external, and not in the interest of its parent company. BT's separation was driven by the need to incorporate EU directives into the UK regulatory framework, the failure of competition to develop as anticipated in the UK, and the establishment of Ofcom in 2003, which undertook a strategic review of the telecommunications market in order to examine the level of competition of the telecommunications industry in the UK and identify the regulatory options available. Ofcom concluded that deregulation was not possible because sector-specific regulation was faster and more precise than the alternatives – competition. Ofcom opted for a functional separation called 'real equality of access' under which independent purchasers of BT's wholesale products could buy these products under the same terms as BT's own retail operations. In practice, the 'real equality of access' took the form of equivalence of outcome and equivalence of input. Under the former, wholesale customers receive products that are comparable to those offered to BT's own retail operations, but the underlying processes are different. Under the latter, wholesale customers receive the same products as BT's own retail operations using the same set of underlying processes. Despite the delays and the definition problems, the period subsequent to the adoption of functional separation has seen the emergence of significant broadband competitors. LLU played a central role in the strategies of entrants and offered incentives to other service providers to invest in other parts of the 'ladder of investment'. The emergence of LLU as a vehicle for the deployment of broadband services lifted the importance of BT in the market and made sure that the relationship between BT and those companies using its network was functioning as planned.

Toshiya Jitsuzumi is the author of the twelfth chapter. This chapter provides a theoretical justification and an econometric analysis of the hypothesis that LLU is contributing to the deployment of new technologies such as FFTx. He emphasizes the need to take into account social and economic aspects of the issues arising from the network neutrality debate, and he concentrates his analysis on the solutions that are most efficient from an economic point of view in the short term and long term. In the short term, it is assumed that entry does not occur and the market is served by the existing network operators. In the long term, entry occurs, competition is more intense and network congestion becomes a less acute issue. The proposed solutions are dependent on the assumptions made.

For instance, when the network capacity is fixed in the short run, the problem becomes one of static efficiency maximization. In that case network operators have market power and regulators have to find efficient solutions to discipline incumbents who control bottleneck facilities. But given that content providers depend on the presence of other firms in the industry in the provision of their own content, it is unlikely that they will exercise their market power and foreclose the market from competition. Indeed, in this industry, a firm is interested in the internalization of complementary efficiencies (ICE) arising from applications created by others. This behavioral characteristic – the ICE in conjunction with the costs associated with regulation and information asymmetry – makes the competition solution a more desirable solution than regulation. Knowing that a profit-seeking bottleneck monopolist acts to maximize its efficiency, regulators should not use rules-based regulation but rather principles-based, unless there is pressure arising from the exercise of significant market power (SMP) by some service providers. ICE makes regulation less desirable and government intervention is kept to a minimum.

In the long run, the issues become more challenging than in the short run. In the long run, market conditions must be such that incumbents and new entrants have the appropriate incentives to invest in new technologies and maximize dynamic efficiency. Investments, being in the form of virtual capacity, better protocols – peer to peer (P2P), proactive network provider participation for P2P (P4P) – and/or better network management, must be financed but the problem is that, as yet, there is no any sound business model that provides incentives to stakeholders to invest in infrastructure. Network neutrality proponents suggest the use of a subscription model – that is, through additional monthly subscription revenues such as quality-of-service (QoS) surcharge from end-users – while the opponents suggest a business model according to which investment in infrastructure is financed by charging content and application providers. Neither of these models can assure the collection of sufficient revenues for capacity expansion or the quality of the transmission of the content. These solutions may be utility-decreasing and unsustainable in the long term.

In order to verify whether the subscription model is sustainable in the long run, Jitsuzumi conducted a survey of Japanese broadband users using an e-mail and web-based system. His data and econometric analysis indicate that such a business model is not sustainable in the long run unless there are positive expectations concerning the future technological developments and the existence of a fund allocation mechanism. His results are quite informative and useful for other countries.

Bruno Basalisco, Andy Reid and Paul Richards contribute the thirteenth chapter, delving into the interesting question about the effects of regulation on innovation and on evolving technologies, most of which emerge outside the domain of the telecommunications industry. Departing from the self-evident fact that the main objective of innovators is the commercialization of their innovations, they argue that such commercialization is more successful when various parties across sectors coordinate their activities. The converging nature of new technologies requires a keen interplay between innovation and regulation and makes the latter less desirable, particularly when the boundaries of the industries are merging.

In an increasingly competitive world, erstwhile competitors realize that innovation requires more than large market shares and market dominance, it needs 'co-opetition'. The latter is used to create a competitive advantage in innovation processes involving all 'co-opetitors', that is, their network of suppliers, users and customers. Collaborative ventures with other industry stakeholders lead to greater coordination of the innovative activities and reduce the uncertainty associated with technological changes across sectors. Innovation interdependence delivers maximum benefits to their participants. The standardization of technologies through collaborative ventures and interconnectivity minimize the level of risk of the entire value chain and not just one part of it. But regulatory decisions may contribute negatively to the tendency of collaborative innovative agreements and thwart the appearance of new business models. The presence of network spillover effects in the various processes of innovation provide incentives for different networks to interconnect. Further, the presence of path-dependence in innovations implies that successful innovation by one or more player(s) is likely to influence future technological choice across the industry. Co-opetition, path-dependence, spillover effects and standardization add considerable complexity to the exploitation of network-based innovations, but reduce the risks associated with innovations in converging industries. Basalisco, Reid and Richards argue that the regulatory frameworks should be operated in such a manner that different risks associated with different innovations are treated differently in regulatory terms. Failing to recognize this means the current regulation may provide disincentives to firms to commit resources to invest in innovative activities and infrastructure.

The penultimate, fourteenth, chapter by Claudio Feijóo, Sergio Ramos and José-Luis Gómez-Barroso adopts an interesting approach to examine the impact of regulation on the pace of deployment of the next generation mobile networks (NGMNs). They argue that regulators may adopt either a stable and coherent framework which provides incentives to investment in broadband technologies, or a framework which may retard the deployment of these new technologies by amplifying all techno-economic uncertainties. According to the authors, spectrum management is one of the most important areas where regulation may have an immediate impact. Referring to the regulation of spectrum in the EU, the authors urge regulators to make spectrum management more flexible in order to be able to accommodate the next generation mobile networks (NGMNs) and other fast-evolving technologies. They also suggest a better harmonization of spectrum management mechanisms across member states for easier deployment of ubiquitous broadband infrastructure and faster realization of the benefits of the NGMNs. Drawing from the US experience and the relative failure of the Universal Mobile Telecommunications System (UMTS) as opposed to the great success of Global System for Mobile Communications (GSM) in Europe, the authors argue that harmonization should be subtler and focus mostly on the new conditions for use of spectrum and particularly on the 'converging competition' which emerges from the fixed–mobile convergence of NGNs. Uncertainties caused by an ill-conceived regulatory framework which does not take into account the conditions for investment in NGMNs, or the conditions for competition and the conditions for innovation, will retard the NGMN deployment and it will result in a loss of the EU's competitiveness. Regulation has an important role to play in the intensity and speed of the NGMN deployment and can contribute significantly to the creation of value through the arrival of new applications and services. To do so, regulation must be swift and adaptive to the requirements the new technologies bring about.

In the final chapter, Chapter 15, Morten Falch, Anders Henten and Karsten Vandrup address the deployment of mobile data in Europe, East Asia and North America in order to identify reasons that may explain the difference in performance among these regions. The chapter focuses on the market conditions that prevail in each geographical market but it also addresses the role of policy in the deployment of data and mobile Internet. It is argued that to promote mobile data one needs to encompass a wide variety of policy areas since the development of mobile data depends on structural factors in the markets, particularly those pertaining to the supply side of the industry and the take-up factors like the general e-readiness of the potential users on the demand side.

The chapter notices and documents that the East Asian countries, Japan and Korea, are ahead in the global development of mobile data. While Europe took the lead in the 1990s with the second generation GSM system, the East Asian countries have been the front-runners with respect to mobile data, first on 2.5G platforms and later on 3G and 3.5G platforms. The US has been trailing somewhat behind in the 2G development but seems to catch up regarding mobile data. The question is what explanations there are for this development and what one can learn from it.

In order to examine this question, the chapter first discusses the reasons for the East Asian lead as compared to Europe and North America. The chapter includes explanations of a structural kind on the supply side as well as diffusion issues on the demand side. Theoretically, the chapter therefore takes its points of departure in theory on innovation systems and theory on diffusion. Secondly, the chapter examines empirically the development of mobile communications more generally, and mobile data more specifically, in the East Asian countries, Europe and North America. Focus is on the terminal markets, network infrastructures, and services and content. The numbers show that the US is forging ahead regarding mobile data and that the European lead over the US in mobile communications is disappearing.

The prime reason put forward in the chapter for the North American catch-up is the position of the US in the information technology (IT) area, hardware as well as software. The US is positioned very strongly regarding Internet technologies and services, and the hypothesis is that the US can leverage this position onto the mobile field. The Internet innovation system and the mobile communications innovation system have to a large extent been separate. The East Asian countries have managed to merge them to a certain degree. However, the competences on the supply as well as the demand side in the US with respect to Internet technologies constitute a strong point of departure for developing mobile Internet in the US.

In sum, Regulation and the Evolution of the Global Telecommunications Industry is a collection of 15 chapters that bring a variety of theoretical perspectives and empirical evidence to the question of how regulation could be applied (even eliminated) to the deployment of BB technologies and NGMNs in an era of dramatic changes in the structure and performance of the global telecommunications industry. The material well illustrates the diversity of thoughts and research that characterize this important area of academic and business research. We hope that this volume will spur others on to research this challenging topic.

PART I

Regulation versus investment: the balance
between static and dynamic efficiencies and
the main issues of regulatory policy

1. Investment in broadband technologies and the role of regulation[1]

Anastassios Gentzoglanis and Elias Aravantinos

1.1 INTRODUCTION

In the two decades from the mid-1980s to the mid-2000s network industries have undergone dramatic changes in structure and regulation. The thrusting changes have initially occurred in the technologies used by these industries to produce, transmit and distribute essential services to consumers. Technological changes have not uniformly affected all segments of the production processes. Few segments still remain natural monopolies while others are more suitable for competition. This uneven impact of technological changes on the cost functions has created new challenges to regulators and the industry. Both still strive to find a 'new' business model and new regulatory frameworks which will allow more investment in infrastructure in general and in broadband (BB) technologies in particular.

Some countries have applied 'light-handed' regulation while some others have even decided to go further and liberalize entirely a few segments of the telecommunications industry by introducing full competition while keeping others under a regulated monopoly regime. The latter was generally applied to the incumbent while new entrants enjoyed more favorable entry and access regimes. The duality created by this type of regulation has sparked a heated debate as to what is the 'appropriate role' of regulation and its impact on investment and innovation.[2] The main issues concern the capacity of the regulatory agencies to promote investment in infrastructure, especially in broadband technologies, and what are the best mechanisms to increase consumer choice at reasonable prices. This debate has arisen because more and more specialists (Aron and Crandall, 2008) believe that the telecommunications industry has entered into a maturity phase, where competition can work really well. In that case, innovation and dynamic efficiencies are viewed as the outcome of competition. Yet, there are other specialists who believe that dynamic efficiencies can only be achieved through an active use of an incentive regulation (Cave, 2006). In the first case, competition spurs the diffusion of innovations and guarantees the best outcome for customers. In the second case, discriminatory regulation (the one favoring new entrants at the expense of the incumbent) is considered as the most appropriate means to promote BB investments and efficient prices to customers.

But the introduction of competition modifies the vested interests of stakeholders, and regulators have to weigh the interests of entrants, incumbents and customers in the design of their new regulatory framework. If the new regulatory regime encourages entry by allowing access at low prices, competition would increase, but the latter acts as a deterrent to investment in infrastructure since low access prices reduce the net present value (NPV) of the incumbent's investment projects. As a consequence, the conditions to achieve competition and encourage investment in BB are not very clear for the regulatory agencies, and their favored policy to create equal access conditions to new entrants by unbundling incumbent's infrastructure has become a challenging task for regulators.

It is true that in many instances, wireline access is still the key infrastructure to provide BB services.[3] This bottleneck infrastructure is usually controlled by the incumbent and confers on it an undue market power. By regulating the access price, regulators aim at providing incentives to new entrants to use the current technological platform and create competition at the local level, but this may have a negative effect on infrastructure investment. Investments in infrastructure and the diffusion of broadband technologies and services have important externalities and spillover effects and both are increasingly viewed as important factors for economic growth and prosperity (Waverman et al., 2005). It is not surprising therefore to find out that the regulatory systems that allegedly impede investments are under great scrutiny. This debate boils down to questioning the efficiency of regulatory policies and whether they are still justified under the current state of different technological platforms. It is unclear whether they contribute to or inhibit the development of innovation and infrastructure in BB technologies and whether competition can better attain these dynamic efficiencies in the telecommunication industry.

The approaches adopted by the regulatory agencies in Europe and the US to promote the deployment of BB technologies are quite different. In Europe, where traditionally digital subscriber line (DSL) technology was most prominently used, service competition seems to work relatively well and the rate of deployment of BB technologies is considered quite satisfactory, compared to other countries. In North America, service competition has not given the anticipated results and at the time of writing North America trails Europe and Asia. Therefore, the North American regulatory agencies increasingly substitute service competition for facilities-based competition. It is believed that facilities-based competition will attract new entrants whose investments in new infrastructure will fill out the alleged BB gap that at present exists between Europe and North America. But lack of entry and investment in the local exchange has raised doubts about the efficacy of this policy. Critics argue that the regulatory approach which initially favors competition on service via an unbundling mechanism does not necessarily develop to a facilities competition at later stages (the ladder of investment theory). Regulators should consider new frameworks to promote investment and innovation in the industry.

The goal of this chapter is to present the theoretical underpinnings of the ladder of investment theory and to review the recent literature on the effects of the ladder of investment on BB performance of major industrialized countries. By doing so, we aim at shedding more light on the issues concerning the role of regulation in the era of rapidly converging technologies. Our analysis and the review of recent empirical studies reveal that the ladder of investment regulation is not flawed per se, but the way it is applied in various countries may lead to different BB performance (Japan versus Australia, for instance). It seems that the regulatory agencies have to develop more refined techniques and strategies which give incentives to new entrants to become first-movers in the investment race.

Section 1.2 of this chapter presents the BB market and the technological platforms that exist to provide BB services. The institutional settings of the broadband market are examined and this sets the debate as far as investment in BB technologies is concerned. Section 1.3 reviews the main arguments by examining the relationship between regulation and investment in broadband infrastructure. It examines critically the arguments according to which innovation and investment in broadband infrastructure may be promoted within a regulatory environment which favors initially a service-based competition (SBC) and later a facilities-based competition (FBC), the so-called the ladder of investment theory. Section 1.4 reviews the findings of the empirical studies, particularly those dealing with the ladder of investment theory. The purpose of this section is to critically examine the robustness of the results of the most recent empirical studies and their usefulness to regulators and policy-makers in gaining a better understanding of the functioning of the investment behavior under alternative regulatory approaches. Section 1.5 concludes and offers some policy recommendations.

1.2   TECHNOLOGICAL PLATFORMS FOR INNOVATION AND INVESTMENT IN BROADBAND: INSTITUTIONAL SETTINGS AND TECHNOLOGICAL PERSPECTIVES

There are currently several technologies capable of providing access to broadband services each one with different technological characteristics. They can be divided into wired access technologies such as ADSL, ADSL2+, VDSL, hybrid fiber coaxial (HFC) and fiber; and wireless access technologies such as Wi-Fi, WiMAX (Worldwide Interoperability for Microwave Acess) and satellite. From a customer's point of view, the most important characteristic of these technologies is speed. A number of emerging technologies provide high speed broadband of 10Mbps and higher, such as optical fiber, HFC using DOCSIS-2 (Data Over Cable Service Interface Specification) standard, VDSL (very high speed digital subscriber line) and fast wireless broadband. These alternative broadband technologies have been promoted as would-be competitors in different markets. However, competition across platforms is increasingly simmering through the evolving technologies. New entrants and incumbents may use one or more platforms to offer their services but high speed broadband requires upgrade of the final link from the exchange to end-users (the so-called 'last mile'). These upgrades increase the quality of the network and allow the offer of broadband services at speeds well in excess of those offered by existing technologies. The upgrades may be done on the existing copper wires but both incumbents and new entrants may invest important amounts in rolling out FTTH (fiber to the home) networks. The new networks free the 'last mile', that is, the copper part which is still considered as a bottleneck, and offer more bandwidth.

Competition from cable operators is becoming increasingly intense particularly by employing strategies aiming at extending the potential of their HFC networks. The upgrade of their networks to higher bandwidths with the deployment of DOCSIS-3 standards necessitates migration to new topologies and networks. The competition possibilities are quite high but the investments to be realized in line with current network infrastructure and new networks are quite substantial and risky and this makes both entrants and incumbents reluctant to invest. Since current investments in infrastructure are considered inadequate to provide ubiquitous broadband, actual competition is limited and potential competition is latent. In such a context, interplatform competition may have little impact on local markets and the existing institutional frameworks. Regulatory authorities should then work on the elaboration of more appropriate conditions that will make latent competition eventually effective. The current regulatory framework adopted in many countries which favors competition by allowing new entrants to lease the incumbent's bottleneck capacity and offer services to customers has been criticized as rather inefficient (Wallsten, 2006). The so-called local loop unbundling (LLU) regulation is based on the assumption that as competition at the retail level becomes more intense, competition at the wholesale level will eventually increase, provided that incumbents and entrants invest in infrastructure.

With LLU, competition at the retail level has indeed increased in some countries (for instance, the UK; Ofcom, 2006) and became intense in market segments where most services and applications are being developed. To lure customers, both incumbent and new entrants undertook bundling strategies at the retail level (double, triple and quadruple play).[4] To be sure, the regulatory framework of each country plays a significant role in the development of competition among players of converging industries. For instance, due to technological and regulatory imperatives cable operators offered triple play in Europe and the US markets prior to the telecom incumbents and Internet providers (Lee, 2009). This was mainly due to the adoption of the Telecommunications Act of 1996 in the US which gave the possibility to cable operators to establish telecommunications business. Further, Internet providers use bundling as a way to access markets and to gain market shares from their competitors as illustrated by the examples of Fastweb and Free (an Italian and a French Internet provider, respectively) which managed to increase their market shares with triple-play high-quality services.[5]

Moreover, telecom incumbents and mobile operators use bundling in a defensive way, in order to protect their core market. The undertakings of bundling strategies push all players to compete fiercely in several markets which further intensify competitive pressures. Differentiation of service plays an important role in making each company unique in its offerings and this allows it to preserve its market power. Based on evidence from recent studies (Lee, 2009), companies manage to differentiate their packages by offering at least one service in the package which is different from the same service in a competitor's package.[6]

Despite the differentiation strategies, it is generally believed that the intensification of competition at the retail level would lead to more innovation and investment in BB infrastructure[7] (ladder of investment). But such an outcome is not automatic. The experience with this approach throughout the world is quite mixed. In many countries (Canada, Australia, New Zealand, for example), competition at the retail level[8] did not spur adequate competition at the wholesale level and the adoption of the ladder of investment approach left them behind others in terms of BB infrastructure (see OECD, 2009; CRTC, 2009). Some specialists (Waverman et al., 2007) have started questioning the (theoretical) foundations of this approach and its practical applications. Others (Aron and Crandall, 2008) have mostly questioned the ability of the regulatory agencies and the type of regulation they have put in place to spur innovation. This research purports to examine thoroughly the workings of the ladder of investment theory and identify ways which could be used by the regulatory agencies to spur BB investment. The next sections describe the ladder of investment relationship in detail and explore the alternatives available to regulators to prompt more investment in BB infrastructure.

1.3   REGULATION AND INVESTMENT IN BROADBAND INFRASTRUCTURE

The debate whether regulation is conducive to innovation is not new, but recently has revived due to the apparent failure of regulation to provide incentives to incumbents and new entrants to invest in infrastructure, particularly in broadband technologies. After many years of experience with deregulation and the introduction of favorable regulatory conditions for new entrants, which made competition a reality, economists and policymakers do not agree as to the level and kind of competition that is required to induce telecommunication firms to invest in BB infrastructure. Many doubt and some even blame (Aron and Crandall, 2008) the type of regulation, that is, LLU, as the culprit of such performance. LLU and other ambivalent regulatory policies have created an uncertain environment for the incumbents and increased their systematic risk and the cost of capital, as measured by their betas. This makes the conditions for accessing international financial markets more difficult and jeopardizes their capacity to get funding to invest in long-term projects.[9] Recent statistics[10] from the Organisation for Economic Co-operation and Development (OECD, 2009) indicate the performance of major industrialized countries in BB technologies (Figure 1.1).

Economists emphasize more and more the importance of dynamic competition as a criterion of performance. For a long time, though, static competition has been the criterion for performance and policies were adopted aiming at promoting static efficiency. The latter is achieved when firms produce at their minimum long-run average production costs (production efficiencies) and consumers pay the marginal cost of goods and services produced (allocative efficiencies). In the telecommunications industry, regulators have designed various frameworks to achieve static efficiencies. Lately, these frameworks have been largely modified to accommodate dynamic efficiencies. It is believed that static efficiencies

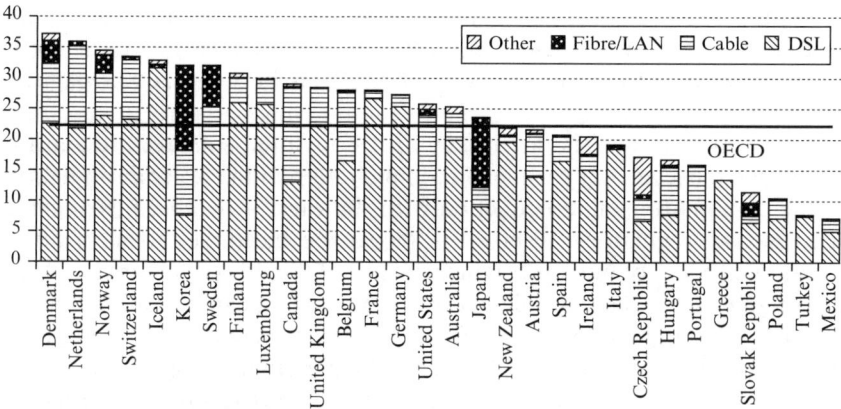

Figure 1.1   OECD broadband subscribers per 100 inhabitants, June 2009

could be achieved by promoting service-based competition. But it is also hoped that service-based competition would eventually lead to facilities-based competition (dynamic efficiency). By allowing new entrants to have equal access to the incumbent's facilities, prices would eventually be driven close to the cost of production and the social welfare would be maximized, achieving thereby the static efficiency objectives. Further, lower prices for the telecommunications services would create sufficient new demand which eventually will be enough to provide incentives to both incumbents and new entrants to invest in infrastructure and satisfy this increasing demand. Thus static efficiency would bring dynamic efficiency which relates mostly to demand creation (new telecommunication products and services) and innovation (new production techniques and alternative technology platforms).[11]

In practice, the relationship between dynamic efficiency and investment seems to be non-linear. The combination of the so-called Schumpeterian and escape effects may result in a reduction of innovation after a certain threshold of competition. In reality, the two effects work in opposite directions and the final outcome depends on the size of each effect. More precisely, as competition increases, it reduces the incentives of firms to invest in new technologies or in new products because of the high risk of post-entry reduction in rents (Schumpeterian effect). Nonetheless, at an initial low level of competition it is possible to increase investment by promoting more competition. As the latter increases so does the investment by incumbents and new entrants. This is so because firms by investing (innovating) manage to escape the negative effects of competition (escape effect). Innovation is thus seen as the means to preserve (or even increase) economic rents. But as competition increases and surpasses a critical threshold, a further increase in competition reduces investment and innovation, the so-called Schumpeterian effect (Breschi et al., 2000; Malerba and Orsenigo, 1996).

Accordingly, the initial level of competition, and its degree after entry, determine the level of investment and innovation in the telecoms industry. These combinations in conjunction with the variety of other instruments used by the regulatory authorities to provide incentives to innovation result in an array of outcomes which become difficult to disentangle. Thus, although it is not easy to isolate them in practice, it can be said, in theory at least, that countries may experience either an increase or a decrease in investment and innovation in broadband technologies, depending on the size of the Schumpeterian and escape effects. This result may be useful to policy-makers and provide an important guidance to regulators in the formulation of their policies. But to make it more operational and to convert it to a powerful decision-making tool, optimality conditions should be developed. It is therefore important to develop realistic models which determine the threshold level of competition and provide answers to the question of whether more competition – be in the form of free or low access price, local loop unbundling, and so on – contributes to investment and innovation in broadband technologies.

Empirical observation shows though that neither entrants nor incumbents have incentives to invest in infrastructure either in established or in new technologies, and this despite the regulatory framework that favors the service-based competition. It is not surprising, therefore, to see that service-based competition is increasingly viewed as a source of dynamic inefficiency and that regulators seek for new ways (total deregulation is one) to bring incentives which would create an industry structure more conducive to dynamic efficiency.[12] In New Zealand, a recent report (Boyle et al., 2008) casts doubts that the telecommunications incumbents are interested in investing in infrastructure, since if they do, their profitability will be reduced. It is stated that: 'leaving the (investment) task to telcos, they will ration technology and bandwidth to stretch out the life of yesterday's investments and maximize their own returns'. According to the same report, governments should, therefore, subsidize the investment in broadband technologies, if an adequate level of broadband deployment is to be achieved more quickly. In the same vein, some authors (Janssen and Mendys-Kamphorts, 2008) go even further by arguing that: 'there may be a role for public investment and ownership of infrastructure, given the seriousness of potential market failures'. Given these conclusions, it becomes important to examine critically the underlying theory used by regulators to formulate their recent regulatory policies and to investigate thoroughly the conditions that lead to dynamic efficiency. This is the subject matter of the next subsection.

1.3.1 The Ladder Investment Theory of Regulation and Dynamic Efficiency

The dynamic impact of regulation on investment has been neatly articulated as what is now called the theory of ladder investment (Cave, 2006). According to the latter, the dynamic efficiencies of competition could be captured through a more or less slow process of deregulation and accommodation of entry. It is argued that entrants cannot invest in infrastructure from the outset unless they get a customer base and familiarize themselves with the costs, demand and other characteristics of the industry. To make the entry attractive, regulators ought to reduce the risks of entry and allow new entrants to realize some economic rents. At this stage, regulation should aim at fixing low access prices so that new entrants could access the incumbent's bottleneck facility. At initial low access prices new entrants find an interest in entering the market and competing head to head with the incumbent on price while both entrants and incumbent use the same infrastructure. At this point, entrants are not interested in investing in new infrastructure but as they get familiar with the market and they realize economic profits, their growth opportunities depend on their capacity to retain proprietary essential assets. These proprietary assets represent a competitive advantage and entrants have an incentive to invest in infrastructure. Initially, service-based investments are complementary to the technologies of the incumbent, but eventually new entrants and the incumbent move up to the ladder of new investments by constructing new facilities.

To attain this level of (facilities-based) competition, regulation plays a significant role. According to the ladder of investment theory, in order to increase investment in infrastructure, regulators, after a certain period of time, have to increase the access price, since its initial low level would simply perpetuate the service-based competition. The higher access fees would lower the revenues of new entrants unless the latter move on the ladder and invest more in infrastructure. As the entrants gain more know-how and master the technologies better, they can climb the ladder even higher by investing more in new facilities. New entrants become full-fledged investors and owners of their infrastructure (Prieger and Heil, 2008), ready to compete with the incumbent not only in terms of price but also in terms of technological and other demand-related characteristics.

Competition which started at the beginning as a service-based competition (SBC) evolves to a facilities-based competition (FBC) where incumbent and entrants vie for more market shares via prices and investments in new technologies (innovation). For regulators, the key to achieve this outcome is to unbundle the local loop and allow new entrants to access the most expensive part of the incumbent's network. The local loop unbundling (LLU) can take three different forms: bitstream access,[13] shared access[14] and full unbundling.[15] The terms and conditions for getting one or another form of access depend on technological, proprietary, institutional and regulatory factors and the preferences of new entrants. For instance, an entrant whose interconnection points with the incumbent's network are rather poor will prefer a bitstream connection, while entrants with a more extensive network will prefer full access. Regulators should then determine the terms and conditions of access by taking into account the technological characteristics of new entrants and the type of competition that they desire to achieve in the telecommunications industry.

In sum, if regulators believe in the ladder of investment theory, they may fix very low access prices at the beginning. This discriminatory regulatory regime provides incentives for further investment in infrastructure. To achieve this goal though, the regulatory authorities have to develop a lot of expertise in order to be able to determine what type of access to allow and the timing of increasing the access fees. Further, regulators have to take into account the strategic interactions of entrants and the incumbent, but this is difficult for them particularly when they have to fine-tune and coordinate competing interactions in a way to achieve the desired results. Moreover, despite the regulators' beliefs that the rise in access tariff will incite new entrants to innovate and develop their own network, it is possible that the risks involved with new investments remain quite high, thereby discouraging firms from delving into such investments. New entrants may prefer to operate as simple resellers of traditional services. This point is quite important because investment decisions are quite sensitive to risks and the latter may be too high to make the investments profitable.

The main risks may be classified as technological and financial, but other risks related to the market and to changes in the regulatory regimes are also important. In a changing environment these risks are difficult to quantify and as competition increases, the probabilities that the new investments will provide economic rents are getting lower. This may be true even though there would be an increase in the size of the market resulting mainly from the introduction of new products and techniques of production. Because competition reduces profit margins, which are important indicators of returns, financiers are willing to provide finance for projects having low risks and generous returns. Lack of finance is an indicator that the investment projects are not highly valued by investors. Because competition is so keen and the risks are high, the telecoms industry has been characterized as a mature, slow-evolving industry. Since the initial development of the theory of the ladder of investment (Cave and Vogelsang, 2003; Cave, 2006, 2007; Bourreau and Dogan, 2006) many researchers have investigated in some detail the conditions under which new entrants will invest in innovation and new facilities. The following subsection presents the main arguments of the models.

1.3.2   The Optimal Timing of Innovation under the Ladder of Investment Framework

The optimal timing of innovation under different market structures and regulatory regimes has been elaborated by Dobbs (2004) but his model does not take into account the complexities of investment decisions when regulators allow local unbundling under an asymmetric[16] regulatory framework. This is a serious drawback of the model because the behavior of the incumbent and new entrants is significantly altered when access is regulated and the effects of this discriminatory regulation on investment and innovation are unclear. A priori, it is expected that investment will increase if new entrants can pre-empt the incumbent's investment, but it may decrease when rival firms cannot or are not allowed to pre-empt the incumbent's investment. This relationship is even less clear if we take into account not only investment in new technologies but also investment in existing technologies. Indeed, if investment decisions are lumped together and no distinction is made between investment in fixed line BB (existing technology) and mobile BB (new technology), the results may be quite different. Given the increasing importance of mobile and other technologies in delivering BB services, the investment decisions of both the incumbent and new entrants may again become biased. It is thus appropriate to see in more detail some of the models which tackle these questions. Their results will be useful to regulators and other policy-makers given that current regulatory policies in both Europe and North America weight the dynamic aspects of competition more than the static ones.

Competition can take various forms but two of them are particularly important for the ladder of investment theory, the SBC and the FBC. Under the ladder of investment theory, the intertemporal variation of access fees by the regulatory agency incentivizes new entrants to invest in infrastructure. When competition is in the form of SBC, the entrants' initial investments are relatively small, since they use the incumbent's network. Nonetheless, they have to make an irreversible investment and sink an initial amount of funds, I_{SBC}, into complementary technology. But as the intensity of competition increases and the regulatory agency increases the interconnection fees, entrants have to make I_{FBC} irreversible investments in new facilities which are much higher than the initial invest-ments in SBC. If the time to make the irreversible investment when the firm is under SBC is t_S, and the time to make the irreversible investment when the firm is under FBC is t_F, where $t_F > t_S$, and if CF_n are the net cash flows of the investment of new entrants, then the expected market value of new entrants will equal the sum of discounted value of the investment and the net cash flows that the investments will bring over time.[17]

The regulated access fee plays a significant role on the timing of investments. On the one hand, investment in SBC is determined by the regulator since it is up to him to specify the type of access that the incumbent must offer to new entrants. On the other hand, the regula-tor, by determining the level of access fees, determines new entrants' net cash flows and their capacity to attract funds to invest in SBC. Since the decision to invest can be viewed as a real option[18] for new rival firms, the latter provides a tool to firms to manage the risk arising from the investment, but this option also provides a bias to the investment decision. Since rivals will invest only if they are sure that their invest-ment will be profitable, it will be more interesting for them to wait and see what happens to the investment realized first by the incumbent. If the incumbent's investment is profitable, rivals will invest too, but if the incumbent's investment is not profitable they will prefer to wait. Thus during good periods investment will increase but this reduces the potential profitability of the incumbent (and new entrants), and during bad periods investments are retarded and all the risks are assumed by the incumbent. Regulation provides an asymmetry in sharing the risks; the incumbent bears the full risk in bad periods but in good times the proceeds are shared by the incumbent and entrants alike. Given that the incumbent's cost of capital is the opportunity cost during good and bad times, the incumbent would not be interested in investing either, or at least its commitment to invest would be relatively low (see Table 1.1).

The situation becomes even more complex when we take into account the level and kind of competition that will prevail on the retail market. The level of competition on the retail market determines the revenues that the incumbent will receive from the wholesale market. If the incumbent is quite aggressive in the retail market this will have an impact on the rev-enues of new entrants and the incumbent alike. If new entrants are losing market shares, the reduction of their revenues has a direct impact on the incumbent's capacity to safeguard a steady cashflow from the business line of new entrants.[19] Low access price and high retail prices make invest-ment in SBC more attractive than in FBC. Entrants delay investments in infrastructure, and competition in facilities does not develop as regulators have originally expected.

Table 1.1　Pay-offs and the timing of investment under the ladder of investment regime

Regulated access fees and timing of adjusting them	I_{SBC} is less inductive to investment		I_{FBC} is more inductive to investment	
Period 1: t_S　Period 2: t_F				
High (H)	N/A	Climb the ladder of investment	Incumbent invests [E(MV)]>0 (expected market value)	entrants invest
Low (L)	Incumbent invests in upgrading	Entrants invest marginally	N/A	N/A

Determining the timing of investment is quite important for regulators and policy-makers alike. Further, the timing depends on the game the incumbent and new entrants play. But the results are sensitive depending on which firm invests first, the incumbent or the new entrants. Stackelberg-type models have been developed to illustrate the intricacies of competition at the retail level and its results on investment. There are undeniable advantages for each firm to be a leader in the industry in terms of investing first in BB technologies. Obviously, the most important first-mover advantage is monopoly profit. The leader reaps monopoly profits as long as a follower does not replicate the leader's investment. But there are benefits associated with the second-mover too. The second-mover advantages may be a better understanding of new technologies, an eventual reduction of the costs of followers, a discovery of new products supplied by the new technologies, and so on. Thus, being second in the race for investment in BB infrastructure can bring some undeniable benefits.

The equilibrium in the market is not unique but depends on the new entrants' preferences concerning their position in the race for investment (being first or second). There are two different types of equilibrium depending whether new entrants prefer the benefits of the second-mover or are indifferent between the advantages of first-mover or second-mover (Keiichi Hori and Mizuno, 2009). In the absence of an obligation to invest within certain time limits, the preferences of new entrants determine the outcome of the game. If new entrants are of a second-mover type, investments will be delayed and the introduction of competition by the regulatory authorities does not have a real impact on the timing decisions to invest by the incumbent.[20] In the opposite case, when new entrants are of first-mover type, investments will be accelerated and the introduction of competition by the regulatory authorities does have a real impact on the incumbent's timing decision to invest. These two equilibria, known as the 'waiting game' and the 'pre-emption game' (Guthrie, 2005), are quite useful for regulatory purposes.

Nonetheless, it is not easy for the regulatory agencies to identify the type of new entrant (first-mover or second-mover), nor the reactions of the incumbent. Given that the models do not provide a clear-cut solution to the problem faced by the regulatory agencies which want to promote dynamic efficiencies, it is important for them to be proactive and develop strategies that make new entrants first-movers in the investment decision process.[21] It is probably the only way to bring more investment in BB technologies and accelerate the growth rates of the economy in general and the telecommunications industry in particular.[22]

1.4   INTERNATIONAL REGULATORY APPROACHES TO INVESTMENT IN BROADBAND INFRASTRUCTURE

Although it is impossible to present each country's regulatory approach to investment in broadband, it is advisable to review some of them in order to get a comparative perspective. It is important to stress from the outset that in some countries facilities-based competition is non-existent and/or quite limited since there is only one platform to provide broadband services, usually through DSL.[23] In other countries, such services can be provided on an inter-platform basis via DSL and cable modems. In either case, an upgrade of the networks is necessary in order to be able to provide broadband services. The old telephone and cable networks have been developed to be shared by a certain number of users and a particular traffic flow (bidirectional in the case of telephone services but unidirectional for cable). Cable companies have to upgrade their network to accept a bidirectional traffic flow and telephone companies have to upgrade their network to increase their transmission performance, especially for the 'last mile' (fiber to the home) for services such as video which requires more bandwidth. Also, to get closer to customers' premises and shorten the length of the local loop, the telephone companies need to increase the number of central offices. This is also a way to strengthen the transmission performance of their DSL modems.

The existence of alternative technologies and vintages has affected both the regulatory and the institutional environments in which the communications incumbents have evolved. In the case of cable-DSL environment, the regulation has been asymmetric, in the sense that cable companies enjoyed less stringent regulations than telecommunication firms because the former were more competitive due to the existence of alternative platforms such as satellite and terrestrial transmission which were judged sufficient to keep monopoly power at bay. By contrast, the telecommunications industry has always had a more rigid regulatory environment because of the existence of one technology platform (copper and coaxial cables) which was considered a source of 'market power'. Incumbents could use it to reap consumer surpluses. From the consumers' interest perspective, regulatory approval of retail prices and control of entry were thus judged necessary. This historical evolution and country particularities are reflected in the diversity of approaches adopted by various countries to promote investment in BB infrastructure. But behind each approach there is a common hypothesis: regulatory policies which initially focus on fostering competition at the service level will eventually provide incentives to both entrants and incumbents to invest in BB infrastructure. Service-based competition moves to a higher level and it becomes facilities-based (ladder theory).

Many countries have adopted different regulatory frameworks. As a consequence, their ranking in terms of BB penetration differs significantly (OECD, 2009). The differences in performance may be explained by many factors,[24] but the most recent studies have focused on the institutional setting which prevails in each country, particularly access regulation and LLU. To offer guidance to policy-makers and regulatory authorities, we briefly summarize the main findings of these studies. Most of the studies have been realized on an aggregate level. They examined the relationship between access regulation and investment in BB infrastructure in order to offer guidance concerning the use of the best international practices to achieve the objective of dynamic efficiencies.

For instance, an OECD (2007) study examined empirically the effects of unbundling on the roll-out of BB technologies. This was the first systematic empirical attempt to test this relationship using data for 2002 and 2005 from a number of OECD countries. The conclusions were very powerful and quite straightforward. It appeared that local loop unbundling was a catalytic factor for the deployment of BB in the OECD countries examined. The study concludes that: 'unbundling . . . is currently more significant than platform competition in explaining broadband penetration' (p. 20). The author of the OECD study underlines the important role that the national regulatory authorities (NRAs) can play in determining an appropriate structure of access fees for the unbundled local loop infrastructure and in using subsidies for more BB penetration.

These results were in sharp contrast to those obtained in previous empirical studies. For instance, a group of studies (Distaso et al., 2006; Cava-Ferrer and Alabau-Munoz, 2006; Kim et al., 2003) found that the effects of unbundling on BB roll-out, although positive, were very small but statistically insignificant. Another group of studies (Denni and Gruber, 2005) found that the effects were rather transitory and very small, while others (Wallsten, 2006) found that these effects were not always positive or statistically significant. This discrepancy in the empirical results led Boyle et al. (2008) to investigate further both the data used in the OECD study and the econometric techniques.

Indeed, Boyle et al. (2008) vehemently criticized the OECD results on both economic and statistical grounds. First, they argue that unbundling is just one single factor but there are many others which may affect the BB uptake. Singling out the LLU factor ignores other important factors which may contribute to the explanation of the relationship between access regulation and BB investment. Second, once they correct for the estimation procedure used in the OECD study, the 'statistically significant relationship between local loop unbundling and broadband uptake disappears' (p. 4). The authors conclude that there are neither economic reasons nor statistical ones to justify the strong relationship between unbundling and BB roll-out. This refuting of the OECD results brings the debate back to its fundamental question concerning the role of access regulation in the deployment of BB technologies.

In another study, Gruber (2007) examined the relationship between regulation and investment in the EU and he found that new entrants have a tendency to use a single technological platform and also to avoid investing in infrastructure. The author also investigated the behavior of entrants and incumbents in mobile technology. He asserts that the behavior of entrants and incumbents in this sector is in stark contrast with that observed in fixed lines infrastructure. In the mobile sector, competition on alternative technological platforms is already quite widespread and access regulation is less prevalent. Investments in BB infrastructure in the mobile telecommunications sector are more aligned with the market shares of new entrants and incumbents. In order to favor a greater expansion of the BB technologies, he proposes a new regulatory framework which puts greater emphasis on platform competition through incentives and special access regulation provisions. His ad hoc analysis shows that the investment behavior in the mobile sector is indeed aligned to the ladder of investment theory, but the fact that the behavior of incumbents and entrants in the fixed line sector is quite different than expected reveals that the origin of the investment problem resides in access regulation. He cites the FCC's (Federal Communication Commission) successful shift of focus from the regulation of broadband access and wholesale markets towards more infrastructure competition.[25] All in all, he concludes by stating that: 'the underlying hypothesis of the current regulatory regime, i.e., appropriate access prices lead to facility based competition, still awaits confirmation'. He emphasizes the need to adopt 'new measures which could be much more conducive to new infrastructure investment' (Gruber, 2007, p. 23).

In the first exhaustive empirical study concerning the link between dynamic efficiencies, regulation and investment in BB in Europe, Distaso et al. (2009), examined the approaches adopted by 12 European regulatory authorities to increase investment in BB infrastructure. They claim that the approaches adopted by national regulatory authorities (NRAs) were mostly consistent with the ladder of investment theory, but given the complexities of national markets, the results could not be interpreted properly. By dividing the sample into three subgroups,[26] they were able to make comparisons among individual countries within each category and identify the ones which have adopted the highest level of competition and diffusion of BB services. They found that although considerable progress has been realized, the adoption of more dynamic pricing schemes at the wholesale level and/or commitments by the regulatory agencies stipulating the phase-out period of asymmetric regulation is still so slow that it makes it difficult to predict 'when and in which way regulators will introduce the full set of ladder of investment policies' (Distaso et al., 2009, p. 13). This study is the first one to examine empirically the above-stated relationship but it falls short of giving more guidance to regulators in terms of the elaboration and adoption of a better regulatory platform to increase investment in BB infrastructure. It is true that the limited sample and incomplete data delimit the scope of the analysis. Obviously, additional detailed data are needed to get more convincing results.

Sutherland (2007) analyzes the effects of unbundling on BB deployment in several countries and he concludes that countries which have followed the successful pattern of infrastructure competition perform better than those which have chosen the service-based competition. Like the previous authors, he does not distinguish between investments in fixed or mobile networks and therefore he cannot make a precise statement about the effectiveness of the regulatory regime which puts the unbundling at the center of BB deployment.

All in all, the theoretical underpinnings of the ladder of the investment theory are quite solid but the empirical evidence does not show clearly that it can work in practice as the theory predicts. Despite the limitations of the empirical studies, it can be said that there are some indications that the existence of alternative platforms and the competition that they entail are more conducive to investment in BB infrastructure than the service-based competition. Nonetheless, in both cases, regulation plays a significant role because it defines the appropriate conditions for both types of competition to develop. The regulators' role is thus enhanced during this phase of deregulation and reliance on market forces.

1.5 CONCLUSIONS AND POLICY RECOMMENDATIONS

The level of broadband investment is not the same across countries. Given the importance of BB to economic growth and development, various regulators have adopted original regulatory frameworks to promote investment in BB infrastructure and safeguard the benefits of dynamic competition. The ladder of investment theory has contributed to a great extent to the development of these new regulatory frameworks. According to this theory, regulators by favoring LLU and other types of asymmetrical regulation (which favors new entrants rather than incumbents), are able to reap initially the static benefits of competition and eventually the dynamic ones.

To be achieved, the regulatory agency has to gradually adjust the access fee entrants pay to get connected to the local loop. The increase of access fees makes service-based competition less interesting in terms of profits and this gives incentives to entrants to make investments in new facilities. According to this theory, it is possible to achieve the desired dynamic efficiencies by using a regulatory framework which appropriately regulates the access fees. Our theoretical analysis reveals that the ladder of investment regulation is not flawed per se, but the way it is applied in various countries may lead to different BB performance as the cases of Japan and Australia make clear. The empirical studies reviewed in this chapter demonstrate that the application of the ladder of investment theory in practice did not bring the anticipated benefits of dynamic efficiency. New entrants have either been passive in terms of investment in new infrastructure or their investments fell short of expectations. Incumbents too have limited their investments in BB technologies by making small upgrades just to keep up with the evolving demand for bandwidth. It seems that the regulatory agencies have to develop more refined techniques and strategies that give incentives to new entrants to become first-movers in the investment race. By this process, the competition for infrastructure will bring the benefits associated with dynamic efficiencies. Thus, the role of regulation may be even more important now that the telecommunications industry has became more competitive than it used to be, when competition was rather limited.

NOTES

1. We would like to thank the anonymous referees for their valuable comments and suggestions. We also thank Stephen Schmidt and Anders Henten for reading the chapter and making very insightful comments. All errors are ours.
2. Investment is not necessarily synonymous with innovation. Since our emphasis here is on broadband technologies which are new innovative technologies, we use both terms interchangeably.
3. Wireline access is not necessary to offer mobile broadband, which may be done on a full-fledged facilities-based competition.
4. At the retail level, bundling refers to the practice of selling two or more differentiated products as a package.
5. Free's monthly fee, €29.99, has never increased since 2002, and its 'triple-play' services include very high-speed broadband Internet access (with downloads up to 24 megabits per second and uploads up to 1 Mbps), unlimited calls (to France and 70 countries abroad) and more than 100 digital television channels. Its decision to adopt point-to-point FTTH will allow Free to deploy in a cost-effective way fiber optical links to each home and enable it to offer 1 Gbps and even 10 Gbps connections over the same fiber (Cisco, 2008).
6. In the case of cable and telephone companies, their differences in the triple-play bundle were mostly on the video service category while voice and data was rather similar (Lee, 2009).
7. Ofcom has stated that: 'Sustainable competition is only one element necessary for an effective broadband market . . . However, we recognize that competition is not the only issue for broadband customers and are working to address other important issues relating to broadband separately' (Ofcom, 2006).
8. In some countries, like New Zealand, it was the uncertainties concerning whether mandatory LLU would have been adopted by the regulatory authorities that left this country behind its major trading partners in terms of BB infrastructure.
9. Aron and Crandall (2008) have estimated the betas (systematic risk) of major incumbent local exchange carrier (ILECs) and found that the most recent two-year betas (2006–08) are higher than the five-year betas (2003–08) except for France Telecom.
10. These statistics have been the subject of heated debate lately because they fail apparently to measure accurately the BB infrastructure and by doing so they result in a biased ranking of the countries concerned.
11. The static–dynamic efficiency is at the forefront of the new European regulatory framework for electronic communications services. It is stated explicitly there that regulation should 'promote competition' and 'encourage efficient investment in infrastructures and to promote innovation' (European Parliament and Council of the European Union (2002), p. 38.
12. Some economists (Wallsten, 2006) do not see it as a failure of competition but, on the contrary, as a failure of asymmetrical regulation which favors new entrants at the expense of incumbents.
13. Bitstream access is the most basic form of access to the incumbent's network. Under this type of access, entrants cannot modify any of the technical characteristics of the technology used by the incumbent by adding devices or other equipment, and they are de facto restricted to the function of reselling. Since new entrants have a very limited infrastructure themselves and quite a few interconnection points with the incumbent's

network, it is not surprising to see that the best strategy for new entrants is to rent at a bulk price the incumbent's services and resell them to final customers.

14. Shared access allows entrants to offer services to their customers other than voice by leasing the high frequency non-voice spectrum of copper wire from the incumbent. The latter owns and controls the entire line and offers voice services normally by using the low-frequency voice spectrum.

15. With fully unbundled access, entrants get hooked to the incumbent's capacity by gaining full control of the copper line and this allows them to offer a full gamut of services to their customers competing head to head with the incumbent.

16. The asymmetries arise because the regulator favors new entrants by fixing low-access fees while regulating the incumbent on a number of monopoly and potentially competitive segments of the market.

17. $$E[MV(t_F, t_S)] = E\left[-\frac{I_{SBC}}{(1+r)^{t_S}} + \sum_{t=t_S+1}^{t_F} \frac{CF_n}{(1+r)^t} - \frac{I_{FBC}}{(1+r)^{t_F}} + \sum_{t=t_F+1}^{\infty} \frac{CF_n}{(1+r)^t} \right].$$

This equation does not take into account any interactions among incumbents and new entrants and ignores their strategic behavior.

18. Real options have been studied extensively in the literature but this approach has still not managed to give more precise answers to the questions raised here (Smit and Trigeorgis, 2004).

19. Of course, this problem may be avoided or at least become less acute in a growing industry.

20. Obviously, the incumbent's cashflows are affected by the new entrants but its decision to invest sooner or later is not affected by entry.

21. Waverman and Dasgupta (2007), believe that: 'the incumbent's local loop infrastructure will remain an enduring bottleneck . . . even if the current copper infrastructure is replaced by the Next Generation access network (NGNs)'. This enduring bottleneck warrants an 'enduring regulation of the access network' (ibid. p. 7).

22. Functional separation has been proposed as an alternative means to foster competition and investment in BB technologies. There are considerable doubts concerning the effectiveness of functional separation to bring the desired dynamic efficiencies in the industry (Waverman and Dasgupta, 2007). Public–private partnerships (PPPs) have also been suggested as a means to increase investment in BB infrastructure.

23. This is the case for many European countries. For instance, Greece relies exclusively on DSLs to provide Internet and other communications services.

24. There is ample literature identifying various factors in explaining country performance in BB penetration. For a review of the literature see Gentzoglanis and Aravantinos (2008).

25. In 2003, the FCC (2003, 2005) shifted focus and phased out the obligation for unbundling because one of the incumbent regional telecommunications carriers won a legal case for refusal to unbundle.

26. The first group of countries included those with high rates of bitstream and resale services and low investments in LLU and proprietary infrastructure. The second group included countries with high levels of LLU and shared access but low levels of proprietary infrastructures. The third group included countries with high levels of proprietary infrastructures.

REFERENCES

Aron, D.J. and R.W. Crandall (2008). Investment in next generation networks and wholesale telecommunications regulation. White Paper, 15 September.

Bourreau, M. and P. Dogan (2006). 'Build or Buy' Strategy in the local loop. *American Economic Review*, **96**(2), 72–6, May.

Boyle, G., B. Howell and W. Zhang (2008). Catching up in broadband regressions: does local loop unbundling really lead to material increases in OECD Broadband Uptake? NZ Institute for the Study of Competition and Regulation, July.

Breschi, S., F. Malerba and L. Orsenigo (2000). Technological regimes and Schumpeterian patterns of innovation. *Economic Journal*, 110, 388–410.

Cava-Ferreruela, I. and A. Alabau-Munoz (2006). Broadband policy assessment: a cross national empirical analysis. *Telecommunications Policy*, 30, 445–63.

Cave, M. (2006). Encouraging infrastructure competition via the ladder of investment. *Telecommunications Policy*, 30(3–4), 223–37.

Cave, M. (2007). The regulation of access in telecommunications: a European perspective. Working Paper, Warwick Business School, University of Warwick, UK.

Cave, M. and I. Vogelsang (2003). How access pricing and entry interact. *Telecommunications Policy*, 27(10–11), 717–27.

Cisco (2008). French 'triple-play' service provider deploys fiber to the home. Customer case study. http://www.cisco.com/en/US/solutions/collateral/ns341/ns524/ns562/ns577/case_study_C36-454892_ns577_Networking_Solutions_Case_Study.html.

CRTC Communications (2009). *Monitoring Report*, Ottawa, http://www.crtc.gc.ca.

Denni, M. and H. Gruber (2005). The diffusion of broadband telecommunications: the role of competition. Paper presented at the International Communications Society conference, Ponte Verda, Spain.

Distaso, W., P. Lupi and F.M. Manenti (2006). Platform competition and broadband uptake: theory and empirical evidence from the European Union. *Information Economics and Policy*, 18(1), 87–106.

Distaso, W., P. Lupi and F.M. Manenti (2009). Static and dynamic efficiency in the European telecommunications market: the role of regulation on the incentives to invest and the ladder of investment. University of Padova.

Dobbs, I.M. (2004). Intertemporal price cap regulation under uncertainty. *Economic Journal*, 114, 421–40.

European Parliament and Council of the European Union (2002). Directive 2002/21/EC of March 7 2002 on a common regulatory framework for electronic communications networks and services (Framework Directive) (*Official Journal* (OJ) L108, 24.04.2002, pp. 33–50). Brussels.

FCC (2003). In the Matter of Review of the Section 251 Unbundling Obligations of Incumbent Local Exchange Carriers; Implementation of the Local Competition Provisions of the Telecommunications Act of 1996; Deployment of Wireline Services Offering Advanced Telecommunications Capability, CC Dockets Nos. 01-338, 96-98, 98-14, Report and Order and Order on Remand and Further Notice of Proposed Rulemaking, Federal Communications Commission, (Triennial Review Order).

FCC (2005). In the Matter of Unbundled Access to Network Elements; Review of the Section 251 Unbundling Obligations of Incumbent Local Exchange Carriers, WC Docket No. 04-313 and CC Docket No. 01-338 (Triennial Review Remand Order).

Gentzoglanis, A. and E. Aravantinos (2008). Forecast models of broadband diffusion and other information technologies. *Communications and Strategies*, Special issue (Nov), 73–98.

Gruber, H. (2007). European sector regulation and investment incentives for broadband communication networks. *EIB* Working Paper.

Guthrie, G. (2005). Regulating infrastructure: the impact on risk and investment. *Journal of Economic Literature*, **44**(4), 925–72.

Janssen, M. and E. Mendys-Kamphorts (2008). Triple play: how do we secure the benefits? *Telecommunications Policy*, **32**, 699–700.

Keiichi Hori, K. and K. Mizuno (2009). Competition schemes and investment in network infrastructure under uncertainty. *Journal of Regulatory Economics*, **35**(2), 179–200.

Kim, H.S. Hong, J. Hee Kim, G. Il Yoo and W. Ha Kim (2003). Emerging Broadband Access and IP Multimedia Architecture of KT. Technical Report, Korea Telecom 2003. Available at http://users.ece.utexas.edu/~hkim4/index_files/commag2003.pdf.

Lee, S. (2009). The triple-play bundle strategy of cable and telephone companies in the current US telecommunications market. *International Journal on Media Management*, **11**(2), 61–71.

Malerba, F. and L. Orsenigo (1996). Schumpeterian patterns of innovation are technology-specific. *Research Policy*, **25**(3), 451–78.

OECD (Organisation for Economic Co-operation and Development) (2007). Catching-up in broadband – what will it take? Communication Infrastructures and Services Policy Paper DSTI/ICCP/CISP(2007)8/FINAL, OECD, Paris. http://www.oecd.org.

OECD (2009). Organisation for Economic Co-operation and Development, Directorate for Science, Technology and Industry, OECD Broadband Portal. Available at http://www.oecd.org/document/54/0,3343,en_2649_34225_386901 02_1_1_1_1,00.html.

Ofcom (2006). Review of the wholesale broadband access markets 2006/07. 21 November. http://www.ofcom.org.uk/consult/condocs/wbamr/summary/.

Prieger, J.E. and D. Heil (2008). Is regulation a roadblock on the information highway? In I. Lee (ed.), *Handbook of Research on Telecommunications Planning and Management for Business* (pp. 15–32). Hershey, PA: IGI Global.

Smit, H.T.J. and L. Trigeorgis (2004). *Strategic Investment: Real Options and Games*, Princeton, NJ: Princeton University Press.

Sutherland, E. (2007). Unbundling local loops: global experiences. Working paper, Link Centre.

Wallsten, S. (2006). Broadband and unbundling regulations in OECD countries. Working Paper 06-16, June, AEI–Brookings JointCenter. Available at http://ssrn.com/abstract=906865.

Waverman, L. and K. Dasgupta (2007). Mandated functional separation: act in haste, repent at leisure? Unpublished paper, November.

Waverman, L., M. Meschi and M. Fuss (2005). The impact of telecoms on economic growth in developing countries. Unpublished paper.

Waverman, L., M. Meschi, B. Reillier and K. Dasgupta (2007). Access regulation and infrastructure investment in the telecommunications sector: an empirical investigation. Unpublished paper.

2. Intermodal telecommunications competition: implications for regulation of wholesale services

William E. Taylor

2.1 INTRODUCTION

The historical paradigm for regulation of wireline telecommunications services has been to regulate both wholesale and retail services of incumbents with the expectation that continued regulation of wholesale services – mandatory unbundling and provision of essential wholesale services at regulated prices – might someday encourage entry and investment in facilities that would permit deregulation of retail services. Regulation of wholesale services would remain a necessary feature of the landscape, unless competition develops for wholesale services. As it happens, intermodal competition for retail telecommunications services has turned this expectation on its head. Paradoxically, intermodal competition for retail services makes regulation of wholesale services unnecessary.

Intermodal competitors for retail services such as cable, wireless and Voice over Internet Protocol (VoIP) suppliers serve their customers without requiring use of incumbent local exchange carrier (ILEC) facilities, and retail telecommunications markets are often effectively competitive. In this case, there is no economic rationale to regulate wholesale services. The demand for wholesale services is a derived demand, derived from the demand for retail telecommunications services. And if dependent wireline competitive local exchange carriers (CLECs) can earn no supracompetitive returns in the retail market, so an ILEC monopolist of wholesale wireline services cannot increase its profits by charging a supracompetitive price for its essential wholesale services.

Where intermodal competition, by itself, is insufficient to constrain retail prices but requires competition from dependent wireline CLEC competitors, an unregulated ILEC might be able to exercise market power over wholesale services. In this case, the wholesale service might be considered 'essential' in the sense that competitors require it to compete in the downstream market and there is some likelihood of ILEC anticompetitive conduct in the retail market. Even in this case, however, *ex ante* regulation in the wholesale market would likely be self-defeating. First, without a market trial, we would never know what market-based wholesale prices might be and what competition would emerge at those rates and become sustainable in the retail markets. Second, the mandatory unbundling of ILEC facilities at regulated rates would reduce the incentive of retail wireline competitors to invest in their own network infrastructures and compete on an end-to-end basis. At the same time, the requirement that ILEC facilities be shared with competitors reduces the ILEC's incentives to introduce and roll out that infrastructure, particularly for services associated with investment that will be – but is not currently – sunk. Third, price regulation at multiple stages of production – wholesale and retail – is particularly problematic, inviting inefficient arbitrage and unintended substitution.

The institutional and market setting of this chapter is the US. However, the ubiquity of technology – particularly the growth of intermodal alternatives to traditional wireline telephony – means that these conclusions are applicable more generally.

2.2   BACKGROUND

Stemming from the assumption that parts of the telephone network are natural monopolies or essential facilities, regulation of wholesale services can make competition possible for retail services. Terms and conditions and prices of wholesale services are pervasively regulated, as there are not generally multiple suppliers of these facilities. To be clear, retail services are the familiar telecommunications services that residence and business customers buy: access to the network, usage, and 'vertical' services such as call answering or automatic call forwarding. Wholesale services are sold by one carrier to another, generally to permit the purchaser to supply telecom services to retail customers. There are four kinds of players to keep track of in these markets:

1.   ILECs, which supply wholesale and retail services using traditional wireline technologies. Examples include Verizon, AT&T and Qwest in the US.

Competitive local exchange carriers (CLECs), which use traditional wireline technologies[1] and come in two relevant flavors:

2. dependent CLECs, which use ILEC facilities to serve their customers; and
3. facilities-based CLECs, which build or buy (from non-ILEC carriers) their own facilities.

Obviously, many CLECs serve customers with their own facilities where such investment is profitable and with ILEC facilities where it is not. Finally,

4. Intermodal carriers that do not use ordinary wireline technologies to reach their customers. Examples include cable companies (Comcast, Cox), wireless carriers (Verizon Wireless, Sprint) and VoIP carriers (Vonage).[2]

2.2.1 Historical Retail and Wholesale Regulation

Privatization of telecommunications companies since the mid-twentieth century has generated a wealth of experience in regulating and deregulating retail telecom markets in the US, the UK, the EU and elsewhere. Pervasive cost-of-service regulation has evolved through price regulation to deregulation or pricing flexibility regimes as competitive circumstances in the retail markets warranted. In this process, there has been general agreement that absence of market power in the retail market is the relevant trigger for reduced regulation.

For wholesale regulation, the history is shorter and the guiding principles ambiguous. US policy since the breakup of the Bell System in 1984 assumed a single vertically integrated ILEC network in each geographic market with dependent competitors that required interconnection and access to customers. This model has produced two radically different regulatory regimes. For long-distance carrier access services, regulation by the Federal Communications Commission (FCC) (and state regulatory authorities) has evolved along traditional patterns: cost-of-service regulation replaced by price caps with some very limited pricing flexibility for interstate special access services. In contrast, for access to the local network to provide local services, regulation under the Telecommunications Act of 1996 has followed a different path. ILEC provision of unbundled network elements for local service is required wherever CLECs would be impaired without them, and wholesale prices are set at the long-run incremental cost of a hypothetical efficient supplier of network elements.

For both paradigms, with the exception of pricing flexibility for special access services, there has been little consideration of deregulating prices or mandatory provision of these wholesale services.

The purpose of this wholesale regulation was not simply to control market power but also to induce self-sustaining competitive entry into the retail telecommunications markets.[3] In theory, new entrants would evolve from dependent to facilities-based competitors by investing in their own facilities as their customer bases grew.[4] If such an evolution took place, then CLECs would no longer be 'impaired' without access to ILEC unbundled elements at regulated prices.[5] Thus, mandatory supply and regulation of ILEC wholesale services would be expected to diminish as the ability of CLECs to compete without using the ILEC's network elements increased.

Similar models apply outside the US. In many countries,[6] traditional monopoly wireline ILECs are required to supply wholesale telecommunications services to their competitors, and terms and conditions (including mandatory access and pricing) for such wholesale services are pervasively regulated.[7] Moreover, in determining whether regulation of wholesale services is necessary, many regulators have defined separate wholesale and retail markets and have examined market power separately in the wholesale market.[8]

Two surprises occurred. First, the 'ladder of investment' theory failed to predict market outcomes. Few wireline CLECs evolved into full facilities-based carriers, particularly for mass-market residential and small business customers. CLECs grew quickly as long as the ILECs' unbundled network element platform (UNE-P) wholesale service was available at TELRIC-based prices.[9] However, once these services were no longer available, CLEC growth stopped, and by June 2008, the majority of facilities-based CLEC access lines belonged to cable companies – which never used ILEC facilities – rather than to the traditional wireline CLECs. Second, retail markets became more competitive due to intermodal carriers that made no use of ILEC facilities. The ILEC and CLEC combined share of local access connections fell steadily as demand shifted to intermodal suppliers such as wireless, cable and VoIP suppliers.

2.2.2 Future Regulation

In light of these surprises, it is timely to ask whether continued regulation of wholesale services is necessary and what standards we should use to assess the wisdom of deregulation. At issue are the circumstances under which *ex ante* regulation of unbundling requirements and pricing of wholesale services is preferable to *ex post* regulation, in which the ILEC would be under no explicit obligation to provide wholesale services or price them at some regulated rate, leaving regulatory or competition authorities to respond *ex post* to specific complaints of anticompetitive behavior.

Regulation entails costs and benefits. The potential economic benefits of *ex ante* economic regulation of wholesale and retail services are fundamentally different in economic theory. As discussed below, economic welfare effects are measured in the markets for final goods – the retail markets – and it is only the downstream effect of regulation at the (upstream) wholesale level that matters. As for costs, the cost consequences of regulating wholesale markets are more complex.

First, regulation at successive stages of production (that is, of wholesale and retail markets) can be inconsistent and a source of important unintended consequences in competitive markets. A modest example from the US: the FCC required that wholesale unbundled loops be priced at a version of incremental cost, which was disaggregated into at least three different geographic areas to reflect cost differences among urban and rural exchanges. On the retail side, many states required that the price of basic exchange service reflect 'value-of-service', so that retail prices were higher in urban exchanges (where there were more customers in the local calling area) than in rural exchanges. The resulting regulation of both wholesale and retail prices led to inefficiently large margins in urban areas and inefficiently small (and negative) margins in rural areas.[10]

Second, regulation at the wholesale level inevitably induces distortions in the retail markets because some network platforms are regulated and others are not.[11] Such regulatory disparity is not a simple squabble over rents; it does not merely transfer wealth among competing carriers. Rather, because telecom markets are characterized by rapid technological change and competing platforms that are subject to lock-in or path-dependence, regulatory disparities can have large and irreversible welfare effects on consumers. Platforms currently in play include fiber to the home, fiber to the node, coaxial cable, various incompatible mobile wireless platforms and various fixed wireless alternatives. Regulatory distortions here are inevitable, if only because the regulators' jurisdiction differs across the wireline, cable, wireless and broadband platforms. And distorting a competitive market outcome here could drive the market to adopt an inefficient platform or technology which could then persist for years.

2.3 INTERMODAL COMPETITION

Historically, different networks were designed and deployed to carry different types of traffic. The wireline public switched telephone network and mobile telephone networks were optimized to transport basic voice communications, cable networks were engineered to transport video, and the Internet was designed to transport packet-based data traffic. Today, these technologies have converged so that providers can offer multiple types of services over a single network. With convergence, the same services are provided (and marketed) over different network platforms (for example, traditional cable systems, wireline telephone networks and wireless mobile networks).

Three fundamental factors have driven this convergence: (1) technological change (such as the advent of two-way, digital, broadband networks and Internet Protocol – IP – technology) which has allowed all kinds of wired and wireless networks to be used for any kind of service; (2) consumer demand for bundled services; and (3) competition among providers seeking gains from improved efficiency (economies of scale and scope), and the promise of increased revenues and lower churn rates. Because convergence enables different types of platforms to provide increasingly similar bundles of services, traditional wireline carriers must now compete with multiple platforms, including Internet and broadband services, cable companies that have made substantial investments in their networks to provide video, data and voice services, wireless services providers, VoIP providers and other carriers using emerging technologies. These industry developments have resulted in dramatic line losses to wireline local exchange carriers and have made retail markets for telecommunications services effectively competitive.

The growth of intermodal competition has been extensively documented in many forums.[12] Here, I give some simple examples from US data. Figure 2.1 shows dramatically the effect of intermodal competition on local exchange access for US mass market telecommunications connections. In June 2000, wireline access exceeded intermodal (wireless and broadband) access by about 50 million lines; by June 2007, there were approximately 200 million fewer wireline (ILEC plus CLEC) access lines than intermodal connections.

Cable telephony is widely available across the US. By 2005, essentially 100 percent of US households had cable service[13] and cable penetration – by some measures – was 58 percent of homes passed.[14] Comcast, for example, reports that of its homes passed, broadband services were deployed to 99 percent and telephone was enabled to 92 percent. Current penetration of available homes was 30 percent and 13 percent for broadband and telephony respectively.[15]

Mobile wireless services are available throughout the US. Measured on a census block basis, approximately 96 percent of the US population has access to three or more wireless providers and more than half (57 percent) have access to five or more.[16] Wireless subscribership approximately doubled between December 2001 and December 2007, reaching about 250 million subscribers at the end of 2007. According to the FCC, about 187

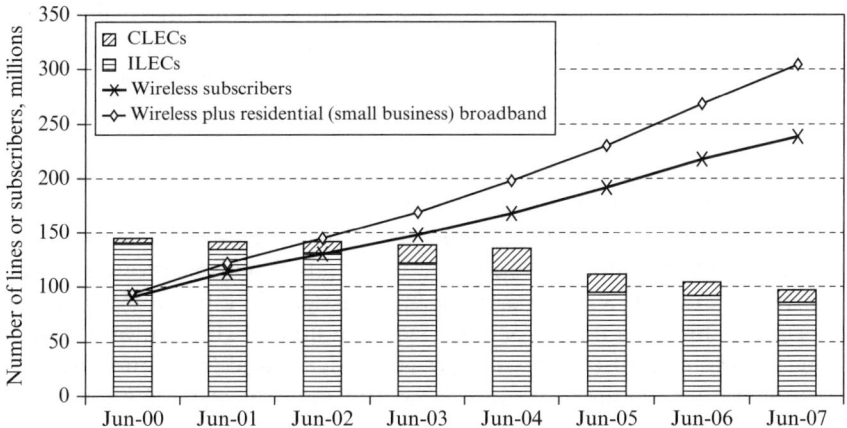

Source: US FCC (2007), US FCC (2009), Tables 1, 14, US FCC (2008b), Table 15.

Figure 2.1 Intermodal competition and number of lines or subscribers

million of these subscriptions were residential,[17] so that residential wireless subscriptions were approximately double the number of wireline subscriptions in December 2007.[18] That wireless service actually substitutes for wireline service is implied by the growth of 8.4 million wireless-only households since 2005; the latest National Health Institute Survey data track the significant growth in the percentage of adults living in wireless-only households, a figure that reaches 17.5 percent of total households as of June 2008 and has increased consistently and significantly over time across all age groups.[19]

Finally, for VoIP availability, high-speed Internet service is now available throughout the US. As of June 2007, about 96 percent of US Zip Codes have three or more high-speed providers with lines in service, and 89 percent of all Zip Codes have four or more such providers.[20] Digital subscriber line (DSL) and cable broadband are both widespread: high-speed DSL connections were available to 82 percent of the US households where ILECs can provide local telephone service, while high-speed cable modem service was available to 96 percent of the households where cable system operators can provide cable TV service.[21] Broadband subscription rates have grown as well, reaching approximately 55 percent of US households in early 2008.[22]

A significant proportion of US households use their broadband connections to make VoIP telephone calls. According to a HarrisInteractive survey conducted between October 1997 and January 2008, about 15 percent of US adults used VoIP to make telephone calls. The largest US VoIP supplier was Vonage, which reported about 2.6 million access lines in August 2008,[23] followed by carriers such as Skype, MSN, Yahoo! and Google. According to InStat, total subscribership for these VoIP services amounted to about 5 million in 2008.[24]

In sum, the assumption that ILEC network elements and services are essential facilities for the supply of retail telecommunications services is unlikely to hold. Retail customers can buy services from wireline and intermodal providers, and facilities that a potential competitor requires to be a wireline CLEC and which must be purchased from on ILEC are nonetheless not required for some competitors – using some platforms – to provide competitive retail telecommunications services.

2.4   WHOLESALE PRICE REGULATION: THEORY

In economic theory, assessing the need for *ex ante* price regulation of a wholesale market is a very different exercise from its (familiar) counterpart for retail markets.[25] It is only to the extent that a market failure in the wholesale market impacts prices in the downstream retail markets that there would be any effect on retail customers and thus any economic efficiency consequences to ameliorate with a wholesale regulatory intervention. Said another way, demand for wholesale services is a derived demand, like the demand for labor services or raw materials. Thus, if the exercise of market power or the abuse of market dominance for a wholesale service has no effect on downstream retail market prices, customers would not be harmed, and there would be no likely circumstances under which *ex ante* economic regulation would lead to an increase in social welfare or an efficiency justification for economic regulation of the wholesale service.

Conventional regulatory practice ignores this fact when it assesses the need for wholesale regulation by examining available substitutes for the wholesale service. In practice, the attempted exercise of market power for a wholesale service frequently can have no effect on retail market prices. Consider some simple cases where the wholesale service in question is an essential facility and the ILEC is the only supplier.

2.4.1   Dependent Competition

In Figure 2.2, the only suppliers are wireline carriers: one ILEC and four dependent CLECs, each of which must purchase wholesale services – say

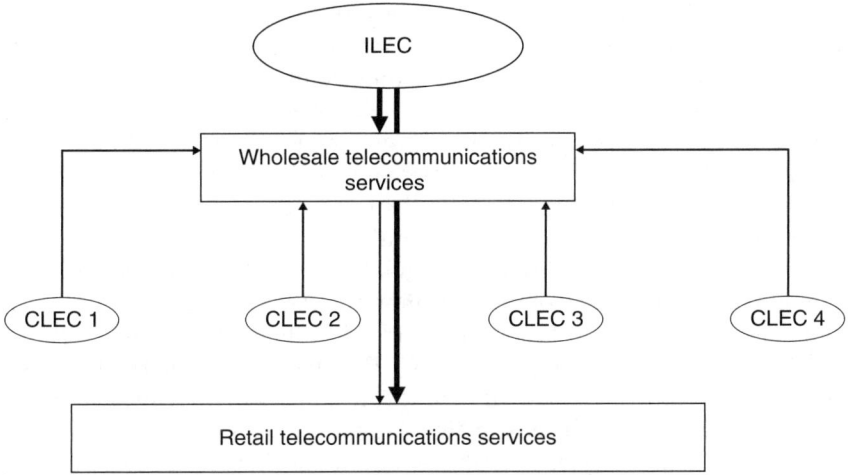

Figure 2.2   Dependent competition: monopoly ILEC and oligopoly CLEC

local loops – from the ILEC. If the CLECs were absent from the retail market, the ILEC would have a retail monopoly and, if unregulated in the retail market, would price those services at the profit-maximizing monopoly level.

Suppose first that the dependent CLECs effectively act as nothing more than additional distribution channels for the ILEC's retail service. In particular, assume that: (1) the CLECs' incremental costs for the retailing function are the same as those of the ILEC and (2) the presence of CLECs in the retail market does not increase total retail demand. If the ILEC and dependent CLECs pursued a cooperative strategy, their combined profits would be the same as the monopoly profits, and the joint-profit-maximizing price would be the same as the monopoly price.

In the absence of collusion or wholesale regulation, the ILEC's profit-maximizing wholesale price is the avoided-cost (or efficient component) price:

$$P_W = P_R - [IC_R - IC_W] \tag{2.1}$$

where (P_W, P_R) and (IC_R, IC_W) represent the wholesale and retail prices and incremental costs, respectively.[26] To see this formally, it can be shown that the ILEC's profit-maximizing wholesale price is a Ramsey-price markup of the avoided-cost wholesale price:[27]

$$P^*_W = [\eta_w / (1 + \eta_w)] \times [P_R - (IC_R - IC_W)] \tag{2.2}$$

where η_w is the wholesale price elasticity of demand. At P_W, wholesale demand is infinitely elastic (so that $P^*_W = P_W$) because no CLEC could pay a wholesale price higher than P_W and compete profitably against the retail price P_R. Intuitively, P_W maximizes ILEC profits in a non-cooperative setting, because P_W permits the ILEC to extract all of the joint monopoly profits for itself. Perhaps more intuitively, P_W and P_R maximize ILEC profits because at those prices, the ILEC is indifferent between serving customers through its retail and wholesale channels.

Unfortunately, in the absence of regulation in the wholesale market, the unregulated retail price would be the monopoly price in this case, and consumers would realize no benefit from the additional retail competition.

Alternatively, if we suppose that CLECs have lower marketing costs or that they introduce new and innovative services that expand retail demand, the joint-profit-maximizing profits for a collusion of the ILEC and CLECs can be higher than the monopoly profits of the ILEC alone.[28] And in the absence of wholesale regulation, the ILEC can extract some of that additional profit by pricing the wholesale service at P^*_W, which would exceed P_W because CLECs could compete profitably against P_R at wholesale prices greater than P_W.[29]

In either case, regulating ILEC wholesale services could reduce the retail price below the monopoly level and increase consumer welfare. For example, if the ILEC's wholesale service were priced at IC_W, there would be five independent, equally efficient competitors in the retail market instead of one. What was previously a monopoly would become a more competitive five-firm oligopoly. In the case of Cournot competition, wholesale regulation would cause the Lerner index of market power for retail services to fall by a factor of 5, and the retail price would approach the competitive market price.[30] In theory, regulating the wholesale price makes sense in this setting because every firm in the retail market (except the ILEC) must pay this price, and in the absence of regulation, the ILEC can use the wholesale price to force the retail price to the monopoly level and extract the full amount of monopoly profits from its retail and wholesale customers.

2.4.2   Intermodal Competition

In Figure 2.3, the ILEC–CLEC nexus is identical but we assume the retail market is also supplied by firms that make no use of the ILEC's wholesale services. As before, dependent CLECs must purchase network access services from the ILEC, but wireless, cable and VoIP suppliers use their own facilities (or, in the case of VoIP, a broadband connection supplied by the end-user). Suppose the market for retail telecommunications services

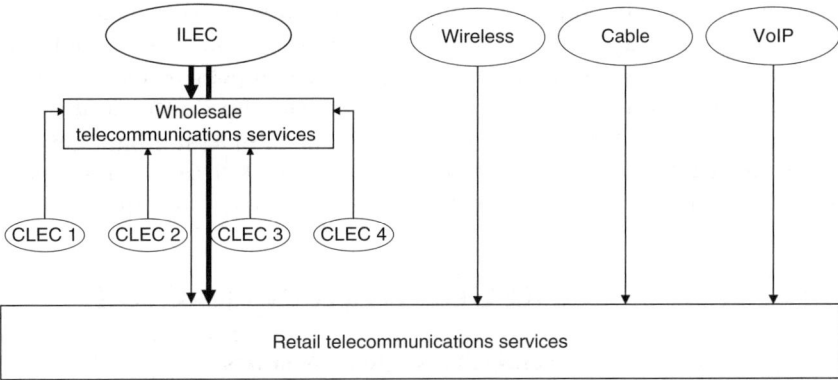

*Figure 2.3   Intermodal competition: ILEC and dependent and independent
          CLEC*

is effectively competitive, due to the presence of intermodal competitors.
Then, even though the ILEC is, by assumption, the sole supplier of wireline network access, it cannot generate any monopoly profits from those
services because its CLEC customers could not profitably pay an excessive
wholesale price and compete against the competitive retail market price.
The demand for wholesale services is derived from the demand for retail
services, and in this case (unlike Figure 2.2), the retail price is unaffected
by the price of the wholesale service.

Intuitively, dependent CLECs earn normal profits at competitive market
wholesale prices. If the ILEC increases the wholesale price, those CLECs
cannot profitably continue to compete with cable, wireless and VoIP retail
service, which reduces the demand for the ILEC wholesale service. By
assumption, no monopoly profits are earned in the retail market; hence,
a wholesale monopolist cannot extract monopoly profits from a subset of
the providers of the retail service.

Formally, this situation is illustrated in the expression (2.2) for the
profit-maximizing wholesale price and is identical to that in Figure 2.2,
except for the level of the retail price. Here, P_R is not controlled by the
ILEC and is equal to the competitive market price. As before, at P_W, the
demand for wholesale services is perfectly elastic because no CLEC could
compete at a higher price. Thus the profit-maximizing wholesale price for
the ILEC is P_W, the competitive market retail price less the avoided-cost
discount. Again, at these prices, the ILEC is indifferent between serving
customers through its retail or wholesale channels.

Thus when the retail market is effectively competitive, *ex ante* price
regulation of the wholesale service is unnecessary because the ILEC is

unable to extract monopoly profits from its (essential) wholesale service. In fact, its unregulated, profit-maximizing wholesale price would be an incremental-cost-based discount from the competitive retail market service price. In addition, the ILEC would have no direct incentive to initiate a price squeeze.[31] At its unregulated, profit-maximizing wholesale price, the regulated retail price would be consistent with (would just equal) the relevant price floor that defines an anticompetitive price squeeze.

2.5   WHOLESALE REGULATION IN PRACTICE

2.5.1   The Level of Competition in the Retail Markets

It is useful to consider three cases.

Case 1: Retail market is effectively competitive
First, suppose the downstream market is effectively competitive due to the presence of facilities-based competitors that do not use or require the ILEC's facilities. This corresponds to Figure 2.3. ILEC wholesale prices are constrained by competitively determined retail prices, whether or not there is a separate market for wholesale facilities. Even though the ILEC is assumed to be the sole supplier of wholesale wireline services, there is no reason to subject wholesale services to *ex ante* economic regulation.

In practice, where current (regulated) wholesale prices are not necessarily set at competitive market levels, deregulation in Case 1 could lead to an increase in wholesale prices. But such a price increase would not constitute an exercise of market power, which requires a profitable (significant and non-transitory) increase in price above the competitive level.[32] It would also not be necessarily anticompetitive: pricing that drives efficient competitors out of an effectively competitive market does not reduce social welfare and is not anticompetitive in economics unless it leads to the acquisition or retention of significant market power.

Note in this case that the criteria for deregulation may differ in practice between wholesale and retail services. In Figure 2.3, an ILEC with no wholesale competitors may have no market power in the wholesale market. 'Defining the relevant market for a wholesale facility' may not involve any substitute wholesale services. In a recent case, the Canadian regulator overlooked this point:

37. With regard to future applications to consider the essentiality of a non-mandated service, the definition will read as follows: To be essential, a facility, function, or service must satisfy all of the following conditions: (i) The facility

is required as an input by competitors to provide telecommunications services in a relevant downstream market; (ii) The facility is controlled by a firm that possesses upstream market power such that denying access to the facility would likely result in a substantial lessening or prevention of competition in the relevant downstream market; and (iii) It is not practical or feasible for competitors to duplicate the functionality of the facility.

43. The Commission considers that determining duplicability is comparable to defining the relevant market for a wholesale facility in the sense that both exercises require the identification of potential substitutes, either through existing or potential alternatives.[33]

Case 2: Retail market is conditionally competitive

The downstream market is effectively competitive but that competition is due, at least in part, to dependent competitors (that is, to dependent CLECs that use – or require – ILEC facilities). To distinguish Case 2 from Case 1, the presence of dependent competitors is assumed to be necessary for competition, that is, that without these carriers, the ILEC would otherwise possess significant market power in the retail market. Here the ILEC and the dependent CLECs combined effectively possess market power in the downstream market, in the sense that a hypothetical coordinated increase in retail prices above competitive market levels could be profitably sustained. In the absence of price regulation in either market, the ILEC could then profitably hold the wholesale price above competitive levels and capture all of the potential supracompetitive profits.

Even though there is no apparent retail market failure in Case 2, the fact that the necessary competition comes from competitors that are (by assumption) dependent on the wholesale service of the ILEC means that the ILEC wholesale service technically fits the definition of an essential facility. However, there is nothing in economic theory or in current experience in telecommunications that suggests this outcome is anything but a temporary market disequilibrium that is actively evolving towards full Case 1 facilities-based competition.

Ironically, a major deterrent in this process would be the classification of wholesale services as essential and *ex ante* regulation of their price. Such regulation would deter investment in facilities by the dependent CLECs and would place intermodal competitors at an artificial competitive disadvantage in the retail market. In addition, requiring ILEC to give competitors access to their network elements at incremental-cost-based rates distorts the ILEC's incentive to invest in new facilities and technologies: incremental costs do not account for either the risk and uncertainty in investment in new technology or the option value to the CLEC of using rather than owning facilities. The net effect would harm consumers in two ways. Wholesale regulation would become a self-fulfilling prophesy in which competition would never emerge to replace it. And investment by CLECs and ILECs in this rapidly-developing high-tech industry would be artificially diminished.

As a result, *ex post* regulation – market observation tempered by competition law principles – of wholesale services would better serve telecommunications customers than *ex ante* price regulation of wholesale services.

Case 3: Retail markets are not competitive
Some downstream markets are not effectively competitive, for example, rural or high-cost areas where the ILEC is the only current supplier of retail services. However, in these markets, ILEC retail services generally remain subject to economic regulation, so consumers would derive little direct benefit from *ex ante* wholesale regulation. In particular, regulation of the wholesale price would be – at best – redundant. An ILEC could not increase its profits by charging wholesale prices in excess of:

$$P_W = P_R - [IC_R - IC_W]$$

where P_R is the regulated retail price because no equally efficient CLEC could pay that price and compete profitably.

Indeed, welfare gains from additional retail competition in these markets may be problematic. It may be the case that entry in retail markets would be possible in these areas if there were profitable opportunities to exploit. For example, regulation may have capped prices in rural areas below a competitive market level, and were the ILEC to price at or above the competitive level in these areas, entry would more likely occur. In general, telecommunications networks are deployed incrementally over time, targeting the most profitable areas – areas having a high concentration of likely customers – first because higher teledensity areas have more customers and lower costs of serving those customers. Rural and high-cost areas may be targeted in later stages so that the current lack of competitors may not necessarily reflect a natural monopoly condition for which *ex ante* regulation might be appropriate.

2.5.2 Summary

In current telecommunications markets, retail services are provided over at least three distinct platforms – traditional narrowband wireline, wireless and cable – and VoIP suppliers provide service over unaffiliated broadband access facilities. Where retail markets are competitive and consumers can choose among multiple platforms, there is no market failure for regulation to redress. Thus in Case 1, *ex ante* regulation of wholesale services on one of those platforms – the wireline platform – cannot improve consumer welfare and the inherently asymmetric regulation of one platform relative to others would likely reduce consumer welfare. For Case 3, regulation of some retail services remains in place and, as discussed, there is no need for additional price regulation of wholesale services. Efficient regulation of wholesale services in Case 2 depends critically upon the facts in the market. If there is uncertainty as to whether the ILEC wholesale service is essential, a reasonable policy would be to undertake a market-based experiment: permit but not require the unbundling of the wholesale service at regulated prices and allow the market to correct itself from the distortions caused by current overinclusive wholesale regulation. Observe the course of competition in the retail market to determine *ex post* whether the wholesale service is truly essential. Given the ubiquity of facilities-based competition (cable, wireless and VoIP providers) in telecommunications markets, it would be surprising to find – even in markets dominated today by an ILEC and dependent CLECs – that the retail market was inherently dependent on access to the ILEC's wireline facilities.

That experiment has been undertaken in the US, where unbundling requirements on ILECs have been relaxed and, in particular, the combined loop and switch (UNE-P) mandatory offering has been replaced with negotiated commercial contracts. All indications suggest that local exchange competition remains robust, as ILEC customer counts have fallen, while customer counts of cable companies, wireless carriers and VoIP suppliers have grown.

2.5.3   Ex post Regulation of Anticompetitive Conduct

Generally speaking, throughout the economy, anticompetitive conduct and abuse of dominant position are regulated and controlled on an *ex post* rather than *ex ante* basis through the enforcement of competition laws. Advantages of *ex post* regulation of anticompetitive conduct with respect to essential facilities include the reduced risk of economic distortions (for example, changing the basis of negotiation for use of facilities) and the efficiency gain from intruding only in cases where anticompetitive behavior actually occurs. Thus, across many different jurisdictions, *ex post* regulation of provision and pricing of essential facilities is the norm for other industries, and the network externalities and network effects that distinguish telecommunications from many other industries do not justify industry-specific *ex ante* economic regulation.

For telecommunications services, what is the potential effect in retail telecommunications markets from the anticompetitive denial or pricing of access to essential wholesale services that *ex post* regulation would be expected to control? In economics, such behavior is termed a price or margin squeeze, where a vertically integrated firm (for example, an ILEC) prices its wholesale service at a level that prevents an otherwise efficient dependent competitor from competing against the price of the ILEC's retail service.[34] A firm that sets a wholesale price for an essential facility that (together with the retail market price) entails an anticompetitive price squeeze necessarily sacrifices profits, at least in the short run because it loses profits on every retail service it sells. A component of the cost to an ILEC of supplying the retail service is its opportunity cost – the contribution (wholesale price less wholesale incremental cost) from the wholesale service – that is forgone when the ILEC serves the retail customer.

Currently in telecommunications markets, there is little likelihood for such behavior to happen because the conditions necessary for a price squeeze to be profitable do not generally hold. Such a sacrifice of short-run profits is unlikely to be profitable in the long run in telecommunications markets because of the difficulty of recoupment – of eliminating competitors and then raising prices above a competitive market level. Characteristics of telecommunications markets that make recoupment difficult include: (1) the absence of retail barriers to entry; (2) the presence of competitors (for example, cable and wireless companies) that do not depend on the putative essential facility, that have sunk costs and that provide retail services in markets other than retail telecommunications; (3) the pace of technological change; and (4) the rates of growth of markets for retail telecommunications services. Like predatory pricing, denial or anticompetitive pricing of an essential facility is a logical possibility in telecommunications markets but is sufficiently unlikely that it makes little sense to regulate access and pricing of essential facilities on an *ex ante* basis.

Finally, the determination that an ILEC facility is essential and must be provided on an unbundled basis ought not to be an *ex ante* exercise. At market-determined prices, some ILEC services may generate substitutes that are not viable at regulated prices or prices determined by mandatory arbitration. A less intrusive and more market-consistent approach would permit market forces to set wholesale prices and would classify a service or facility as essential only on an *ex post* basis after complaint and a finding by the regulatory or competition authority. Only after a finding of essentiality would the commercially negotiated prices (and other terms and conditions) for those facilities determined *ex post* to be essential be required. And only in the case that the parties could not agree to prices, terms and conditions would regulatory or competition authorities' step in on an *ex post* basis to resolve the differences.[35]

2.6 CONCLUSIONS

In many telecommunications markets, *ex ante* economic regulation of both wholesale and retail services is generally unwarranted, mutually inconsistent and rife with inefficient, unintended consequences. Where retail markets are effectively competitive due to intermodal competition, regulation of wholesale services serves no function and is unnecessary even though there may be no substitutes for ILEC wholesale services. Even where retail markets are not effectively competitive, regulation of wholesale services may not be warranted: substitutes for the ILEC's wholesale facilities may be forthcoming at competitive market rather than regulated prices. In both cases, the fact that wholesale regulation cannot be consistent across intermodal platforms is a well-known recipe for inefficient outcomes.[36]

Ex post regulation of wholesale services through application of antitrust and competition law can be applied at parity across these different platforms and would avoid many of these problems of *ex ante* regulation.

NOTES

1. The term 'CLEC' is often used to refer to non-incumbent local exchange carriers generically, not necessarily restricted to carriers using traditional wireline technology. Thus, in many states in the US, cable companies are considered CLECs for some regulatory purposes. For simplicity, I use the term 'CLEC' to mean 'wireline CLEC' here, and use 'intermodal carrier' to include non-wireline competitors that may or may not be CLECs.
2. Note that VoIP suppliers, unlike cable companies and wireless carriers, do not generally provide their own network infrastructure but rather supply their services as applications carried over broadband facilities supplied by another carrier.
3. See, for example, US FCC 1996.
4. Propounded by Martin Cave (2006), this 'ladder of investment' theory has been widely discussed as an underpinning to regulatory policy at the EU and elsewhere.
5. In the US, Section 251(3)(2)(B) of the Telecommunications Act of 1996 requires ILECs to unbundle network elements and provide them at cost-based prices whenever competitors would be 'impaired' without access to them.
6. Countries requiring some form of local loop unbundling include the US, Canada, Australia, New Zealand, and the European Union member countries, Hong Kong and South Africa.
7. See, for example, EU Access (European Parliament and Council of the EU, 2002a).
8. See, for example, in Canada, Telecom Decision CRTC 2008-17 (CRTC, 2008); or in Europe, EU SMP Guidelines (European Parliament and the Council of the EU, 2002b).
9. The unbundled network element platform (UNE-P) was a combination of unbundled local loop and switching which was essentially a form of very cheap resale. Total element long-run incremental cost (TELRIC) is the cost standard in the US for unbundled network elements. Depending upon the regulators' interpretation, this standard resembles the economist's total service long-run incremental cost.

10. See for example, J.R. Abel and V. Watkind-Davis (2000), pp. 14–15.
11. For example, cheap wireline access (UNE-P in the US) gave wireline CLECs an uneconomic advantage to compete successfully for a period against cable and wireless suppliers.
12. See, for example, US Department of Justice (2008).
13. US FCC (2006).
14. NCTA (n.d.).
15. Comcast Press Release – (2008).
16. US FCC Twelfth CMRS Report (US FCC, 2008a).
17. '25 percent of wireless users were business customers, with the remaining 75 percent being ordinary consumers.' (US FCC, 2008a), footnote 633.
18. There were approximately 94 million residential ILEC plus CLEC switched access lines in December 2007. US FCC, 2009 Local Competition Report, Table 2.
19. All of the growth in wireless-only households may not reflect substitution because there has also been an increase in the number of households, some of which may never have had a wireline phone. However, the relevant numbers are small. About 1.5 million households were added on average over the last five years according to US Census data. Assuming that 21 percent of these households (that is, the average percentage of 'wireless-only' 18-29-year-old adults) decided not to install a wireline phone implies that only about 320 000 new households per year have never purchased wireline access. Over that same period, the number of wireless-only households has grown by 4.2 million per year. Thus, the number of households that substituted wireless for wireline service has averaged about 3.8 million per year or 92 percent of the total gain in the wireless-only count.
20. US FCC (2008b).
21. Ibid., Table 14.
22. Ibid., Table 14.
23. Press release at http://www.marketwatch.com/news/story/vonage-cuts-quarterly-loss-only/story.aspx?guid=%7BFE7B3F24%2DFAC6%2D4300%2D8867%2D76728F3DB6B3%7D&dateid=39667.3494664699-935212285.
24. http://www.instat.com/promos/08/dl/IN0804301WWI_as78sa.PDF .
25. For a retail market, economists would generally advise forbearance when no carrier possessed market power as indicated by such structural market characteristics as the presence of alternative suppliers and low barriers to entry and expansion. See for example, the EU SMP Guidelines (European Parliament and Council of the EU, 2002b, p. 5).
26. Note that IC_R is the incremental cost of the entire retail service. $IC_R - IC_W$ can be thought of as the ILEC's incremental cost of the retailing function, assuming that the ILEC's cost of using the wholesale service itself is the same as the cost of providing that service to CLECs.
27. See Taylor (2008), p. 18.
28. Intuitively, CLEC demand is higher at the monopoly price and CLEC incremental costs are lower.
29. This is an interesting special case because the ILEC finds it profitable to set a wholesale price above its avoided-cost discount. In economics, such wholesale and retail prices constitute a vertical price squeeze, and an open question in antitrust enforcement and competition policy is whether such prices necessarily entail a reduction in the ILEC's profits.
30. See, for example, Carlton and Perloff (2005), pp. 194–7.
31. There may still be an indirect incentive to price squeeze, analogous to predatory pricing, in which current profits are sacrificed in the expectation of driving rivals from the market and recouping lost profits in the future. While a logical possibility, the pre-conditions for such a strategy to be successful do not often hold in telecommunications markets. In addition, as above, if CLECs were more efficient than the ILEC in providing the retail function, the profit-maximizing wholesale price could exceed P_W because

CLECs could profitably compete – at least as well as the ILEC – at a wholesale price above P_W, so that wholesale demand would still have finite elasticity at P_W and P^*_W would exceed P_W.

32. As defined by the US DOJ Merger Guidelines, 'market power is the ability profitably to maintain prices above competitive levels for a significant period of time' (US Department of Justice and Federal Trade Commission, 1992, § 0.1).

33. Telecom Decision (CRTC, 2008).

34. For this discussion, denial of access is envisioned (constructively) as simply pricing access too high to permit any efficient firm to compete.

35. It does not follow that the price and other terms and conditions should be set on an *ex ante* basis and tariffed. Costs and market conditions for essential facilities can differ significantly across potential customers, so that economic efficiency would be sacrificed by an *ex ante* regulatory determination of prices and terms and conditions irrespective of market circumstances.

36. The classic example of these distortions is US regulation of surface transportation in the 1950–80 time period and the resulting inefficient mix of railroad, truck and barge infrastructure investment and, ultimately, traffic.

REFERENCES

Abel, J.R. and V. Watkins-Davis (2000). Geographic deaveraging of wholesale prices for local telephone service in the United States: some guidelines for state commissions. National Regulatory Research Institute. April.

Canadian Radio-television and Telecommunications Commission (CRTC) (2008). *Canada Telecom Decision* (CRTC, 2008-17), 3 March.

Carlton, Dennis W. and Jeffrey M. Perloff (2005). *Modern Industrial Organization*, 4th edn (pp. 194–7). Boston, MA: Pearson Education.

Cave, M. (2006). Encouraging infrastructure competition via the ladder of investment. *Telecommunications Policy*, **30**, 223–37.

Comcast (2008). Press release: third quarter 2008 results. Table 6. Retrieved from http://media.corporateir.net/media_files/irol/11/118591/Earnings_3Q08/3Q08PR.pdf.

European Parliament and the Council of the European Union (2002a). Directive 2002/19/Ec on access to, and interconnection of, electronic communications networks and associated facilities. EU Access Directive. Retrieved from http://www.ictregulationtoolkit.org/en/Pusblication.1496.html. 7 March.

European Parliament and the Council of the European Union (2002b). European Commission Guidelines on Market Analysis and the Assessment of Significant Market Power under the Community Regulatory Framework for Electronic Communications Networks and Services (2002/C 165/03) ('EU SMP Guidelines'). Retrieved from http://www.ictregulationtoolkit.org//en/Publication.2589.html.

National Cable and Telecommunications Association (NCTA) (n.d.). Industry data. Retrieved from http://www.ncta.com/Statistic/Statistic/Statistics.aspx.

Taylor, W. (2008). Intermodal telecommunications competition: implications for regulation. International Telecommunications Society. Retrieved June 2008 from http://www.canavents.com/its2008/abstracts/27.pdf.

Telecommunications Act of 1996, Pub. L. 104-104, 110 Stat. 56 (1996). Section 251(3)(2)(B).

US Department of Justice (DOJ) and Federal Trade Commission (FTC) (1992). Horizontal Merger Guidelines, April ('DOJ Merger Guidelines') § 0.1.

US Department of Justice (DOJ) (2008). Voice, video and broadband: the changing competitive landscape and its impact on consumers. November. Retrieved from http://www.usdoj.gov/atr/public/reports/239284.pdf.

US Federal Communications Commission (FCC) (1996). Implementation of the Local Competition Provisions of the Telecommunications Act of 1996, CC Docket Nos. 96-98, 95-185, First Report and Order ('Local Competition Order') 11 FCC Rcd 15499, 231–232.

US Federal Communications Commission (FCC) (2006). In the Matter of Annual Assessment of the Status of Competition in the Market for the Delivery of Video Programming (MB Docket No. 05-255). Twelfth Annual Report. Released: 3 March, Table 1.

US Federal Communications Commission (FCC) (2007). Local Telephone Competition: Status as of June 30, 2006 ('2007 Local Competition Report'), released January 2007, Tables 1, 14.

US Federal Communications Commission (FCC) (2008a). Annual Report and analysis of Competitive Market Conditions with Respect to Commercial Mobile Services, Twelfth Report ('Twelfth CMRS Report'), FCC 08-28, released 4 February, Table 1.

US Federal Communications Commission (FCC) (2008b). High-Speed Services for Internet Access: Status as of June 30, 2007 ('2008 High-Speed Internet Report') released March, Table 15.

US Federal Communications Commission (FCC) (2009). Local Telephone Competition: Status as of June 30, 2008 ('2009 Local Competition Report'), released July 2009, Tables 1, 2, 14.

3. Access regulation versus infrastructure investment: important lessons from Australia[1]

Martyn Taylor

3.1 INTRODUCTION

A key focus of increasing concern in telecommunications regulatory policy is the impact of access regulation on the long-term incentives for new infrastructure investment. This chapter analyses the theory and practice behind the regulation versus investment debate and makes a series of key policy recommendations. Importantly, many jurisdictions have not yet struck an appropriate balance, particularly in circumstances where substantial investment in next generation networks is now required.

3.2 DETERMINANTS OF TELECOMS INFRASTRUCTURE INVESTMENT

The Internet is one of the most significant innovations of the twentieth century and is profoundly shaping modern society in the twenty-first century. Governments around the world are increasingly recognizing the value of Internet access, including via positive externalities that spill over into economic growth. A common policy vision among advanced industrialized nations is ubiquitous broadband Internet access.

A key constraint to the supply of high-speed broadband access into the home in most countries is legacy wireline 'last mile' customer access network infrastructure. With the exception of new housing developments, such infrastructure in most nations typically comprises either a twisted copper pair, often many decades old and prone to quality issues; or, less commonly, a separate coaxial cable, normally as part of a hybrid fiber coaxial network initially deployed to provide cable television services.

Broadband Internet access is normally provided over such infrastructure via xDSL (digital subscriber line) and DOCSIS (Data Over Cable

Service Interface Specification) technologies, respectively. However, both technologies have inherent limitations borne from the quality of the underlying infrastructure. The policy vision of ubiquitous broadband Internet access is therefore underpinned by the roll-out of so-called next generation networks (NGNs) that upgrade legacy infrastructure by extending high-capacity fiber optic cable towards the consumer.

3.2.1   Realizing the Vision: the Economics of Infrastructure Investment

The roll-out of NGN infrastructure involves significant labor and capital sunk costs. The investment is essentially irreversible: once deployed, fiber cannot easily be redeployed. As the firm's ability to redeploy capital is limited, its potential losses from poor investment decisions are commensurately greater.

The investment also involves economies of scope and scale and network effects that make it optimal to roll out the network rapidly on a holistic rather than piecemeal basis. As a consequence, the investment is 'lumpy' and most costs are borne immediately rather than spread over the network life. In contrast, revenues are normally recovered over the substantial lifetime of the network assets, creating a significant timing mismatch between outgoing cost cashflows and incoming revenue cashflows. Financing arrangements must address this mismatch. Given such issues and the magnitude of the costs involved, NGN investment decisions are serious matters for industry participants.

Generally, a firm should only invest in telecommunications infrastructure if it can achieve aggregate revenue cashflows over the project life that exceeds its aggregate cost cashflows, adjusting for timing, financing and opportunity costs. Under a standard net present value (NPV) analysis, this is assessed by discounting the incremental net ungeared after-tax cash flows of the project at an appropriate weighted average cost of capital (WACC). The cashflows are probability-weighted to adjust for project-specific risks.

Any investments achieving an average return greater than the WACC create value and are represented by a positive NPV. Conversely, any investments achieving an average return below the WACC are value destructive and are represented by a negative NPV. Most firms choose between investment projects based on their NPV ranking.

The WACC is the financing cost of the project, calculated as the weighted average of the cost of debt and equity. The precise weighting is determined by the capital structure of the project: debt increases leverage while providing a benefit via tax-deductible interest payments. The cost of debt depends on the credit rating of the borrower. The cost of equity is usually determined by the capital asset pricing model (CAPM) as the risk-free interest rate, plus a beta-adjusted market risk premium.

3.2.2   The Impact of Risk on Project Decision-Making

Bearing this decision-making framework in mind, the concept of 'risk' affects investment decision-making in two principal ways:

- Symmetric risks: risks with similar upside and downside impacts (for example, changes in interest rates) increase the volatility of project cashflows. Such generic market risks are considered 'undiversifiable' for equity investors under modern portfolio theory so are factored into the cost of equity, leading to a reduced NPV. The project must therefore earn a higher rate of return before it proceeds relative to competing projects.
- Asymmetric risks: risks with asymmetric upside and downside impacts (for example, price regulation) are considered 'project-specific' and therefore 'diversifiable' for equity investors under modern portfolio theory so are not factored into the cost of equity. Rather, such risks are weighted directly into the project cashflows and hence similarly reduce NPV. Again, the project would need to earn a higher rate of return before it proceeds relative to competing projects.

However, the level of return obtainable from telecommunications infrastructure is constrained by market conditions. Project revenue is a function of price and the volume of services supplied. The price is constrained by consumers' willingness to pay, competition from substitutable services (for example, wireless broadband), and regulatory constraints. Volume is determined by price under a market demand curve.

An NGN investment will not therefore proceed unless and until a firm has reasonable certainty (in light of forecast long-term market conditions and project-specific risks) that it will earn a return that exceeds its cost of its capital over the long-term project life, including a market risk-adjusted return on that capital.

3.3   THE IMPACT OF REGULATION ON INVESTMENT

There are a variety of circumstances beyond generic competition law in which policy-makers have intervened to regulate markets in the public interest. Sectoral telecommunications regulation is an important example.

However, well-intended regulation can have unintended adverse effects. In essence, regulation can harm investment incentives by increasing the risk faced by investors without providing scope for offsetting higher returns. To illustrate this effect it is useful to consider the impact on telecommunications infrastructure investment of network unbundling and cost-based access regulation.

3.3.1 Excessive Regulation as a Deterrent to Infrastructure Investment

Mandated network unbundling has long been justified as promoting allocative efficiency by increasing the level of competition to which incumbent network owners are subject. Market entrants can selectively bypass existing infrastructure via their own network elements while obtaining exclusive use rights to remaining network elements at cost-based access prices. Unbundling is intended to lower barriers to market entry while enabling competitors to aggregate sufficient customers to enable them to finance a more extensive infrastructure roll-out.

While unbundling undeniably stimulates short-term competition, this may often be at the expense of long-term investment incentives. The international trend to balance this trade-off has been to roll back unbundling obligations as competition develops. The trend towards such regulatory 'forbearance' is exemplified by the following comment by the United States Federal Communications Commission (FCC) in 2003 in conjunction with its subsequent decisions:

> While unbundling can serve to bring competition to markets faster than it might otherwise develop, we are very aware that excessive network bundling requirements tend to undermine the incentives of both incumbent LECs [local exchange carriers] and new entrants to invest in new facilities and deploy new technology. (FCC, 2003a, para 3)

The rationale for such forbearance is well explained in the literature. Professor Sidak commented in a 2006 journal article that the FCC implementation of unbundling led to: 'tens of billions of dollars of investment flowing into business models that were neither particularly innovative nor sustainable in the absence of regulatory distortions in their favour. That distortion of investment represented a staggering destruction of wealth' (Sidak, 2006, p. 55).

Professors Gayle and Weisman concluded in a 2007 article that United States competitive local exchange carriers 'proceeded to lose billions of dollars when the FCC's introduction of [unbundling] along with artificially

low prices for network elements resulted in what was ultra-free-entry',
and, 'policies that reward imitation rather than innovation will attract
those market entrants adept at innovation, predominantly arbitragers,
while driving away genuine innovators' (Gayle and Weisman, 2007, p.
321). Importantly, the genesis of the concerns with unbundling involves
the price at which access is provided to unbundled network elements.

3.3.2 Criticisms of Cost-Based Incentive Pricing: TELRIC and its Variants

The impact of unbundling on investment is a function of the price at which
access is provided to unbundled infrastructure. The selection of that price
involves a balancing exercise between promoting allocative and dynamic
efficiency:

- If the access price is set too low, a firm will have a reduced incentive to invest in new infrastructure: hence dynamic efficiency will be impeded.
- If the access price is set too high, a firm will face less competition and will have greater market power to set prices at levels that maximize its profits: hence allocative efficiency will be impeded.

Most nations have sought to balance allocative and dynamic efficiency
by adopting so-called 'incentive regulation'. Such regulation is intended to
mimic competitive markets by constraining prices at a level consistent with
that which would otherwise be imposed by competition.

The United States has historically used a form of incentive regulation
known as 'total element long-run incremental cost' (TELRIC). Other
jurisdictions have used variants of this, including total service long-run
incremental cost (TSLRIC). TELRIC and its variants determine prices
based on the incremental costs faced by an efficient cost-minimizing firm
with an optimally configured network that uses the best available current
technology. The network elements in the hypothetical network are normally assumed to be located at the same geographic nodes as the regulated
firm's existing network.

In this manner, the price of access is determined not by the regulated
firm's need to achieve a return on its historic capital costs, but rather by
a forward-looking model in which the only costs recoverable are those
incurred by a hypothetical efficient firm. The policy rationale for this
approach is that a network provider in a competitive market would be
constrained in its pricing by the price of its most efficient competitor.
Under the TELRIC philosophy, that competitor is assumed to have

the most efficient network and aggressively undercut the incumbent's pricing.

In theory, if the TELRIC approach were applied accurately, any investment in network infrastructure would realize an NPV near zero. The firm would not earn above the project WACC. Professor Pindyck explained this in a statutory declaration in the following terms (FCC, 2003a):

> The TELRIC pricing methodology relies on the simple Net Present Value (NPV) investment rule. The NPV rule states that a firm should invest in a project if the sum of the discounted cash flows (the NPV) from the project is positive. TELRIC is designed to produce prices that price an ILEC with a competitive return, i.e., with no profits in excess of those that would arise in a competitive market. In other words, under TELRIC the expected NPV of the included costs at any given discount rate is zero. The theory behind the TELRIC methodology is that, if the NPV was greater than zero, additional firms would enter until excess profits were driven to zero.

TELRIC pricing creates a clear disincentive towards new infrastructure investment. By investing in a project subject to TELRIC pricing, the firm is exposed to little upside to create shareholder value. Rather, the firm is exposed to significant downside that could destroy shareholder value. In such circumstances, a rational firm would choose to invest its money in alternative projects that do create shareholder value via a positive NPV.

Beyond this immediate NPV = 0 issue, there are a range of other difficulties with the application of TELRIC pricing. These difficulties create significant asymmetric downside risk and therefore further reduce the project NPV.

Firstly, rapid technological innovation and regular cost reviews create difficulties. Continued innovation usually results in the price of modern telecommunications technology declining over time for a given functionality. If TELRIC pricing is regularly reassessed, such declining prices will mean that a firm will never recover the initial cost of its investment. The firm will only be permitted to recover its historic costs to the level of the best available technology at the time of the review.

Generally, the shorter the TELRIC review period and the faster the rate of technological innovation, the greater the risk to the firm that it will not recover its costs. Professors Mandy and Sharkey (2003) have identified that the magnitude of this loss is considerable: if asset prices fall 11 percent per year, TELRIC pricing leads to an under-recovery of around 50 percent (Mandy and Sharkey, 2003, p. 437).

Secondly, the regulator always has the benefit of perfect hindsight. In investment decision-making, the firm will have imperfect information regarding future demand. The higher standard deviation of such forecasts may lead to a firm to overdimension its network to address high-demand scenarios.

In contrast, the regulator has perfect hindsight. At the time it makes its regulatory decisions, it will know actual traffic patterns and can ascertain the optimal network to meet those patterns. This difference in information quality means that regulators are often skeptical of network dimensioning and have a propensity to disallow costs.

Thirdly, competitors share in rewards but not risks. In making an investment, the firm is exposed to the market risk of incorrect demand forecasts. If the network is a failure, the firm alone will bear the cost of that failure. However, if the investment is a success, access regulation will mean the firm is subject to access requests from competitors ('access seekers') wishing to share in that success at cost-based access prices. In this manner, regulation exposes the firm to an asymmetric project return profile with an unlimited downside risk but truncated upside benefit. Such an asymmetric profile will negatively skew the NPV calculation and is a disincentive to investment.

Furthermore, the disincentive to invest is even more acute when considered under 'real option' theory. The regulated firm must make a long-term irreversible investment, so is committed to its substantial investment for the network life. The firm foregoes the value of an option to wait for lower-cost technology and is exposed to the downside risk that forecast demand does not materialize. In contrast, the access seeker retains flexibility and has a valuable option to enter or exit the industry at will, or to roll out its own infrastructure with a cheaper technology should demand forecasts be correct.

Fourthly, free-riding and disincentives for competitors to invest cause concerns. Not only will TELRIC pricing create a disincentive for the regulated firm to invest, but the pricing will also create a significant disincentive for access seekers to invest. Access seekers will rank the NPV of investment options when making their investment decisions. A rational access seeker will prefer the investment delivering the highest NPV. It will not incur the significant capital cost and long-term risk involved in rolling out infrastructure, if it can obtain immediate access to such infrastructure at a TELRIC price consistent with that of the best technology available. The access seeker would prefer to free-ride on the investment by the regulated firm rather than rolling out its own competing infrastructure.

Fifth, and finally, is regulatory error. Regulators are not perfect and may exhibit a natural bias in favor of the market entrant or may be prone to error. The scope for regulatory error is particularly high in the context of highly complex telecommunications pricing models given the sheer range of interacting assumptions, their limited verifiability and associated measurement complexity.

Given these concerns, it is not surprising that Professors Crandall and Sidak commented in a June 2007 review of the literature that: 'the net effect of mandatory unbundling on dynamic efficiency (investment by both entrants and incumbents) is therefore unequivocally negative' (Crandall and Sidak, 2007, p. 10).

3.3.3   Conclusions on the Impact of Regulation on Investment Incentives

For existing (legacy) telecommunications infrastructure, there is little risk to regulators that such infrastructure will be removed if excessively low access prices are imposed. The cost of such infrastructure is sunk. For existing infrastructure owners, low access prices will largely transfer value from the infrastructure owner's shareholders to the shareholders of access seekers. Any impact on investment is limited to ongoing maintenance of the network.

However, the situation is very different for new (greenfield) infrastructure. If excessively low access prices are imposed, the infrastructure owner has the ability to avoid incurring the sunk cost by not undertaking the investment. In this manner, low access prices will likely result in the infrastructure not being built. Low access pricing therefore has a direct and disproportionate impact on the roll-out of new infrastructure.

Bearing this important distinction in mind, international best practice in telecommunications regulation has historically favored a high degree of network unbundling of legacy network infrastructure with the associated application of cost-based access regulation. While such regulation has promoted market entry and services-based competition, it has impeded infrastructure investment and longer-term facilities-based competition. Specifically, it has deterred investment in greenfield infrastructure and the upgrading of legacy networks. Given such issues, the optimal regulatory approach should be to wind back such regulation as competition develops so as to preserve long-term network investment incentives.

The unbundling experience demonstrates that care is required in the application of telecommunications price regulation to ensure that an optimal balance is achieved between allocative efficiency and dynamic efficiency. While TELRIC pricing is intended to achieve this balance, the current application of TELRIC has seriously impeded dynamic efficiency by deterring infrastructure investment. Serious policy concerns therefore arise regarding the continued appropriateness of applying TELRIC pricing to greenfield telecommunications infrastructure investments or network upgrades.

3.4 AN AUSTRALIAN CASE STUDY IN EXCESSIVE REGULATION

As a geographically vast country with a highly dispersed population, located at a great distance from its trading partners, Australia depends heavily on the quality, efficiency and innovativeness of its telecommunications system. However, attracting sufficient funds for investment in Australian fixed telecommunications infrastructure has been a significant challenge.

3.4.1 Adoption of a Negotiate–Arbitrate Model Underpinned by TSLRIC Pricing

Australia was one of the earliest countries to liberalize its telecommunications sector, commencing an interim phase of liberalization from 1991. Australia subsequently implemented full liberalization under legislative reforms in 1997. A sectoral telecommunications access regime known as 'Part XIC' was enacted into the Trade Practices Act 1974 (Cth) based on a negotiate–arbitrate model.

Part XIC enables the Australian Competition and Consumer Commission (ACCC) to declare various wholesale telecommunications infrastructure and carriage services after a formal public inquiry on the basis that the declaration is in the 'long-term interests of end-users'. Following declaration, various standard access obligations apply that require an access provider to supply a declared service to access seekers on request on a non-discriminatory basis. There are currently 9 declared services in Australia and they include unbundled network elements.

A consequence of declaration under Part XIC is that, if the access provider and access seeker cannot agree on any aspect of the supply of a declared service, either party may notify the ACCC and seek arbitration. Most access arbitrations are lodged by access seekers and relate to the price at which the declared service is supplied. The ACCC has published non-binding pricing principles for each declared service to guide its arbitral determinations. The favored pricing methodology by the ACCC has been TSLRIC pricing, a variant of the TELRIC pricing methodology.

However, the manner of application of TSLRIC pricing by the ACCC has created significant concern in Australia. TSLRIC pricing has deterred investment in infrastructure that could be subject to actual or potential Part XIC declaration. In this manner, investment flows have been distorted away from regulated (and potentially regulated) services and infrastructure towards unregulated services and infrastructure.

Furthermore, not only has the regulatory regime created a disincentive for investment; but the absence of greenfield investment has been used as political justification to perpetuate and continually strengthen the regulatory regime. This regulation has further exacerbated the risk of investing in greenfield infrastructure, thereby feeding a vicious cycle.

In this manner, Australia has been caught in an upwards spiral of increasing sectoral telecommunications regulation. This spiral has occurred even though the level of competition in Australian telecommunications markets has substantially increased. According to international best practice and economic theory, the level of sectoral regulation should rather have been rolled back as competition developed.

3.4.2   Refinements to the Regulatory Regime to Promote Greater Investment

The potential adverse impact of the Part XIC access regime on investment was recognized in Australia as early as 2001, only four years after the enactment of Part XIC. In a comprehensive review of the telecommunications regulatory regime in 2001, the Australian Productivity Commission commented:

> Mandated access still presents formidable regulatory risks to investors. Telecommunications technology and markets are rapidly moving and very risky . . . For a carrier making a new investment; the risk of future declaration – with regulated access prices – may lead to the delay or termination of the planned investment. This is because access regimes may truncate the returns from risky investments.

The Productivity Commission made a series of recommendations to promote investment, principally the adoption of an 'access holiday' regime.

In December 2002, Australia therefore enacted access holiday provisions into Part XIC, known as 'anticipatory individual exemptions'. A carrier could apply to the ACCC for an exemption if it would promote the long-term interests of end-users, assessed on the basis of promoting competition and encouraging the economically efficient use of, and investment in, infrastructure. The ACCC could grant the exemption subject to any conditions. The explanatory memorandum to the relevant legislation indicated that the purpose of anticipatory individual exemptions was to provide certainty for potential investors in telecommunication infrastructure and services in relation to future access.

However, the exemption mechanism met immediate difficulties the first time it was applied. The suppliers of the Australian analogue (FOXTEL) cable television service sought an anticipatory individual exemption to underpin their upgrade of network assets and equipment to support digital cable television. The ACCC granted the exemption subject to various conditions. The parties proceeded with the upgrade on the basis of the exemption. However, the ACCC's decision was subsequently appealed.

In hearing the appeal, the Australian Competition Tribunal noted that the parties had proceeded with the roll-out of digital cable television before the appeal had been determined. The parties had also made a range of statements in the public domain regarding their intention to roll out digital television without qualifying these by the need for the appeal to be determined. The Tribunal therefore held that the network upgrade would have proceeded regardless of the exemption; the exemption was unnecessary for the investment to proceed.

Even though the digital upgrade had already occurred on the basis of the exemption, the Tribunal overturned the exemption, leaving the parties exposed to the risk of access regulation. Ironically, the Productivity Commission had used the roll-out of digital cable television as a key example of the need for the 'access holiday' mechanism in the first place.

Following the Tribunal decision, anticipatory individual exemptions would only be effective in Australia if an access provider delayed its investment until the exemption had been granted and any appeals of that decision were resolved. This could take at least two years and potentially significantly longer given the potential for further appeals to higher courts.

Not surprisingly, no further applications were made to the ACCC for anticipatory exemptions. Ironically, the mechanism intended by the Australian government to facilitate investment had the unfortunate converse impact of further impeding it.

3.4.3   The Saga of Australia's Fiber to the Node Network

Issues with the application of Part XIC to new infrastructure investments came to a head between Australia's incumbent telecommunications carrier, Telstra Corporation Limited, and the Australian government during 2006 and 2007 in the context of the proposed upgrade of Telstra's network to implement fiber to the node (FTTN).

Given the perceived difficulties with Part XIC, Telstra indicated that it was not intending to seek an anticipatory individual exemption from the ACCC. Rather, Telstra sought to enter into negotiations directly with the government with a view to obtaining direct legislative relief from the Part XIC access regime. Negotiations between Telstra and the government occurred throughout much of 2006. Telstra's requested the government to enact legislation that disapplied Part XIC to Telstra's network upgrade, effectively to provide a long-term access holiday by way of specific legislation.

After many months, it became clear that the government and Telstra were unable to reach agreement. In August 2006, Telstra announced that it would not continue with its proposed network upgrade as it had not reached agreement on pricing issues. In a share offer prospectus released in 2006, Telstra commented:

> Telstra seeks a competitive rate of return when it invests its capital. If Telstra cannot be confident that ACCC regulation of prices for competitor access to a new network will allow a competitive rate of return, Telstra will not invest in the network. This year, Telstra planned to start building a $3 billion FTTN network. However, Telstra disagreed with the ACCC on the price its competitors should pay for access to the network and, as a result, Telstra decided not to build the network.

Meanwhile, a consortium comprised of Telstra's competitors (known as 'G9') announced a rival proposal to build its own FTTN network. The G9 proposal required the government to enact legislation that prohibited Telstra overbuild and required the forced cutover of existing Telstra copper infrastructure to the new network at cost-based rates. By requiring a prohibition on competitive overbuild, the G9 proposal effectively sought to create a monopoly franchise in its favor. The G9 proposal was also conditional on agreement of an undertaking with the ACCC on access prices to reduce regulatory risk.

During 2007, the FTTN network became a significant issue in Australia's federal election. The Minister for Communications announced that a tender process would be established. Few parties subsequently expressed serious interest, leaving G9 and Telstra as the only potential bidders. Some notable investors cited excessive regulation as a deterrent to investment.

The Liberal government lost Australia's December 2007 federal election and a new Labor government came to power. The new minister terminated the previous FTTN tender procedure and substituted a new procedure. The new tender sought proposals to roll out an Australian national broadband network (as a NGN) that would cover 98 percent of the population with a minimum bandwidth of 12Mbit/s into the home. The government indicated that it would make a funding contribution of up to AU$4.7 billion to the total estimated cost of AU$15 billion.

In November 2008, six parties submitted proposals to government for the NGN. Telstra refused to submit its full proposal on the basis that the government had not provided sufficient regulatory certainty. Controversially, Telstra's proposal was disqualified in December 2008 due to non-compliance with tender requirements. The government is now building its own fiber to the premises network at very substantial cost, effectively replicating Telstra's existing access network.

3.4.4   Insights from the Australian Case Study

In Australia, regulation has persisted, even increased in some cases, as competition has increased. Such regulation has deterred investment by the incumbent. Such regulation has also deterred long-term infrastructure investment by market entrants by enabling them to obtain access to existing infrastructure at a materially lower cost than if they rolled out infrastructure themselves. While this has promoted short-term market entry, it has created a market dependence on continued access regulation at the expense of the development of long-term infrastructure-based competition.

Australia recognized the problems that Part XIC created for investment at an early stage and sought to create an access holiday regime. However, investment is sensitive to uncertainty and time delays and the new regime suffered both problems. As a consequence, Telstra sought to bypass Part XIC by negotiating directly with the government when upgrading its infrastructure. However, the continued emphasis by the ACCC on TSLRIC has led to an impasse and no network roll-out.

Ultimately, the issue became highly politicized. Infrastructure rollout was proposed via a public–private partnership (PPP) supported by partial government funding and amendments to the current regulatory settings. However, following a further impasse on these issues, Telstra was disqualified from the NGN tender procedure. As Telstra was widely viewed as the only entity capable of rolling out the NGN to the government's specifications, the likelihood of an NGN roll-out in Australia remains uncertain.

Given this outcome, the existing Australian regulatory regime has undeniably failed to achieve one of its stated objectives of 'achieving economically efficient investment in infrastructure' (Trade Practices Act 1974 (Cth), s 152AB(2)(e)).

3.5   TECHNIQUES TO PROMOTE INFRASTRUCTURE INVESTMENT

Appropriate incentives can be created to encourage large-scale infrastructure investment by reducing the level of regulatory risk. Of the various mechanisms adopted in other industries, the most common are:

- the sharing of regulatory risk between the public and private sector via binding governmental commitments (enforced by compensation), as frequently occurs in public–private partnerships; and

- access holidays, typically providing certainty that a network owner will earn a positive NPV sufficient to justify the infrastructure roll-out.

Both of these techniques are considered in further detail below. Project financing structures also utilize techniques such as third party insurance as a means of mitigating regulatory risk for financiers. Such insurance is usually known as 'political risk insurance' and is normally directed at the most extreme forms of regulatory risk, such as expropriation of assets and currency inconvertibility.

3.5.1   The Public–Private Partnership Model

Large-scale infrastructure projects involve significant risks. A wide range of stakeholders are involved in such projects, whether by contributing debt or equity, or by contributing goods or services. An underlying principle guiding the negotiation of contracts between these parties is that the project risks should be allocated to those parties best placed to control or manage them. In this manner, the level of risk should be reduced and the NPV of the project should increase.

A recent legal innovation in many countries has been the formal development of integrated PPPs to reduce the financial burden on the public sector while increasing the efficiency of service delivery. A PPP is best described as a set of contractual arrangements between public and private sector entities to deliver cost-effective and high-quality services to the public over an extended time period. Each party contributes its skills and expertise and shares in the risks and rewards.

Importantly, PPPs seek to implement a key principle of risk redistribution. Participating entities are allocated those risks that they are best placed to control or manage. Under this principle, a party will bear a risk where it is within that party's control and/or which it can most efficiently manage relative to other parties (including by spreading the risk to third parties). In effect, project risks are allocated so that each party bears those risks that it can best control and/or manage and at the least overall cost to the project.

Relevantly, the government normally bears a substantial proportion of the regulatory risk associated with large PPP projects under the terms of the relevant 'concession agreement'. This is primarily because:

- Regulatory risks arise as a result of governmental action, hence the government will normally have the political and legal power to prevent adverse regulatory events occurring.

- If adverse regulatory events do occur, the government can most efficiently bear such risks by pooling and spreading the cost of compensation across taxpayers. Such spreading of risk to the public is appropriate, for example, if the government has determined that regulation is required in the public interest.

The key insight from PPPs is that it is possible to reduce regulatory risk via a binding contract between the public and private sector. The private sector will agree to roll out infrastructure in the public interest with agreed pricing that provides a positive NPV. In consideration, the government will forbear imposing certain forms of regulation on that network (for example, price controls) and will bind itself to pay compensation if this commitment is breached.

A key characteristic of NGN telecommunications projects is their high degree of exposure to regulatory risk. As evidenced by the Australian experience, such risks can be sufficiently great as to render an NGN project uneconomic. By reducing the regulatory risk via a PPP arrangement, the NPV of the project will increase and it will be more likely to proceed.

In essence, an NGN project will be more likely to proceed if the private sector can obtain a meaningful binding commitment (enforced via compensation) from government that it will not be subjected to value-destructive regulation that will result in a negative NPV.

3.5.2 Access Holidays

A specific form of binding commitment given by the government to the private sector to promote investment is an 'access holiday'. An access holiday is a defined period of time within which new infrastructure investment is exempted from some or all access regulation. Typically, the access holiday mechanism is enacted into law as a statutory exemption to an existing legislated access regime.

The benefits of an access holiday were summarized in the 2003 article by Professors Gans and King in the following terms:

> If regulators are expected to set low access prices ex post but, at the same time, there is a commitment that any new essential facility will not be subject to infrastructure access for a significant period of time, then this will raise investor incentives. It is this ex ante commitment to delay access ex post that is the basis of an access holiday.
>
> In this sense, access holidays play a role similar to a patent in innovative activity. Patents encourage innovations by conferring on the inventor temporary monopoly profits. Similarly, access holidays encourage infrastructure investment by allowing investors to temporarily exploit any market power

associated with their facility. Both patents and access holidays are second-best solutions in that they impose a temporary monopoly cost. Both an optimal patent and access holiday needs to be designed to trade-off this temporary loss with increased incentives to invest. (Gans and King, 2003, p. 163)

The Gans and King analogy of a patent can be extended:

- A policy rationale for granting exclusive rights to patent holders is that the patented innovation may have spillover positive externality benefits for wider society. In this manner, the patent holder is, in effect, being allocated (via a monopoly right) a proportion of the wider utility benefit that accrues to wider society.
- The same can be said in relation to access holidays for new telecommunications networks. Such networks have similar positive spillover externalities for wider society, explaining why governments are so keen on promoting them. It is therefore appropriate that network owners receive an additional reward in order to stimulate such investment.

As indicated by the patent analogy, the central policy concern in conferring an access holiday is the ability of the infrastructure owner to utilize any pre-existing substantial market power that it may possess. Specifically, how should policy-makers balance the promotion of infrastructure investment against the prevention of misuses of market power? This issue of an optimal access holiday has been debated in Australia in the context of investments in natural gas pipeline infrastructure.

In August 2004, the Australian Productivity Commission released a final report on its review of Australia's national third party access regime for gas pipelines. The Productivity Commission identified that access pricing for gas pipelines was impeding investment and innovation. It reasoned that the existing regime was excessively focused on cost-based price regulation and this caused a truncation of returns that deterred investment. For gas pipeline infrastructure that did not confer substantial market power, the Productivity Commission recommended the introduction of a 15-year access holiday regime.

However, the Productivity Commission did not recommend an access holiday recommendation for gas pipeline infrastructure that did confer substantial market power. Following significant resulting concern expressed by pipeline owners, the Australian Ministerial Council on Energy proposed a compromise: a partial access holiday would apply to pipelines with substantial market power in which access prices would be exempted from regulation but all non-price terms would remain regulated.

Following significant concern expressed by pipeline users, a negotiated political compromise was ultimately achieved that was supported by both pipeline owners and pipeline users:

- All new pipelines without market power were permitted to apply for a 15-year holiday from all access regulation.
- New international pipelines with market power were permitted to apply for a 15-year holiday from price regulation, but would remain subject to non-price regulation.
- New domestic pipelines with market power would have no right to an access holiday, but would instead be subject to price regulation triggered by a higher regulatory threshold.

This compromise was subsequently enacted into Australian law. Interestingly, this compromise may still deter investment in those domestic pipelines that would clearly meet the higher regulatory threshold. The Gans and King solution would suggest that such pipelines should be guaranteed some kind of positive NPV return sufficient to create an incentive for investment and reward the investor.

3.5.3 Policy Recommendations to Promote Greenfield Investment

Several potential policy recommendations for NGN infrastructure can therefore be drawn from the analysis set out in this chapter. Firstly it is potentially beneficial for government to grant an access holiday to encourage investment in new telecommunications infrastructure that will not confer substantial market power on the network owner. This may be the case, for example, if the NGN would not be the only means of delivering broadband Internet access into the home due to intermodal competition.

Secondly, if new telecommunications infrastructure would confer substantial market power on the network owner, it is still potentially beneficial to provide a more limited access holiday arrangement. Under this arrangement, the network owner would not be subjected to TELRIC pricing (and its disincentives) but would instead earn a regulated return that generated a value-accretive positive NPV sufficient to incentivize the investment. In this manner, any regulated access pricing should be increased well above the TELRIC levels that have traditionally applied.

Thirdly, particular jurisdictions may be susceptible to changes in policy so that 'access holiday' rights could be repealed by a subsequent government. In such circumstances, it may be appropriate for network owners also to consider entering into an agreement with government in which compensation is paid by the government if an access holiday is prematurely removed. However, any such entitlement to compensation would likely be conditional on the network achieving certain quality standards, possibly under a PPP arrangement.

3.6   CONCLUSIONS AND RECOMMENDATIONS

A key focus of telecommunications regulatory policy at present concerns the impact of regulation on the long-term incentives for new infrastructure investment. This chapter has analyzed the theory and practice behind the regulation versus investment debate and made a series of key policy recommendations.

In summary, legacy customer access network infrastructure has its technological limitations and therefore substantial investment in NGN infrastructure will be required. However, investment in such infrastructure is inherently risky because of sunk costs and long asset lifetimes. Under modern finance theory, such investments will only proceed if the network owner can make a return on its investment that exceeds its cost of capital over the project life. The rate of return will depend on the risk profile of a project. In the case of an NGN project, the most critical project risks are often regulatory risks.

Against this context, international best practice in telecommunications regulation has historically favored a high degree of network unbundling and the application of cost-based access regulation. While such regulation has promoted market entry and services-based competition, it has impeded infrastructure investment and the development of long-term facilities-based competition. The optimal regulatory approach should be to wind back sectoral regulation as competition develops so as to preserve long-term network investment incentives. In the absence of such wind-back, low access prices may result in telecommunications infrastructure not being built.

The genesis of the network unbundling concerns the price at which access is provided to unbundled network elements. Incentive-based TELRIC pricing was intended to balance allocative and dynamic efficiency considerations, but it has clearly had a disproportionate adverse effect on dynamic efficiency. Generally, such pricing has capped the upside of telecommunications investment while exacerbating the downside risks. Furthermore, competitors can free-ride on the investor's infrastructure at a lower cost than rolling out their own infrastructure. Serious policy questions therefore arise regarding the appropriateness of applying TELRIC pricing to greenfield investments and network upgrades.

A case study of the proposed roll-out of NGN infrastructure in

Australia provides some practical insights into these issues and suggests some policy solutions. Generally, regulation has persisted in Australia, even increased, as competition has developed. Such regulation has had an adverse effect on investment. Australia sought to correct the adverse incentive effects of access regulation at an early stage by introducing an access holiday regime. However, for various reasons, this attempt was unsuccessful. Given the extent to which the access regime has deterred the roll-out of the Australian NGN, the Australian government intervened by offering partial government funding; in effect, proposing a PPP arrangement. Ultimately, the government has been required to make the investment itself.

As evidenced by the attempts in Australia, the general policy solution to the investment problem is to generate sufficient and certain returns for investors that they are guaranteed a value-creative positive NPV over the long lifetime of their investment. However, if the infrastructure investment also confers substantial market power on the network owner, it will also be necessary to implement a policy solution that prevents the misuse of that market power over the investment lifetime.

This chapter concludes that the appropriate pricing methodology to be applied in that scenario is not an incentive-based TELRIC price. Rather, the price methodology should provide a return to the investor in the nature of a patent: the investor should earn sufficiently above its WACC that the investment becomes highly desirable. Such higher returns to investors are justifiable given that telecommunications networks have positive spillover externalities for wider society.

In summary, nations need to adopt a more appropriate regulatory balance that appropriately rewards investors for investment in new telecommunications infrastructure. Only if a better balance is achieved will we achieve a vision for the year 2015 of ubiquitous broadband Internet access.

NOTE

1. The views in this chapter reflect the personal views of the author and not necessarily the views of any of his current or previous clients.

REFERENCES

Australian Productivity Commission (2001). Telecommunications competition regulation: Productivity Commission inquiry report. Retrieved 10

March 2009 from http://www.pc.gov.au/inquiry/telecommunications/docs/finalreportcfm?abstract_id=996065.

Crandall, R. and G. Sidak (2007). Is mandatory unbundling the key to increasing broadband penetration in Mexico? A survey of international evidence. Social Science Research Network, Retrieved 10 March 2009 from http://papers.ssrn.com/sol3/papers.

FCC (Federal Communications Commission) (2003a). In the Matter of Review of the Section 251 Unbundling Obligations of Incumbent Local Exchange Carriers (CC Docket No. 01-338). Report and order on demand and further notice of proposed rulemaking, released 21 August.

FCC (Federal Communications Commission) (2003b). In the Matter of the Commission's Rules Regarding the Pricing of Unbundled Network Elements and the Resale of Service by Incumbent Local Exchange Carriers (WC Docket No. 03-173). Notice of proposed rulemaking, released 15 September (*Triennial Review Order*).

Gans, J. and S. King (2003). Access holidays for network infrastructure investment. *Agenda*, **10**(2), 163.

Gayle, P. and D. Weisman (2007). Efficiency trade-offs in the design of competition policy for the telecommunications industry. *Review of Network Economics*, **321**(6), 321–41.

Mandy, D. and W. Sharkey (2003). Dynamic pricing and investment from static proxy models. *Review of Network Economics*, **4**(2), 403–37.

Sidak, G. (2006). A consumer-welfare approach to network neutrality regulation of the Internet. *Journal of Competition Law and Economics*, **4**(55), 55–99.

Telstra Corporation Ltd. (2006). Telstra 3 Share Offer: share in the future. Prospectus, 9 October. Retrieved 10 March 2009 from http://www.telstra.com.au/abouttelstra/investor/prospectus.cfm.

4. Behavioral economics and telecommunications policy

Patrick Xavier and Dimitri Ypsilanti

4.1 INTRODUCTION

Regulation in the telecommunications sector has focused mainly on the supply side of the market including, for example, market entry and licensing, access to and use of networks, interconnection, control over retail and/or wholesale pricing. This emphasis on the supply side has been appropriate because the task was to install effectively competing alternative suppliers in former monopoly telecommunication markets. As competition has developed in telecommunication markets and users have a wider choice of service providers, there has been increased attention by some regulators on the demand side. For instance, a demand-side measure introduced in many countries is the requirement for 'number portability' aimed at facilitating consumer 'switching' in the fixed line and mobile markets.

Such attention to the consumer demand side is timely, because informed consumers prepared to choose between competing suppliers are necessary to stimulate firms to innovate, improve quality and compete in terms of prices. Indeed, in making well-informed choices between suppliers, consumers not only benefit from competition, but they exert the sustained pressure for providers to compete for their customer. Conversely, where consumers have too little information, poor-quality information, or misinformation, they may end up misled and confused by the choices on offer, may pay too much or may buy the wrong service. This may, in turn, inhibit and dampen the competitive process. Moreover, if suppliers can exploit consumers this could at some point lead to 'reputational damage' and disillusionment with the competitive process as a whole. Hence consumer protection is critically important from an economic as well as social rationale.

4.1.1   Telecommunication Policy and Regulation in the Interests of the Consumer

Policy and regulation in the telecommunications sector has long been said to be concerned with the consumer interest (including consumer protection and empowerment). Legislation, policy and regulatory statements repeatedly claim that the welfare of consumers is a, if not the, primary objective of policy and regulation. But, despite this rhetoric, there has been relatively little analysis of the extent to which consumers have in fact benefited and become more empowered. Competition among suppliers to attract and retain consumers was supposed to result in suppliers that serve consumers well prospering relative to those that do not.

Now with market liberalization and alternative telecommunications suppliers becoming firmly installed in many countries, questions are emerging about whether consumers have in fact become empowered and 'switched' to alternative suppliers as much as had been expected. If not, why not? Should policy-makers and regulators intervene? And if so, how? In seeking to answer these questions, important information can be gleaned from analysis of the demand side, including new insights from behavioral economics into actual (as distinct from normative) consumer behavior.

4.1.2   Demand-side analysis, including behavioral economics

Conventional economics starts with a number of assumptions about consumer behavior, including that:

- Consumers approach markets with a set and stable set of preferences.
- Consumers are concerned only with their own welfare.
- In aggregate at least, consumers can rationally use available information to make optimal decisions.

Demand-side analysis, including so-called behavioral economics, goes beyond these assumptions. It argues that in some situations consumers consistently depart from behavior predicted by these assumptions. In other words, consumer choice is biased away from that which would occur if behavior conformed to the assumption of 'rationality'. Behavioral economics accords particular reference to the observed, as opposed to theorized, behavior of consumers (Pesendorfer, 2006). A key issue for a demand-side analysis is examining 'what is actually going on in terms of consumer outcomes' (Sylvan, 2006b). This involves examining the product features as well as consumer choice issues and how the product or service is being offered or presented. An important insight of behavioral economics is that it is not only the product that matters, but the context as well.

The conventional economic approach recognizes that information asymmetry and information failure may lead to suboptimal consumer outcomes. It also recognizes that consumers face a 'bounded rationality' (for example, in the form of various costs of acquiring and processing information) and, as a result, rely on 'heuristics' (such as reliance on a firm's reputation and other price and quality signals) in decision-making. Consumer policy has therefore focused on removing those market failures that prevent consumers who, even though they behave rationally, are nevertheless prevented from making optimal choices that maximize their welfare.

A substantial amount of regulation has been directed at these problems, including implementing misleading-conduct laws (for example, prohibition on fine print disclaimers, requirement for plain language contracts); disclosure requirements; product regulation; and regulation aimed at allowing consumers to switch conveniently between suppliers – such as the development of interoperability standards and number portability (as discussed later) in the telecommunications industry. Other potentially useful tools include the provision of price and quality comparison data for consumers, and the use of calculators (technical tools usually on websites) allowing consumers to enter data to enable them to make price comparisons.

Behavioral economics challenges some of the assumptions of conventional economics that consumers make their choices coherently and rationally given their preferences and the constraints upon them. Behavioral economics involves the study of actual consumer behavior (by contrast with how consumers should behave) (OECD, 2008). It argues that consumers often fail to act in their own best interests due to behavioral traits such as failure to process information objectively or misevaluations about the costs and benefits of prospective decisions. For example, research conducted on switching activity by low-income households in UK electricity markets found that on the whole these consumers did not switch provider in a way that could be explained by any rational set of criteria. Only 7 percent of consumers chose the cheapest option and, indeed, 32 percent changed to a supplier that was more expensive (Waddams-Price and Wilson, 2005).

Among the biases identified by behavioral economics, the following may be particularly relevant to the telecommunications market and may help to explain how, even where there is adequate information, consumers may be making seemingly irrational decisions in choosing an operator or service package (OECD, 2006).

Choice overload

Consumers having too many products or features to compare and may experience increased anxiety about the possibility of making a bad choice. This can lead to random choice, or failure to make any choice, resulting in missed opportunities for buyers and sellers. A type of 'analysis paralysis' can take hold when information and choice become very complex.

Endowment

Consumers may be reluctant to give up what they have, even though they would not buy such goods or services if they did not already have them (for example, consumers may stay with the incumbent fixed line provider because of misplaced loyalty, a failure to acknowledge poor choices in the past, or an irrational consideration of sunk costs).

Defaults

The ordering of options, particularly in markets where a choice must be made, influences choice. Default bias means the decision to opt-in or opt-out (for example, of extending a mobile phone contract) is not the same decision for all people. Also, consumers may tend to take a path of least resistance, particularly if they feel that there is a 'normal' option (for example, people may buy 'standard' bundles offered by telecommunications suppliers, even if they do not want the whole telecommunications bundle).

Hyperbolic discounting

Consumers tend to be short-sighted when making decisions with immediate costs or benefits to be weighed against future costs or benefits (for example, consumers may enter long-term telecommunications contracts because they place more value on the immediate benefits of the offer, such as a free or heavily subsidized handset or a reduced first month rate, or free local calls, rather than on the long-term costs of a contract such as high prices for calls exceeding a usage 'bucket', the inability to switch to lower-priced alternatives, and the inability to take advantage of latest technology).

Time-variant preferences

This usually manifests as hyperbolic discounting. Generally, consumers' discount rate increases as the outstanding time period becomes shorter and this leads to decisions which discount costs (which are incurred over the longer term) more than benefits (which are realized in the short term).

This can lead consumers to sign long-term contracts with mobile phone companies to obtain a free handset, even though this may be more expensive in the long term than buying a handset without being locked in to a long-term contract.

Framing biases
Consumer choice is influenced by the 'frame' in which information is presented. Presentation of the same information in a different frame can lead to a different decision. For instance, cash-back offers can be much more attractive to customers than a similar or even greater discount. For most consumers, 'only 3 percent fat' is likely to be less appealing than '97 percent fat free'.

Loss aversion
The preference for avoiding loss is widely considered to be greater than the preference for gain.

Heuristics
Consumers often take short cuts (for example, by following rules of thumb) when the decision environment is too complex relative to their mental and computational capabilities. These rules of thumb are called 'heuristics' and are often accurate enough to be useful, but may sometimes lead to suboptimal decisions.

4.2   INFORMATION IMPERFECTIONS

Where consumers have too little information or too much information of inconsistent quality, various types of problem can arise, including:

- consumers failing to participate in the market at all, because they have limited awareness of the products and services on offer, or conversely because they are confused by an excess of available information (sometimes referred to as 'information overload');
- consumers paying too much;
- consumers buying the wrong product or service;
- consumer disappointment with the product or service, because it turns out to lack the expected level of quality (Ofcom, 2006a).

Constraints on the ability of consumers to process information can also lead to non-optimal, welfare-reducing decisions even when the information available to them is plentiful and non-deceptive. For example, comparison of prices for telecommunications services offered by different suppliers is complicated by the wide range of possible consumer usage patterns, detailed variations in price levels and price structures, and the large number of possible discount and bundled schemes available. Moreover, operators and service providers regularly adjust their pricing strategies either through changes in components of their tariff structure, or the introduction or withdrawal of various discount schemes and service packages (Miravete, 2007).

Technological advances and market pressures have made telecommunications and information and communication technology (ICT) products and systems increasingly complex and feature-rich. But this has also increased the complexity faced by consumers in determining appropriate products that satisfy consumer needs and in evaluating alternative product offerings by competing providers.

4.2.1   Ability and Willingness to Switch

Barriers to switching service providers can be present due to high switching costs. Switching costs can be defined as the real or perceived costs that are incurred when changing supplier but which are not incurred by remaining with the current supplier.

The ability and willingness of consumers to switch is critically important. If switching is discouraged or impeded this could impact not only on the demand side but also potentially raise supply-side barriers (Barrow, 2007). This is because new entrants could be deterred from entering the market in the belief that it will be difficult to persuade consumers to switch from their existing provider. This could diminish contestability and the effectiveness of competition and limit the benefits that consumers would otherwise derive from it.

In the UK, research conducted by Ofcom (2006b, 2008), the UK National Consumers Council (2006) and others suggests that in the telecommunications sector there are, in fact, a range of important deterrents to switching, including:

- Lengthy and cumbersome switching procedures that make it inconvenient for consumers to switch and can outweigh any potential benefits.
- Early exit charges, imposed by an existing provider, that reduce the benefits of switching.
- Confusing products and non-transparent pricing that makes it difficult or time-consuming to compare deals (as in the case of mobile telephony and the Internet).

- Technical incompatibility of equipment that can make it uneconomical for consumers to switch (for example, if they cannot use a blocked mobile phone with their new provider).
- Long-term deals that lock consumers into lengthy relationships with their providers (as may occur with mobile telephony and Internet contracts) and increase the risk of them being overcharged.

Consumers will not switch to a competing supplier unless the price difference exceeds the switching costs. Where switching costs are high, it is possible for a provider to set very low prices, even prices below cost to attract new consumers, but then subsequently charge these consumers prices well above costs once they are 'locked in'. This 'bargain then rip-off' pricing pattern is a characteristic of many markets with high switching costs (NERA, 2003).

4.3 EVIDENCE OF ACTUAL CONSUMER BEHAVIOR IN TELECOMMUNICATIONS MARKETS

As noted earlier, behavioral economics involves the study of actual consumer behavior as distinct from how consumers should behave. Accordingly, this section examines the available evidence of actual consumer behavior in the telecommunications services sector and influences on such behavior.

4.3.1 Consumer Behavior in the Fixed Line Market

United Kingdom
Ofcom's research (2008) found that in the United Kingdom, one in three (34 percent) of the fixed line consumers surveyed had changed the supplier providing their home fixed-line service between 2003 and 2006. However, two-thirds (66 percent) had not switched in the 2003–06 period and the majority (52 percent) had not even considered doing so. Regardless of whether or not they had switched supplier in the last four years, 34 percent had made some change to their existing service with their current supplier in 2003–06. A similar proportion claimed to keep an eye on the market: around a third (36 percent) agreed that they were always on the look-out for a better deal and a similar proportion (31 percent) agreed that they made a conscious effort to keep up to date with what other providers were offering. In this context, it is notable that according to Ofcom's research, 86 percent of fixed line consumers in the UK were satisfied with their

*Table 4.1   Switching in the UK fixed line communications market during
            2007 and 2008*

	2007 (%)	2008 (%)
Switched supplier	9	12
Actively looking	3	4
Started looking, not switched	4	5
Considered without looking	7	6
None	77	73
	100	100

Source: Ofcom (2008).

overall experience with their current supplier: almost half (46 percent) were very satisfied and a similar proportion (40 percent) were fairly satisfied. Clearly, if consumers are satisfied, there may be no or less reason for them to switch.

The figures in Table 4.1, showing that well over 70 percent of fixed line consumers in the UK did not switch suppliers during 2007 and 2008, are broadly consistent with the data for earlier years.

Consumer responses to Ofcom's questionnaire surveys indicated that the greatest deterrent to shopping for an alternative fixed line supplier is the possibility of getting locked into a contract with a new supplier: two-thirds of consumers (67 percent) agreed that they would be put off by this. The second-greatest barrier to switching was reluctance to leave a known and trusted supplier for one that is unfamiliar – stated by 65 percent of consumers. In the fixed line market, many consumers have been with their supplier for many years. For some consumers, their relationship with their fixed line supplier was perceived as being more important than whether they could get a better deal elsewhere. Because of the 'endowment factor', a factor influencing consumer behavior underlined by behavioral economics, these consumers would only switch if they had experienced a serious betrayal of trust that incites a revenge value to switching. As many as one in two (53 percent) fixed line consumers agreed that they had a strong sense of loyalty to their existing supplier (Ofcom, 2006b). No significant differences were detected between demographic groups. In other words, differences are due to attitudes and behavior rather than demographics.

In summary, Ofcom concludes that in the fixed line market, in the face of an undifferentiated market (perceived or actual), some consumers are adopting fallback, risk-averse strategies and will stay with what they know and trust, even though it might not be the best rational option.

Table 4.2   Degree of satisfaction towards fixed line prices in Portugal

Degree of satisfaction	%
Very satisfied	5.8
Satisfied	58.2
Not satisfied	29.5
Not satisfied at all	4.1
Do not know/Did not respond	2.5

Source: ANACOM (2006).

This appears to support the arguments of behavioral economics. Ofcom's findings suggest that the greatest chance of furthering participation in the fixed line market are efforts to help consumers to overcome their inertia, by allaying fears regarding the potential risks associated with switching services, or by educating them regarding the tangible benefits of any new service over and above their existing arrangement (Ofcom, 2006b).

Portugal
The level of switching in Portugal's fixed line market has also been relatively low. Table 4.2 indicates that about two-thirds (64 percent) of consumers interviewed as part of an Anacom (2006) (the telecommunications regulator) study were satisfied with the prices charged by their fixed network provider. This degree of satisfaction with a key aspect of service would clearly reduce the disposition to switch.

4.3.2   Consumer Behavior in Mobile Telecommunications Markets

United Kingdom
Ofcom found that more than a third (36 percent) of mobile telecommunications consumers in the UK had changed their mobile phone network supplier during 2003–06. The figure rises to 52 percent for those who are on an annual contract and falls to 29 percent for those who are on pre-pay packages. However, two-thirds (66 percent) had not switched during this time period and the majority (53 percent) had not even considered doing so. Even among the 7 percent who had considered switching, not all had actively started looking for an alternative.

The figures in Table 4.3 show that well over 70 percent of mobile consumers in the UK did not switch suppliers during 2007 and 2008, and this is broadly consistent with the data for earlier years.

Ofcom (2006b) pointed out that although the majority of consumers in the mobile phone market have not considered changing their supplier,

Table 4.3   Switching in the UK mobile communications market during 2007 and 2008

	2007 (%)	2008 (%)
Switched supplier	13	12
Actively looking	3	1
Started looking, not switched	6	7
Considered without looking	4	6
None	74	74
	100	100

Source: Ofcom (2008).

they are nevertheless participating in other ways: 32 percent had made some change to their existing service with their current supplier during this time period. While consumers were far more likely to change the tariff or package they were on (31 percent) than ask their supplier to match a better deal they had seen elsewhere (8 percent), mobile phone users seemed aware of their potential to negotiate: 62 percent of those on contracts had changed their existing tariff or package and 16 percent had attempted to renegotiate their package or deal. Around a third of all consumers claimed to keep an eye on the mobile phone market; 32 percent agreed that they were always searching for a better deal and 28 percent agreed that they made a conscious effort to keep up to date with what other providers were offering. In terms of reasons given for not switching, more than half the telecommunications consumers surveyed (54 percent) said that they were very satisfied with their overall experience with their current supplier and another third (36 percent) said that they were fairly satisfied. Post-pay and pre-pay consumers were equally satisfied with their current network supplier.

The research also identified a range of procedural and psychological barriers perceived by mobile telecommunications consumers as Table 4.4 indicates. The greatest deterrent to shopping for an alternative mobile phone network supplier was the possibility of getting locked into a contract with a new company. Around two-thirds of consumers (68 percent) agreed that they would be put off by this.

Reflecting a similar sentiment, the second-greatest barrier to shopping was reluctance to leave a known and trusted supplier for one that was unfamiliar, an issue for 64 percent of consumers. This refers to the extent to which consumers value their relationships with their mobile phone network supplier. More than one in two (55 percent) agreed that they had

Table 4.4 Perceived barriers to shopping and/or switching mobile phone suppliers in the UK

Response to question (with prompted reasons): 'I did not switch my mobile phone supplier because . . .'

Reason	Agree (%)	Strongly agree (%)	Total agree (%)	Post-pay (%)	Pre-pay (%)
Don't want to get locked into contract with new provider	47	21	68	47	52
Reluctant to leave provider I trust for one I don't know	47	17	64	52	59
Strong sense of loyalty towards current provider	36	19	55	No difference	
Shopping for new provider too much of a hassle/chore	41	12	53	41	49
Only short term gain as providers follow each other	43	9	52	No difference	
Difficult to make comparisons b/w providers	38	9	47	35	53
Don't have time to research options	38	9	47	34	54
Don't want to lose current deal/package	32	14	42	No difference	
Big risk that something will go wrong in transition	30	7	37	No difference	
Don't know enough to make right choice	28	9	37	27	42
Bound to feel stupid/out-of-date when I talk to sales staff	21	12	33	24	37
No difference in cost of supplier	28	4	32	27	33
No difference in quality of supplier	26	3	29	23	32
Don't know where to find trusted info about options	21	5	26	No difference	

Note: Base: all mobile (500). Percentages represent share of total sample.

Source: Ofcom (2006b).

a strong sense of loyalty to their existing network supplier. This was particularly the case for inactive consumers of whom 39 percent were likely to be relational people (with a preference for using trusted brands known or recommended to them) and only 14 percent were likely to be transactional people (willing to consider unfamiliar brands if they offer a good deal). These manifestations of the endowment factor and/or irrational fear or risk-averseness relating to a change to a new provider are consistent with the arguments of behavioral economics.

More regulatory attention might be directed to this 'barrier to switching'. For instance, the cost of breaking a contract might be lowered where a consumer can demonstrate that conditions of the contract were not being delivered by the supplier.

The consumer traits of lack of confidence, heuristics and susceptibility to information overload emphasized by behavioral economics also appeared to play a role in decision-making. Among so-called inactive consumers, 48 percent did not feel they knew enough to make the right choice and 42 percent expressed concern about appearing stupid in front of sales staff, while 44 percent were willing to accept a solution that they felt was good enough rather than investigate all options to find the best one.

In this market, consumers on contracts also appear to be bound to their suppliers as a result of having negotiated or been given special deals: two-fifths (42 percent) of consumers expressed concern about losing the package or deal they were on. Early-exit penalties could also apply here.

The process of shopping was itself likely to discourage around half of the consumers surveyed: 53 percent agreed that shopping for a new supplier was too much of a hassle; 47 percent agreed that it was difficult to make comparisons between suppliers; and 47 percent agreed that they did not have enough time to research the options. Moreover, many (52 percent) perceived that the gain would be short term because all the suppliers follow each other. As Table 4.4 indicates, this is particularly the case among pre-pay consumers, who perceived higher barriers to switching than those on contracts: 82 percent of pre-pay users agreed they did not want to be locked into contracts with a new provider. They were also significantly more likely to regard shopping for a new network supplier as an onerous process, approaching it with lower levels of interest and confidence than contract users.

In summary, Ofcom concluded that the evidence suggests that in the UK mobile telecommunications market, in the face of complexity and lack of market differentiation (perceived or actual), some consumers will stay with what they know, even though it might not be the 'best' option. Ofcom concluded that the greatest chance of furthering consumer participation in the mobile phone market is through efforts made to influence these key drivers, for example by educating them in regard to the tangible benefits of any new service over and above their existing set-up, by making the switching process easier or by helping consumers to overcome inertia.

Portugal

In Portugal, Table 4.5 indicates that of those mobile telecommunications consumers responding to a survey conducted for Anacom about 19 percent have switched. Of these, about one-third switched because most of their contacts were clients of the new operator. Presumably there was some advantage with being on the same network such as network coverage and/or discounted or free calls to consumers on the same network. Another one-third of those who switched did so because they were unhappy with the prices of their old operator; while 13.7 percent of those who switched did so because they were unhappy with the quality of service provided by their old operator, and 8.6 percent because of an offer from the new operator.

Of the 80.7 percent who had never changed operator, 66 percent were satisfied with their current operator and 31.2 percent because most of their contacts are clients of their current operator. Only about 10 percent of

Table 4.5 Switching mobile telecommunications operator in Portugal

Reasons to change operator	%
Have already changed operator	19.3
Most of my contacts are clients of new operator	34.6
Unhappy with prices	33.3
Unhappy with quality of service	13.7
Offer from new operator	8.6
Other	20.6
Never changed operator	80.7
Satisfied with current operator	66.2
Most of my contacts are clients of current operator	31.2
Offers the best prices	6.6
Switching is too complicated / inertia	3.2
Decision of another person	2.8
More / better network coverage	0.8
Keep the same number	0.4
Other	4.0
Do not know / Did not respond	2.3

Source: ANACOM (2006).

consumers seemed focused on price as a driver of switching and even less (about 4 percent) on quality of service.

Australia

The results of a questionnaire survey in Australia for ACMA (Australian Communications and Media Authority – the telecommunications regulator), summarized in Table 4.6, provide some additional information on consumer satisfaction.

During the July 2005 to June 2006 period, satisfaction levels (defined

Table 4.6   Consumer satisfaction with telecommunications services in Australia, 2004 to 2006

Respondents were asked: 'In the last 12 months, how well have each of the following service providers met your overall expectations?' (Excludes 'no answer' and 'doesn't apply')

	Exceeded my expectations		Mostly met my expectations		Sometimes met my expectations		Rarely met my expectations	
	July 2004 – June 2005 (%)	July 2005 – June 2006 (%)	July 2004 – June 2005 (%)	July 2005 – June 2006 (%)	July 2004 – June 2005 (%)	July 2005 – June 2006 (%)	July 2004 – June 2005 (%)	July 2005 – June 2006 (%)
Local telephone company	7.8	8.2	73.9	73.3	13.3	13.5	5.0	4.9
STD telephone company	7.0	7.5	75.4	74.7	12.9	13.2	4.7	4.6
International telephone company	7.4	7.1	73.5	74.3	13.7	13.4	5.5	5.3
Mobile phone service provider	9.4	9.9	72.5	72.1	13.1	13.4	5.0	4.6
Internet service provider	10.2	11.2	68.6	68.9	15.6	15.3	5.5	4.6

Source: Roy Morgan Research, questionnaire survey sample of approximately 23 000 people aged 14+ years, cited in ACMA (2006).

as 'exceeded' or 'mostly met' expectations) were more than 80 percent for mobile telecommunications providers as well as fixed line and Internet service providers. The highest satisfaction levels were for long-distance fixed line telephone companies (82.2 percent). Internet service providers were the providers who most exceeded expectations (11.2 percent).

Consumer expressions of the likelihood of them switching service provider offer another measure of consumer dissatisfaction with their current provider. The Australian survey also provides some information in this regard, summarized in Table 4.7. The likelihood of consumers switching service provider ranged from: about 13.5 percent for fixed line providers, and 14 percent for mobile telecommunications providers, to 16.5 percent for Internet service providers during July 2005 to June 2006. The remaining consumers said they were either 'unlikely' or 'neither likely nor unlikely' to switch providers.

Table 4.7   Consumer likelihood to switch providers in Australia, 2004–06

Respondents were asked: 'How likely would you be to switch companies, if you were able to buy that service from another company?' (Excludes 'no answer' and 'doesn't apply')

	Very or fairly likely to switch		Neither likely nor unlikely to switch		Very or fairly unlikely to switch	
	July 2004 – June 2005 (%)	July 2005 – June 2006 (%)	July 2004 – June 2005 (%)	July 2005 – June 2006 (%)	July 2004 – June 2005 (%)	July 2005 – June 2006 (%)
Local telephone company	13.5	13.2	29.2	27.9	57.3	58.8
STD telephone company	13.0	12.8	29.8	28.3	57.2	58.9
International telephone company	11.8	11.5	30.7	29.1	57.5	59.5
Mobile phone service provider	14.5	13.9	29.2	28.2	56.4	57.9
Internet service provider	16.8	16.5	30.6	28.3	52.5	55.2

Source: Roy Morgan Research, questionnaire survey sample of approximately 23 000 people aged 14+ years. Cited in ACMA (2006).

Table 4.8   Early termination fees of major US mobile telecommunications service providers, July 2005

Company	Early termination fee
Cingular	US$150
Nextel	US$200
Sprint	US$150
T-Mobile	US$200
Verizon	US$175

Note: These operators control about 80% of the US mobile telecommunications market.

Source: CALPIRG Education Fund (2005).

United States

A report based on responses to a questionnaire survey of 1000 households conducted in the United States in 2005 found that 36 percent of respondents replied that early termination fees ranging from US$150 to US$240 (as shown in Table 4.8) had prevented them from switching (CALPIRG Education Fund, 2005).

In fact 89 percent of mobile telecommunications consumers considered that early termination fees are designed to prevent consumers from switching. They disagreed with the mobile telecommunications operators' claim that the termination fees are a necessary part of the rate structure (to enable them to recover the costs of subsidizing handsets, 'buckets' of free calls, and so on).

4.3.3   Consumer Behavior in Internet Markets

United Kingdom

According to Ofcom's research (2006b), awareness of alternative suppliers is lowest in the Internet market with around 20 percent of Internet consumers in the UK unable to name spontaneously any narrowband or broadband Internet service providers (ISPs) in their area. However, half of Internet subscribers were spontaneously aware of two or more narrowband suppliers, and this rises to 74 percent when prompted (Ofcom, 2006b).

During 2003–06, just over one quarter (28 percent) of UK consumers switched their ISP supplier. However, about three-quarters (72 percent) had not switched during this time period and the majority (46 percent) had not even considered doing so. Even among the 13 percent who had considered switching, not all had actively started looking for an alternative.

Table 4.9   *Switching in the UK broadband communications market during 2007 and 2008*

	2007 (%)	2008 (%)
Switched supplier	11	10
Actively looking	6	9
Started looking, not switched	10	11
Considered without looking	9	8
None	65	62
	100	100

Source: Ofcom (2008).

The figures in Table 4.9 indicate that only about 10 percent of broadband consumers in the UK switched suppliers during 2007 and 2008, although about another 20 percent had considered doing so. Another 8 percent had considered switching, but had not actively started looking for an alternative. However, more than 60 percent had not even considered doing so.

Ofcom found that regardless of whether or not they had switched ISP in the last four years, almost three-quarters (72 percent) had made some change to their existing service with their current ISP in this time period. This was mainly the result of consumers switching connection type (62 percent had done this) and/or switching tariff, including upgrades to connection speed (53 percent had done this). A few (10 percent) had renegotiated their deal (that is, asked their current ISP to match a better deal they had seen elsewhere). Thus, even though the research shows that many consumers in the Internet market have not considered changing their ISP, competition had allowed them to 'participate' in other ways. Compared with broadband users, narrowband users were significantly less likely to have made any changes to their existing package: 24 percent had changed connection type and 25 percent had changed their tariff or package.

According to Ofcom's research (2006b), the greatest deterrents to active participation in the Internet market include the possibility of getting locked into a contract with a new supplier (68 percent of consumers); reluctance to leave a known and trusted ISP for one that was unfamiliar (63 percent of consumers); and the perceived effort of shopping around, including difficulty in making comparisons between ISPs (44 percent) and not enough time to research all the options (40 percent). The main drivers of participation include interest in technology, and desire for low cost and/or willingness to consider unfamiliar brands if offered a good deal. There were few demographic differences, suggesting participation is influenced by attitudes and behavioral factors rather than by differences in demographics.

This information suggests that in the face of complexity (perceived or actual), inactive consumers are adopting fallback, risk-averse strategies and will stay with what they know, even though it might not be the best option. The information is consistent with the argument of behavioral economics that an endowment factor serves to influence decisions in favor of the present provider.

4.4   IMPLICATIONS FOR POLICY AND REGULATION

4.4.1   Regulation of Information Disclosure

More information is usually seen to be desirable. A demand-side behavioral perspective warns that if consumers have limited cognitive abilities, either generally or in a particular situation, then adding more information may result in information overload and hence in worse decision-making (Camerer, 2003). Excessive disclosure can confuse consumers (as evidenced in the case of mobile phone and Internet tariff options) and can also discourage firms from providing useful information through their advertising. In these cases, the need is not for more but for better (perhaps less) information in a structured, easily comprehensible format (Gans, 2005a).

Typically, a mandated disclosure is intended to improve the information received by the less-informed buyers to a greater extent than by those more sophisticated consumers who may already possess and be able to process the requisite information. But the reverse can also be the case where it is the more sophisticated (and more cognitively capable) consumers who can make more use of the information contained in a mandated disclosure. For example, mandating information on broadband download speeds and caps can be of far more use to those consumers who have at least a basic knowledge of technical features of broadband capacity (for example, bits per second).

Differences in the incidence of benefits from mandated disclosure means that any costs could tend to be different as well. Some consumers may be provided with information they cannot use, some with information they do not need, and some others will be overloaded with information, but all will have to pay for the compliance and related costs (OECD, 2006).

This cautionary note aside, there are many situations where an information disclosure remedy is necessary and appropriate. Even in competitive markets, suppliers may possess market power if consumers are not well informed about products, supply conditions and/or alternatives and feel unable or unwilling to switch between these alternatives (ERG, 2009). Providing consumers with more and better information, for example through mandatory disclosure or through third party certification, may facilitate more active participation in the market and wiser decision-making (Better Regulation Executive, 2007).

Just as informational problems may be multifaceted, so too may problems that result from behavioral biases. It is important for policy and regulation to recognize these biases and develop a fuller understanding of the needs and motivations underlying consumer behavior in telecommunications markets. Not all behavioral biases lead to consumer detriment. For the most part they do not lead consumers to depart significantly from optimal decisions. Public policy should be concerned only with those biases that lead to significant detriment. When markets fail because of such costly biases, remedies should be shaped accordingly. For example, a situation of choice or information overload could be aggravated by a requirement for more information disclosure. Rather, the appropriate intervention may involve reframing the information that is available to consumers in a way that makes choice easier (OECD, 2007).

The policy implications of demand-side analysis in some respects are more challenging than those derived from the more traditional approach to consumer protection and empowerment. For instance, behavioral economics predicts that for various reasons some consumers (or consumers in some circumstances) may act in ways that are inconsistent with their *ex ante* preferences. Consumers may use information in ways not predicted by neoclassical theory or they may, for various reasons, not use available information. Thus, while in some cases providing more information or providing information in a different form may remove or reduce the risk to consumers, this will not always be the case. If it is to be effective, an effort to inform consumers must appreciate how people actually think. The behavioral response to identical pieces of information will depend on how they are presented and framed.

Thus, in circumstances where the conduct of suppliers alters the preference set of consumers and hence their choices, resulting in an inferior outcome for those consumers, the solution may lie in regulatory intervention that aims 'to steer people's choices in welfare-promoting directions without eliminating freedom of choice' (OECD, 2006).

Resetting defaults
Options can be presented in ways that lead the consumer to gravitate towards certain choices that are in their interests. For example, to overcome the biases of hyperbolic discounting and of default inertia, renewal of a telecommunications contract can be presented with 'opting out' as a default. Policy-makers and regulators could consider the use of 'light' interventions such as a mandatory 'opt-in' default provision (rather than an opt-out default) as part of arrangements for extending a telecommunications contract.

Reframing
Suppliers can be required to present information in a variety of frames, or in specific frames that may guide sound consumer choice. The policy solution under a demand-side behavioral approach is that rather than requiring that the consumer read ever more complex contracts, the government can mandate standard form clauses or even standard form contracts. This recognizes that consumers probably will not read the contract in detail. But care should be exercised so that in removing ambiguity, the behavioral solution also tries not to limit consumer choice, since this can make consumers worse off if parties are now constrained to government-devised contracts that are inflexible and cannot be altered to fit their personal circumstances.

General de-biasing
Although biases are generally deeply ingrained, there is a role for consumer education, making consumers aware of their biases, and helping them to develop mechanisms to overcome those biases that go against their self-interest. For instance, to counter 'hyperbolic discounting', consumers might be counseled to consider carefully whether a flat rate contract or a 'free' mobile phone handset with a two-year contract really does suit their usage level and pattern. Cooling-off periods can be used to allow consumers to reframe their choices and to give them an opportunity for rational reconsideration to overcome the influence of impulsive choice, such as those resulting from hyperbolic discounting.

Recognizing risk-averseness in the switching process
In addition to making more information available, raising awareness, and addressing some of the behavioral biases that may prevent consumers from actively participating in the market and making decisions to switch where not satisfied, there are some specific measures that can be taken by regulators to reduce practical impediments to consumers switching from one supplier to another. For example, regulators could ensure that the shortest possible time is taken to complete number portability for

consumers switching fixed line and mobile telecommunications providers. Regulators could require that all Internet service providers ensure a simple, costless (or at least cheap) and quick transfer of consumers who choose to switch provider. Regulators could examine the need to limit the 'lock-in' period for mobile phone handsets in order to facilitate switching. Where applicable, the fee for unlocking the handset should be related to the cost involved.

Vulnerable consumers
The circumstances of vulnerable consumers warrant special attention. For instance, Ofcom's research in the UK indicates that while those without Internet access – frequently older and low-income consumers – are less likely to look for information at all, providing price information solely via websites could risk excluding relatively large and vulnerable groups of consumers. There is need for further research to assess:

- Consumers' use of information sources and how consumers use these when making choices.
- Consumer opinion on current information obtained through, for example, supplier websites, the Internet generally and specialist publications.
- The importance of savings in influencing decisions to switch supplier, what trade-offs are made when considering whether to switch, and the key drivers in consumer decisions to search or switch.
- The extent to which misperception, low awareness of achievable savings or ease of the switching process impacts on consumer switching decisions.

There is particular need to assess the needs and motivations of consumers – especially uninvolved and vulnerable consumers – in more detail to ascertain what, if anything, would encourage more participation in telecommunications markets. This would assist consideration of whether, and if so what, regulatory intervention is warranted. All consumers want simply understood information. If businesses and policy-makers design the regulated information with vulnerable consumers in mind, all consumers are likely to benefit.

4.5   CONCLUSION

This chapter distils some basic elements of behavioral economics and demand analysis, examines available evidence of actual consumer behavior and considers implications for telecommunications policy and regulation. The chapter concludes that the insights of a demand-side perspective, including behavioral economics, can serve to alert policy-makers and regulators in the telecommunications sector of the need for a number of measures:

- Policy-makers and regulators should develop a better and fuller understanding of the needs and motivations underlying consumer behavior in telecommunications markets, especially those of vulnerable consumers (such as those in rural areas, the elderly, minors, disabled, those on low incomes, the unemployed).
- Policy-makers and regulators could assist consumer participation in telecommunications markets by educating consumers about their rights, by raising awareness about new services and options offered by the market, and by making the process of switching in the fixed line, mobile and Internet markets easier, cheaper and faster.
- Regulators should consider requiring that all major operators provide complete, comparable, appropriate and accurate information to consumers through different channels (for example, through leaflets, radio, consumer hotline and web-based programs) to enable consumers, especially vulnerable consumers, to identify quickly the most suitable and best-value telecommunications plan (ERG, 2009).
- Regulators could use more effective means of targeting information to vulnerable groups to provide them with practical guidance about how they can get the best deal.
- Regulators could encourage third parties, including consumer organizations, to provide price and service comparison facilities through consumer hotlines, websites, and so on.
- Regulators could work with fixed line (including Internet service providers) and mobile network operators to develop and publicize a set of comparable indicators relating to quality of service.
- Regulators should ensure that the shortest possible time is taken to complete number portability for consumers switching between fixed line and between mobile service providers.
- Regulators should require that all Internet service providers ensure a simple, free (or at least low-cost) and quick transfer of consumers who choose to switch provider.
- Where not already in place, regulators could require truth-in-billing, and restrict harmful business conduct and practices (for example, by prohibiting mis-selling and misleading advertising).

REFERENCES

ANACOM (2006). Electronic communications consumer survey, February 2006 – key findings. May.

Australian Communications and Media Authority (2006). Telecommunications performance report 2005–2006. October.

Barrow, P. (2007). Just enough: empowering fixed-line telecommunications consumers through a quality of service information system. CCP Working Paper 07-2. Retrieved from http://www.ccp.uea.ac.uk/publicfiles/workingpapers/CCP07-2.pdf.

Better Regulation Executive (2007). Government response to the final better regulation executive/National Consumer Council report on consumer information. UK Department for Business Enterprise and Regulatory Reform, London.

CALPIRG (California Public Interest Research Group) Education Fund (2005). Locked in a cell: how cell phone early termination fees hurt consumers.

Camerer, C. (2003). Regulation for conservatives: behavioural economics and the case for asymmetric paternalism. *University of Pennsylvania Law Review*, **151**, 1211–54.

ERG (2009). Report on transparency of tariff information.

Gans, J.S. (2005a). Real consumers and telco choice: the road to confusopoly. Paper presented to the Australian Telecommunications Summit, Sydney. 21 November.

Gans, J.S. (2005b). Protecting consumers by protecting competition: does behavioural economics support this contention? *Competition and Consumer Law Journal*, July, 1–11.

Miravete, E. (2007). The doubtful profitability of foggy pricing. Retrieved from http://www.eco.utexas.edu/facstaff/Miravete/papers/EJM-Foggy.pdf.

NERA (National Economic Research Associates) (2003). Switching costs. A report prepared for the Office of Fair Trading and the Department of Trade and Industry. Economic Discussion Paper 5. April.

OECD (2006). Roundtable on demand-side economics for consumer policy: summary report. Paris. Available at http://www.oecd.org/dataoecd/31/46/36581073.pdf.

OECD (2007). Summary report on the second OECD roundtable on consumer policy. Paris. DSTI/CP (2007)1.

OECD (2008). Enhancing competition in telecommunications: protecting and empowering consumers. DSTI-ICCP-CISP (2007)1/FINAL, May, 2008, Retrieved from http://www.olis.oecd.org/olis/2007doc.nsf/LinkTo/NT00005FB2/$FILE/JT03246386.PDF.

Ofcom (2006a). Ofcom's consumer policy – a consultation. Retrieved 8 February 2006 from http://www.ofcom.org/consult/condocs/ocp/ocp_web.pdf.

Ofcom (2006b). The communications market 2006.

Ofcom (2008). The consumer experience 2008. Telecoms, internet and digital broadcasting. 24 November.

Pesendorfer, W. (2006). Behavioural economics comes of age: a review essay on advances in behavioural economics. *Journal of Economics Literature*, **44** (September), 712–21.

Sylvan, L. (2006a). The interface between consumer policy and competition policy. Department of Consumer Affairs, Victoria. Lecture in honour of Professor Maureen Brunt.

Sylvan, L. (2006b). *Capturing the consumer interest.* Consumer Panel Toolkit launch. Australian Competition & Consumer Commission. 2 February.
UK National Consumers Council (2006). Switched on to switching? A survey of consumer behaviour and attitudes, 2000–2005. April.
Waddams-Price, C. and C. Wilson (2005). Irrationality in consumer switching decisions: when more firms may mean less consumer benefit. Industrial Organization working paper from EconWPA.

PART II

Technology convergence and the future role
of competition and regulation

5. The measure and regulation of competition in telecommunications markets[1]

Marcel Boyer

5.1 INTRODUCTION

Telecommunications and more generally information and communication technologies (ICTs) constitute the backbone of our societies. Nothing is more important for the well-being, present and future, of modern societies than getting an efficient web of telecommunications chains and networks. Such a web is essential for social cohesion, productivity gains, innovation and commercialization, and for reaching the highest level of humanist economic growth.

The development of a telecommunications web is significantly influenced by the regulatory framework put in place to oversee the evolution of this web towards a competitive system through policies aimed to protect efficient newcomers from predatory incumbents and to protect the public against the capacity of large firms to exercise market power. If and when the level of competition is deemed sufficient, specific economic regulation is expected to disappear in favour of general competition or antitrust policy while technical, architectural or design regulation and coordination will subsist.

When is the level of competition in a national telecommunications industry sufficient? When it is not, then when will it be: that is, what value of which indicators will tell us that its time has come? The answers to those questions are vital for any society since too much economic regulation for too long as well as too little economic regulation for too short a time generate important social costs in terms of consumer welfare, productivity, innovation and growth.

The specific objectives of this chapter are, first, to develop a methodological framework, which will allow a proper characterization of the level of competition in the residential local access market; and second, to recommend a new approach to regulation. It addresses major concerns of the regulators regarding two 'factual measurements': first, the small number of competitors in the residential local access; and second, the high market shares captured by the incumbent firms, even after a decade of favourable policies aimed at 'more competition' in that market. The situation is basically the same in all regional residential local access markets.

The above 'factual measurements' are poor indicators of the level of competitive pressures in those markets. Indeed, those facts may be credible indicators that competitive pressures are relatively intense, so much so that entry strategies by newcomers are not profitable. The traditional measures of competition based on market shares are inadequate because this industry has more characteristics of an emerging industry than of a mature industry.

The current approach to economic regulation, mainly based on the tight control of incumbent local exchange carriers (ILECs) and light-handed surveillance of competitive local exchange carriers (CLECs), has run its course as market delineations become rapidly blurred.

The new regulatory framework should rest on three specific principles of economic efficiency:

- The pursuit of a dynamic regulatory approach based on implementing proper competition processes and information systems.
- The promotion of competition through proper incentives ensuring:
  - dynamically efficient inter-access prices and conditions;
  - efficient investment programs in network maintenance and development.
- The design of non-predatory pricing rules through full cost sharing, to promote the emergence of a more competitive industry, even if such rules reduce static efficiency.

If the government or regulator wishes to adopt a proactive strategy to favour the emergence of competition, based on the fact or belief that incumbents have an 'unfair' advantage, then the preferred policy is a direct incentive subsidy, such as a generous but deferred investment grant or tax credit to be paid at some time in the future conditional on the entrant's capture of a predetermined market share. Only those potential entrants with superior products or services, and/or superior technology, and/or better-quality consumer service, who believe they can compete with and displace the incumbent one way or another, will enter or consider entering the industry. Inefficient competitors, fly-by-night operators and fast-buck seekers will stay out. Consumers and customers will reap the benefits of efficiency-enhancing competition through proper creative destruction.

5.2   THE RELEVANT MARKET

The telecommunications industry looked very much like a mature industry in the 1970s and even in the 1980s. But over the last 20 years the development of new information and communication technologies made the telecommunications industry appear much more like an industry in constant instability, with new technologies, new products, new competitors, significant level of mergers and acquisitions as well as divestiture activity. These movements were accompanied by significant changes in regulation, from earnings-based or rate-of-return-based cost plus regulation towards a whole array of incentive, price cap, light-handed, competition-based and competition-enhancing regulation.

Regulators in most if not all jurisdictions have forced incumbent firms who own the local loop (the last mile) of the wireline network to give access to their network to CLECs at relatively favourable conditions. In spite of such efforts, the level of access effectively demanded and used remains quite low, raising fears that ILECs may have been able, for some unknown reasons and through some obscure means, to prevent and/or block access, thereby maintaining their market power over the local wireline access services.

However, market shares (whatever way they are measured in practice) may not be the best measure of competitive pressures in a fast-growing and technology-driven industry. The actual and potential competitive products and services in the telecommunications industry are in a constant state of flux. Besides the strictly regulated products and services, we see changes in products and services, appearing in a seemingly unpredictable manner, as well as changes in the prices, price structures (fixed or access charge, variable charge, bundled services discounts, and so on), restricted conditions imposed on different groups of consumers of those products and services, and even their technology delivery platforms, affecting their accessibility and flexibility of access.

The relevant actual market is therefore difficult to determine and the empirical studies that can be conducted can only be incomplete as to what is the real level of competition in the industry. Even the best state-of-the-art econometric studies available are subject not only to explicit caveats but also to a significant level of cavilling given the simplifications and aggregations in product characteristics and in pricing structures that researchers and analysts must inevitably have recourse to in their empirical estimation of demand structures for telecommunications products and services. One must treat these attempts very carefully and with skepticism. Even if they may nevertheless contain some useful information, the use one may make of the results in order to fine-tune the regulatory framework remains very limited.

In most cases, consumers face a menu of pricing structures and it becomes difficult, in empirical statistical or econometric studies, to summarize such pricing structures in a meaningful notion of what is the 'price' of a product or service to be compared to the price of other (substitute) products or services. There are many reasons why telecommunications products and services are priced in such a way.

Menus of pricing structures or calling plans, according to which consumers pay a fixed charge per unit of time (typically a month) and then decreasing or increasing price levels for increasing consumption levels, are or can be implemented when the resale of the product can be controlled or deterred in such a way that the end consumer of the product can be identified by the producer or seller. Menus of contracts or calling plans have three important impacts: first, they promote efficiency and favor a faster development of the industry; second, they generate higher profits through price discrimination; and third, they allow product differentiation thereby softening competitive pressures.

Menus of pricing and contracts promote efficiency insofar as the marginal (last) units of service are sold at or close to marginal cost. Hence at the margin the most important allocative efficiency condition (marginal cost pricing, possibly equal to zero) is satisfied. The welfare value that consumers attach to those marginal units is equal to the cost of producing or generating those marginal units. Hence the efficient social-welfare-maximizing level of production and consumption is achieved; at least once the decision to subscribe is made. However, a surplus may be lost if some of those potential consumers do not subscribe to the service because of the fixed subscription cost, whether a one-time cost or a monthly cost.

Carefully determining the calling plan limits and conditions allows telecommunications firms to differentiate their products and services from what competitors are offering, thereby softening the level of competitive pressures across the industry. Offering the same or very similar calling plans as those offered by another competitor (same limits, conditions and characteristics) would force the two firms to compete on prices only. This latter type of competition is likely to end up in a money-losing price war.

Many factors other than pricing drive the demand(s) for telecommunications products and services. To avoid competing strictly on prices, firms will compete along those other factors: coverage (and dead zones), digital versus analog transmission, physical and design features of phones (such as color, size and weight), battery life, digital subscriber line (DSL) services and cable modem services, household composition, education, employment and occupation, security, commuting or not, and so on.

Consumers are most likely indifferent between the different forms or technologies by which they satisfy their needs for telecommunications services, as long as their demand for connectivity and flexibility, the two most important characteristics that they value, are comparable in terms of quality and affordability. In most cases, consumers are not even aware of the technological characteristics of the platform on which their calls and other services are transmitted. It is more the end product that counts: connectivity, flexibility, safety, dependability, accessibility, capacity (high speed and broadband) and user-friendliness. In that sense, the demand expressed by consumers for different telecommunications devices (wireline, wireless, cellular mobile, satellite-based mobile, Internet Protocol – IP – telephony, and so on) is a derived demand rather than a direct demand. Connectivity refers to the geographic area (or to the set of potential called and calling parties) over which communications can be established; connectivity is a multidimensional characteristic since it often relies on a multilayer pricing plan: different connectivity at different prices. Flexibility refers to the availability of the service under different circumstances, the most important certainly being the availability throughout a given geographic area, as one moves around in the area. Safety refers to the health hazard one may be exposed to in using the service on a regular basis, an example being the concerns for safety which have prompted different jurisdictions to forbid the use while driving of handheld cellphones, and in some cases all uses of cellphones. Dependability refers to the assurance of service of a high quality level, for example free of parasites and free of breach of confidentiality, when one needs it, especially but not only in an emergency situation. Accessibility refers to the availability of the service when one wants to communicate with another party. Capacity refers to the possibility to transmit vast amounts of information, such as large files, and high-definition pictures as well as videos, at a rate high enough to quasi-replicate in situ communications. User-friendliness refers to the ease of use of the technology and service.

The empirical estimates of the level of substitutability (and therefore competition) between wireline and wireless telephony is in terms of the own- and cross-price elasticities of demands for those services. To run such empirical studies, one needs sufficient variability throughout the sample observations. If every observation unit in the empirical sample subscribes to a wireline service, then it is impossible to identify the decision variables behind or justifying the subscription itself. To do so, one needs within the sample a reasonable number of non-subscribers to the different wireline and wireless services, a condition which may not always be satisfied for the basic wireline services. However, it may still be possible to estimate empirically the relative importance of the explanatory variables, drivers or shifters, of the usage level (like minutes) of the services.

In terms of data requirements, it is clearly preferable to have household-level data rather than more aggregated regional data. A study[2] by Rodini et al. (2002) of wireline–wireless communications substitution behavior using household-level data reports own-price elasticities of mobile access demand and cross-price elasticities with relatively high degrees of precision. The results suggest that wireline and wireless telephony may be, for a significant number of consumers, complements rather than substitutes, although the negative (net) cross-price elasticities indicate that on average the substitution effect dominates the complementary effect. The development of a larger number of different telecommunications technologies, each with important specific characteristics (connectivity, flexibility, safety, dependability, accessibility, capacity, friendliness), tends to increase the global usage level of all telecommunications technologies. Computing market shares in terms of telephone lines is not the only way to measure concentration. In fact, it now appears to be an outdated way to do it.

5.3   THE LEVEL OF COMPETITION

Let us first consider the level of substitutability in the local telephone market between wireline and wireless communications. The standard approach of antitrust economics is to consider the level of competition as sufficient to prevent an undue exercise of market power by incumbents if those incumbents would find it unprofitable to implement a small but significant (typically 5 percent) and non-transitory (typically for one year) increase in price (SSNIP) for their services, given the pricing structure of their competitors' products and services. Such a move would be profitable if the reaction of consumers to such a price increase for one year were relatively small and limited, that is, if there were no close substitutes to which they can turn to offset the impact of that price increase.

To do so, one must define the relevant market, the relevant choice set that consumers face regarding the satisfaction of their telecommunications needs, and the relevant set of competitors to determine the level of competition in the telecommunications industry. These questions can be raised in the study of any industry or industrial sector but are particularly important in the context of the telecommunications industry, which has many characteristics of an industry in its early phases of development rather than a mature industry, a rather striking example of the 'back to the future' storyline.

The ever-increasing number of consumers having access to both technologies and having access soon to a third one, the Voice over Internet Protocol (VoIP) technology, continuously reduces the need for protection. The level of market power of the ILEC is seriously diminished up to a point where more regulation may cause more harm than good. Insofar as those consumers who already have the possibility to react to any price increase by switching to alternative modes of telephone communications cannot be isolated from other consumers – that is, cannot be offered a specific pricing plan – the wireline incumbent will have to balance the benefits from an increase in the price with the loss of revenues due to such switching. Hence, on that basis alone, one would be inclined to recommend immediate deregulation even if some consumers have, by choice or not, access to wireline local telephony only.

The potential future competition in products and technologies also plays an important role and defining the set of competitors, actual and potential, is a critical step in determining the level of competition. The identity and role of potential competitors is difficult to determine since many products and services as well as some future providers may simply not be yet identified even though they may be *dans les coulisses* and certainly present in the minds of incumbents, whether they are wireline, wireless or VoIP providers. Any empirical study of the type econometricians usually do cannot avoid neglecting such crucial aspects.

Moreover, telecommunications products and services must be seen as a 'technology' to generate a vector of communications (and entertainment) characteristics that customers value. A consumer is likely to express his or her demand for a given product or service in terms of its contribution to the overall connectivity, flexibility, safety, dependability, accessibility, capacity and friendliness of his or her portfolio of products and services, which are offered or distributed through different technologies (wireline, Internet access, many mobile lines, and so on). It is not the specific stand-alone characteristics of any given service that are likely to count, but their contribution to the individual communications portfolio. This has important implications for our understanding of the proper regulatory approach to competition in telecommunications, in particular in the local wireline access. A product-by-product or market-by-market approach may turn out to be very detrimental to the welfare of consumers because it is likely to miss the interdependencies between products and services and the specific contribution of a product or service to the group or portfolio of telecommunications means, products and services that a customer may have access to.

5.4   THE COMPETITIVE ENVIRONMENT

One must look at local wireline access competition from a 'process' point of view rather than from a 'market share' point of view. This makes the analysis slightly more difficult and complicated but at the same time more transparent, more adequate, and better fine-tuned to the characteristics of the telecommunications industry.

The process approach concentrates on ensuring open access to the existing network facility at properly defined competitive access pricing and conditions, rather than on the number of firms demanding access or the market shares of those firms as compared with the incumbents' market share. In so doing, it avoids opening the doors to inefficient competitors who may benefit from the regulators' overemphasis on market shares at the expense of consumer welfare.

If not properly addressed, these developments may turn out to be value-destroying by reducing the incentives to invest in developing and maintaining infrastructure capacities that are prone to social inclusion, in particular in terms of the obligation to serve. The situation is even more dangerous if potential competitors are in fact unwilling to enter the (residential) local wireline access market at any cost. We will see next that there are good reasons to suspect that this may indeed be the case, and that the regulators are ill advised to concentrate their attention on market shares rather than on the competitive process. The process approach to fostering competition is also likely to ensure a level playing field for all competing players while at the same time recognizing the different responsibilities of those players, especially in developing and maintaining infrastructure capacities.

It is important to understand the competition model relevant for the telecommunications sector, including the pro-competitive or anticompetitive role of the following factors: the number of competitors, the substitutability between the different products and technologies, and the existence of excess capacity. Equally important is the understanding of the entry and exit strategies of competitors (based on '*ex post* entry' market conditions rather than on the 'actual *ex ante* entry' conditions) as well as the dynamic incentives for research and development (R&D) and innovation. Let us consider these in sequence.

Consider an industry with the following three characteristics: firms are engaged predominantly in short-run competition, the products of the different firms are close quasi-homogenous substitutes, the firms have no binding capacity constraint and can produce up to the market demand at a common quasi-constant marginal cost (with some fixed costs). In such an industry, a Bertrand industrial equilibrium will most likely emerge (Boyer et al., 2004b; Boyer et al., 2007). In such an equilibrium, the competition between the firms, which are eager to increase their market share by pricing their product slightly below their competitors' prices, will make prices fall towards the marginal cost level and therefore generate negative profit levels given by the negative of the fixed costs. Such a situation is of course extreme but it is nevertheless quite instructive. It shows that a strategy to enter into an industry characterized by product homogeneity, constant marginal costs (excess capacity) and short-run competition is likely to be a money-losing strategy. Hence, entry in such an industry is very unlikely. Entry by competitors in such an industry cannot be expected unless those Bertrand competitive conditions can be changed or bypassed and controlled in some ways. Hence the importance for the competitors to plan their entry in the industry with a strategy aimed at avoiding the Bertrand competition conditions.

A firm could profitably enter with a differentiated product (possibly and preferably offered through a different technological platform) in terms of those characteristics valued by the consumers, namely connectivity, flexibility, safety, dependability, accessibility, capacity and user-friendliness. Given the diversity of consumers' tastes and needs, there may be a niche in which the firm can enter and make profits. A firm could also profitably enter into the industry with a limited capacity (using a 'judo' economics strategy) such that the incumbent will find it unprofitable to engage in a price war in order to prevent the small and limited loss of customers that the capacity-constrained entrant can attract. The telecommunications frenzy of the 1990s, in terms of both capacity investments and technological developments, led to a 500-fold transmission capacity increase while demand itself increased by a phenomenal fourfold. It is estimated that $200 billion of telecommunications network capacity was built unnecessarily (*ex post*). This building of overcapacity in a 'commodity' market brought the telecommunications industry into the Bertrand competition trap.[3]

A firm could also profitably enter with a set of products and services relatively similar and homogenous to that of the incumbent and with a relatively large capacity, thereby credibly signalling to the incumbent that it is entering for the long term. If the entrant's strategy is sufficiently convincing, the incumbent may find more reasonable and profitable to respond to entry by adopting an accommodation strategy in order to avoid a likely unsuccessful but potentially very costly price war. In the repeated competition game, the firms may be able to find mutually beneficial accommodation strategies avoiding the Bertrand competition outcome.[4]

But entering with such a strategy is quite risky given that the incumbent may not be convinced that it is in its best interest to accommodate entry. Inducing an entry-accommodating reaction by the incumbent will in general require that the entrant either incurs significant investment outlays as sunk irreversible investment costs, or signs contracts with important customers with conditions that are very costly for the provider for renegotiating or breaking the contract, or adopts a bridge-burning (no-escape) strategy to signal clearly and convincingly to the incumbent its forceful intention to remain in the industry (by raising its cost of exiting).

That is how we must understand the competition for the residential local wireline access and why entry in this market has been so weak even after multiple pro-entry decisions and actions by telecommunications regulators since 1997. The residential local wireline market is most likely to be characterized by Bertrand competition conditions: short-run competition (favoured by the CLEC immunity regarding the development and maintenance of the network capability, which makes entry into the industry almost as easy as exit from the industry), homogenous product or service (there is arguably nothing more homogenous in the telecommunications industry than the local wireline services offered by two different providers), constant marginal cost (with this marginal cost being very low), and potentially no binding capacity constraint (by getting access to the whole local wireline network at low cost).

Entry into the local wireline market is clearly not an easy policy to pursue and could potentially be a very risky strategy. One may expect that entrants will prefer to invest in other segments of the telecommunications industry. However, this is not peculiar to the telecommunication industry and markets. Similar difficulties and risks are present in most industries. Indeed, incurring losses during the early phase of entry is quite common for most if not all entrants in any industry. Such a situation would not justify a regulatory intervention to protect entrants from the competitive pressures they will be confronted with. When Toyota decided in the 1950s to challenge the major incumbents of the time in the automobile industry, it came in with a superior technology and eventually a better product. There were no calls for protection from the incumbents through price and marketing controls (except for the legal provisions forbidding predatory pricing). The company made the risky gamble that its value chain model was better, that is, would lead to lower costs and higher quality, than the value chain model of the big three of the day. And that is the way competition serves customers through creative destruction. In the telecommunication industry, such creative destruction could come from incumbents entering into each other's markets or from new entrants capitalizing on their expertise in a related industry (cable or satellite communication services for instance). Many other examples could easily be given.

5.5   A RENEWED PRO-COMPETITION REGULATORY FRAMEWORK

The real conundrum of the regulator is to create and enforce static competitive conditions between the current bundles of products and services or technologies as well as dynamic competitive conditions yielding strong

intensity of incentives for developing new technological platforms, introducing new products and services that could be both cheaper and more
efficient, connectivity-wise and flexibility-wise, and in so doing replace
the current goods and services and possibly their producers and distribu-
tors. To achieve a proper balance between short-run (static) and long-run
(dynamic) goals, regulators must rely on competitive processes, that is,
they must make sure that their interventions are not aimed at micro-
managing prices and quantities but rather at making sure that those prices
and quantities emerge from a competitive environment. Micro-managing
prices and quantities is very much reminiscent of the old regulatory framework (rate-of-return regulation for instance) that led the regulator to
discover through adversarial proceedings the 'true' cost functions of the
firms. We know now that this objective was futile: economic costs are not
accounting costs. Exerting a proper level of competitive pressures is the
only way to induce the firms really to minimize the costs of their portfolio
of goods and services, that is, to make the accounting costs close or equal
to the economic costs.

The competition processes view of regulation developed below stresses
the importance of competition as the most powerful and efficient generator
of social efficiency in the telecommunications industry. Benevolent regulation can, under complete and perfect information, favor the emergence of
an efficient allocation of resources in a partially natural monopoly industry such as telecommunications (proper goods and services, proper prices,
proper investments in R&D, capacity building and network maintenance,
and so on) and in so doing can dominate competition. But in an incomplete information (on technologies, consumers' preferences and needs,
and R&D and investment opportunities) situation such as the one prevailing in the telecommunications industry, it is these competitive processes
and pressures that can generate the best feasible allocation. The proper
reference point here is not the full-information first-best allocation, but
rather the incomplete-information second-best allocation. Many conflicts
regarding the regulators' role, ways and means stem from the ill-conceived
comparison between an infeasible first-best allocation and a feasible
second-best allocation that takes into account explicitly the informational
constraints that the benevolent regulator is facing: the (first-)best is here
clearly the enemy of the good.

The possibility and desirability to inform customers better about the
characteristics, including pricing, of different (packages of) telecommunications goods and services should, in the context of an emerging industry
such as telecommunications, be a well-understood role of the regulator.
This regulation-by-information approach is not new but it is too often
neglected as if customers of complex goods and services were individually

able to make and understand comparisons of pricing and other characteristics such as connectivity, flexibility, safety, dependability, accessibility, capacity and user-friendliness of telecommunications services. In an
industry such as telecommunications, the generation and presentation of
such information in a user-friendly way should be a role and responsibility of the regulator, at least till the customers develop such a capability
of their own. Asking the regulator to assume such a responsibility could
prove very important for efficiency and political acceptability during the
transitory period towards a complete deregulation of the telecommunications industry.

Regulators generally recognize that substituting competition mechanisms for costly imperfect regulation could improve cost-efficiency as well
as allocative efficiency. But for network industries, it is not necessarily
desirable to introduce competition in all segments. Some segments of a
network, which are essential inputs for the potentially competitive segments, may be subject to significant economies of scale. These so-called
essential facilities, specific to a given technology, could remain price
regulated as long as alternative technologies allowing those segments to
be bypassed are not available. For instance, the local wireline may still
represent a non-standard natural monopoly. Even though it is challenged
by competing technologies, it would not make sense to duplicate the last
mile wireline connection to homes and businesses. What is needed is open
(inter-)access to all essential facilities on all technologies at proper prices
and conditions.[5]

To access essential facilities, entrants or competitors pay an access price
(regulated) and the incumbent must grant access to all firms, including its
own divisions, on an equal non-discriminatory basis. Under this regime it
is important to get the access price right. If it is 'too low', then the incumbent has low and inadequate incentives to invest and innovate, a subsidy
is provided to the entrant, and inefficient entry is encouraged. If the access
price is 'too high,' competitors have an incentive to bypass inefficiently
the incumbent's facilities, efficient entrants are discouraged from entering,
and competition is not likely to prevail.

The promotion of efficiency in network development investments is a
role and a responsibility that the telecommunications regulator should
assume. The main reasons are threefold:

● Investments in network development are typically large, long-term
  irreversible investments with significant economies of scale, making
  a given platform a non-standard natural monopoly.
● The different network development strategies offer significantly different levels of embedded managerial flexibility to delay or advance

the development timing (to slow down or speed up the pace of development), to increase or decrease the planned network size and scope, and to raise or reduce the planned quality of the network infrastructure as new information on significantly volatile demands and costs is gathered over time.

- The volatility of demand and costs together with the relative embedded managerial flexibility and the significant irreversibility of investments in network development require new evaluation methodologies (real options valuation), which are still underused by network operators.

In such a context, the regulator must make sure that inter-access to essential facilities is available, at proper non-discriminatory conditions and prices, to foster entry by more efficient providers.[6] The regulator must also recognize that the proper non-discriminatory conditions and prices must account for the real options that are exercised, as the network is being built or developed.[7] Eventually, the network developing firm is stuck with the technological characteristics of its realized network. Exercising those real options can represent a significant cost in network development and must therefore be accounted for in determining the proper non-discriminatory access prices and conditions that will govern the use of the incumbent's network by competitors. Unless a proper account is made of those real options-related costs, the development and maintenance of the telecommunications networks are likely to be inefficient, thereby imposing significant real costs on society.

This leads to two important observations. First, in reviewing pricing rules for inter-access to networks, the regulators must consider not only the development of competing networks on alternative technological platforms but also in each case the implicit costs of the real options exercised in developing the network infrastructures. Second, in reviewing the rules of competition between networks, the regulators should make explicit and transparent the pro-competition special treatment that it wishes to grant to some competitors, rather than directly impose price and technology controls on others.

In a standard natural monopoly industry, profits are likely to rise rapidly when the market expands since revenues then increase linearly or proportionately with the number of subscribers while costs increase typically very slowly due to the presence of economies of scale, economies of scope and network economies. The increase in profits means an increased rate of return on capital, exceeding over time by a significant margin the opportunity cost of capital of the firm. In any normal industry, this increased profitability would favor the entry of additional firms into the industry, but not necessarily in a natural monopoly industry. In this latter type of industry, the potential entrants understand that what is important and crucial for them is not the price they observe or the profits the incumbent firm is presently enjoying, but rather the price and profits that they will be experiencing if and when they enter the market and compete with the incumbent firm.

It is post-entry goods and services prices that preoccupy the potential entrants, not actual pre-entry prices. In particular, if it is understood that regulators are going to control pre-entry prices at a relatively low level but that they will let prices be freely determined once new competitors have entered the market, then observed pre-entry prices will have no effect on the decision to enter or not. Potential entrants will evaluate what competitive post-entry prices are likely to be and then decide if those prices justify entry or not.

If the regulated pre-entry prices in the local wireline services market are fixed at a relatively low level in order to prevent monopoly pricing, it may give the wireline incumbent a hedge insofar as these regulated prices prevent the emergence of alternative bundles of services offered over competing technologies. If those technologies cannot profitably compete with the low prices set by the regulator for the local wireline services market, then consumers will eventually suffer from the reduced technological advances and the reduced variety in products and services. A short-term benefit from lower local wireline prices is obtained at the expense of a long-term cost in terms of reduced choices, reduced innovation efforts and reduced competitive pressures.

If the regulated pre-entry prices in the local wireline services market are fixed at a relatively high level in order to prevent predatory pricing and favour competition (through imposing floor prices on the ILEC without regulating in any way the CLECs' pricing, forbidding win-back actions by the ILEC for a certain period, and other similar policies), it may give the competitors (the CLECs) a hedge insofar as the regulated prices prevent the incumbent from competing efficiently. Such pricing regulation is likely to prevent the emergence of a more efficient portfolio of goods and services offered to customers, at least in the short run.[8]

We need a different, more efficient approach to controlling monopoly pricing and predatory pricing while avoiding the subsidization of inefficient entrants, that is, a more efficient approach to developing a level playing field in short-run competition while at the same time keeping a high intensity of incentives for long-run competition. This more efficient approach rests on unregulated pricing of telecommunication services subject to general pricing policies that the regulator can impose and enforce on all players, ILECs and CLECs.

The current dominant approach to the prevention of monopoly pricing (mainly in wholesale markets) and predatory pricing (mainly in retail services) in networks rests on the notions of marginal, incremental or avoidable (MIA) cost. Although these three concepts of cost are not exactly the same, they share similar foundations. The marginal cost is the additional cost a firm must incur to increase its production of a given service by a small amount; the incremental cost refers to the same concept for more important increases; the avoidable cost refers to the reduction in cost a firm would experience if it were to drop the production of a particular good or service from its current portfolio of products and services (the avoidable cost includes therefore the fixed cost specific to that product or service).

A firm is considered capable of exercising market power if its price for a particular good or service could profitably be set significantly above its MIA cost, including its MIA cost of capital, for a significant period of time. It could be considered guilty of predatory pricing if its price for a particular good or service is below its MIA cost for that good or service, including again its MIA cost of capital. Unfortunately, the use of the concept of MIA cost is a source of major problems and conflicts. The problems and conflicts stem from the fact that the MIA cost of a particular good or service is a concept suffering from five major shortcomings when economies of scale and scope as well as network economies are present, as in telecommunications: it is non additive, order-dependent, unstable, manipulable and horizon-dependent.

Clearly, a proper notion of cost allocation for the purpose of verifying if a firm's pricing behavior violates the non-predatory pricing rules must satisfy the requirements of additivity, order-independence, stability, non-manipulability and horizon-independence. One such cost allocation procedure is the Shapley–Shubik (SS) cost allocation rule (Boyer et al., 2006). In a nutshell, this rule corresponds to the average MIA cost over all possible production orderings of the goods and services offered by a firm. Rather than taking simply the usual MIA cost, the SS procedure takes the average MIA cost over all orderings, thereby making it immune to the shortcomings of the cost allocation rule based on the MIA cost.

Moreover, the SS cost allocation rule has two other important properties, which prove very helpful in practice. First, it is such that the cost shares are invariant to the decomposition of total cost between specific and common or joint costs. This property is important because the decomposition of the total costs between joint costs (to be shared) and specific costs (to be borne by the specific good and service causing them) is very often a source of conflict. The fact that the decomposition is irrelevant is therefore a property that is most welcome. Second, the SS cost-sharing rule satisfies demand monotonicity, which means that the cost allocation of a good or service increases if the demand (production) of that good or service increases while the production of others remain constant, and also that under economies of scale the cost share of a good or service decreases if the production of another good or service increases while its own production remains constant. Again, this property is conducive to an agreement between different stakeholders. More generally, the cost allocations obtained from applying the SS procedure can be shown to be an almost costless surrogate for the cost allocation that would be obtained through lengthy and costly bargaining 'between the different goods and services' (or their demanders/consumers) as to the proper allocation of common costs.

The regulator would consider a firm guilty of predatory pricing if the firm was found to price one (group) of its products or services below the per-unit cost allocation associated to that product or service in the total cost of the firm. Hence, the per-unit cost shares become the price floors under which a firm would not be allowed to price any of its goods and services. Those price floors are determined by the 'real' cost function of the firm itself and not by some argument regarding the necessity to grant some advantage to competitors.

Combining proper access pricing for essential facilities and proper cost allocation rules leads us to the intended result: if a firm is indeed more efficient than another, it should be able to achieve per-unit costs for its goods and services that are in general lower than those of its competitors. In that sense, the SS cost allocation rule is a pro-competitive price floor rule, protecting all firms, incumbents and competitors against predatory pricing. It creates adequate incentives for all firms to reduce the costs of their respective goods and services, thereby developing a level playing field with similar competition rules for everyone. In the presence of economies of scale, economies of scope and network economies, the SS cost allocation rule delivers the proper pricing benchmark compatible with the belief that competitive pressures constitute the most important dynamic driver toward socio-economic static and dynamic efficiency. The experience gathered over time in diverse applications of the SS cost allocation rule[9] makes it a good candidate for the basic pricing rule. This does not mean that the application of the procedure is (always) easy and straightforward, but none of the alternative procedures, and certainly not the one presently used, is easy and straightforward either.

How then can the government or regulator adopt a proactive strategy to favour the emergence of more competition? The answer is simple: by promising entrants a generous but deferred investment grant or tax credit to be paid *n* years after entry conditional on the entrant's acquisition of a market share larger than x percent. Only entrants with superior products or services, and/or superior technology, and/or better-quality consumer service, or who believe that they can beat the incumbent one way or another, will enter the industry.[10]

5.6 CONCLUSION

I have argued in this chapter that the traditional market shares measure of competition is inadequate in the telecommunications industry because this industry has more of the characteristics of an emerging industry than of a mature industry. In such a context, the role of the regulator should be focused on three responsibilities.

Firstly, acting as the trusted generator of information for the consumers on pricing structures and product characteristics, underlying the derived demand for telecommunications goods and services, namely connectivity, flexibility, safety, dependability, accessibility, capacity (high speed and broadband) and user-friendliness.

Secondly, acting as the manager of the level playing field conditions to favor both static efficiency and dynamic efficiency, and enacting indirect policies to control monopoly and predatory pricing, based on full cost sharing between goods and services offered by any given firm (as provided for example by the cooperative game-theoretic SS cost allocation rule).

Thirdly, acting as the promoter of efficient investment programs in network development and maintenance to guarantee the integrity of the global telecommunications network, and designing access pricing rules incorporating all network access costs (including the real options costs embedded in completed networks). Those redesigned responsibilities of the regulator are in fact the end point of the sequence of regulatory reforms that have been implemented since 1997. The telecommunications industry regulator would become a truly pro-competitive watchdog and an integral part of competition policy implementation.

Although the information gathering, the institutional reforms and the change of minds and attitudes required to implement a true pro-competitive regulatory framework for telecommunications, in particular for the local access markets, represent a sizable and significant undertaking, it is no more and probably much less demanding than the prevailing regulatory framework. Since the change in approach will take some time, it is urgent for national authorities to launch a significant program of research, consultation and transformation, aimed at developing the necessary means and procedures to implement eventually a switch to the new pro-competition rules proposed in this chapter.

NOTES

1. This chapter is a revised and condensed version of my paper 'The measure and regulation of telecommunications markets' (CIRANO working paper 2005s-35). I want to express my thanks to the ITS Montreal Conference participants and to the external referees for their comments.
2. The authors used household-level data from the Bill Harvesting dataset from TNS Telecoms ReQuest Market Monitor® along with its survey response (see www.tnstelecoms.com). The estimate for the cross-price elasticity of fixed line telephony (all lines combined) from mobile telephony prices is lower (of the order of +0.06 to +0.08), as expected insofar as the main effect should be on second fixed lines. See also Dzieciolowski and Galbraith (2004) and Zimmerman (2007). More recently, Taylor and Ware (2008, p. 6) estimated an all fixed lines combined cross-price elasticity between wireline and wireless of 1.4.
3. The Economist (2003). The impacts of the emergence of Bertrand competitive conditions in an industry can be quite significant: the Canadian-based Nortel saw its capitalization fall from \$400billion to \$3billion in two financial years 2000–2002, one of the main reasons being the large overcapacity in the industry. Nortel, North America's biggest maker of telephone equipment, has filed for bankruptcy protection under the US Chapter 11 bankruptcy protection law on 14 January 2009 (BBC News).
4. The 2005 Nobel Prize winners in economic science Thomas Schelling and Robert Aumann were pioneers in the development of such competitive modeling.
5. In proceedings under competition law, the definition and characterization of 'essential facilities' remain a hotly debated issue.
6. For a presentation and discussion of access pricing rules, see Boyer and Robert (1998), and Boyer et al. (2008).
7. For a discussion of the real options approach to evaluating investment decisions, see Boyer et al. (2004a). For a discussion of real options and access pricing in network industries, see Haussman and Myers (2002).
8. What is wrong with lower prices? Low prices, or more generally loss leaders, may be detrimental for competition if they are intended to harm competitors. If they are not, for instance if a telecommunications provider announces a sale of telephones at prices below costs in order to reduce costly inventories of slow-moving goods, then such low prices would be considered fair rather than predatory. In boxing, there is a difference between fighting and trying to beat the opponent by throwing hard punches to put him unconscious (legal, fair) and doing it by biting the opponent's ear (illegal, unfair). In tennis, one is allowed to smash the ball as hard as one can, but not to aim at the opponent's head; similarly with fair low prices and predatory low prices.
9. See Young (1994) for discussions of some of those applications.
10. For an application of such principles to spectrum auctions, see Boyer (2007).

REFERENCES

Boyer, M. (2007). Optimal policy relative to spectrum auction. Industry Canada, Government of Canada. Gazette Notice DGTP-002-07, May.

Boyer, M., M. Benitah and S. Weihao (2008). Real options, network development and network access. WIK NETCONOMICA 2008 Conference on Current Issues in Network Economics: Regulatory Risk, Cost of Capital and Investment Incentives. September, Königswinter, Germany.

Boyer, M., P. Christofferson P. Lasserre and A. Pavlov (2004a). Value creation, risk management and real options. *Icf Journal of Management Research*, **3**(10),

42–62; and in CIRANO 2003RB-02 Burgundy Report at http://www.cirano.qc.ca/pdf/publication/2003EB-02.pdf.

Boyer, M., P. Lasserre T. Mariotti and M. Moreaux (2004b). Preemption and rent dissipation under Bertrand Competition. *International Journal of Industrial Organization*, **22**(3), 309–28.

Boyer, M., P. Lasserre and M. Moreaux (2007). *The dynamics of industry investments*. CIRANO working paper 2007s-09.

Boyer, M., M. Moreaux and M. Truchon (2006). Partage des coûts et tarification des infrastructures. CIRANO Monograph 2006MO-01.

Boyer, M. and J. Robert (1998). Competition and access in electricity markets: ECPR, global price cap and auctions. In G. Zaccour (ed.), *Deregulation of Electric Utilities* (pp. 47–74). Norwell, MA: Kluwer Academic Publishers.

Dzieciolowski, K., and J.G. Galbraith (2004). Indicators of wireline/wireless competition in the market for telecommunication services. CIRANO 2004RP-21; Bell Canada, Analysis of Local Residential Voice Network Usage, a detailed study covering the two-month period 1 January to 29 February 2004.

The Economist, (2003). Beyond the bubble. 9 October.

Haussman, J. and S. Myers (2002). Regulating the United States railroads: the effects of sunk costs and asymmetric risk. *Journal of Regulatory Economics*, **22**(3), 287–310.

Rodini, M., M.R. Ward and G.A. Woroch (2002). Going mobile: substitutability between fixed and mobile access. Center for Research on Telecommunications Policy Working Paper CRTP-58, Haas School of Business, University of Berkeley. December.

Taylor, W.E. and H. Ware (2008). The effectiveness of mobile wireless services as a competitive constraint on landline pricing: was the DOJ wrong? NERA Working Paper, December 2008.

Young, H.P. (1994). Cost allocation. In R.J. Aumann and S. Hart (eds), *Handbook of Game Theory* Vol. II (pp. 1191–1235). Amsterdam: North-Holland.

Zimmerman, P. (2007). Recent developments in US wireline telecommunications. *Telecommunications Policy*, **31**(6–7), 419–37.

6. Preventing harm in telecommunications regulation: a new matrix of principles and rules within the *ex ante* versus *ex post* debate[1]

Kenneth Jull and Stephen Schmidt

6.1 INTRODUCTION

A recent tragedy involving the 911 emergency service illustrates the need for regulation in those sectors that impact human health and safety. A family who subscribed to Voice over Internet Protocol service moved from Toronto to Calgary. When their young child became gravely ill, they called 911 and waited for an ambulance that did not arrive. Tragically, the 911 operators dispatched an ambulance to the Toronto address, as it had not been updated to reflect the move to Calgary. Sadly, the young child passed away. It is not our intention to comment on the specifics of this case. We can comment, however, that this case underlines the need for regulation in those sectors where the potential for human harm is the greatest. It is of little comfort to a family who has lost a child to advise that the situation will be remedied in an *ex post* world, if it might have been prevented by *ex ante* measures to ensure comprehensive and accurate 911 coverage across all telecommunications networks. In a six-month investigation of Canada's 911 service, the *Globe and Mail* found that: 'a lack of federal oversight, regulatory loopholes and outdated technology have left this country's emergency dispatchers scrambling to locate callers who dial 911 from cell phones or from Internet phones'.[2]

Most North American regulatory systems, in areas ranging from the environment to securities law, utilize a basic *ex post* model: the regulation or statute sets the standard, and if one breaches the standard, then remedies are imposed after a legal finding of wrongdoing. By contrast, *ex ante* systems operate prospectively, depend on prior approval and impose remedies on an anticipatory basis, prior to any actual breach or wrongdoing. Regulatory systems in the telecommunications industry have traditionally been *ex ante* systems. For example, a review of Canada's Telecommunications Act shows a regulatory scheme that is weighted towards *ex ante* regulation through, among other things, tariff approval requirements, agreement approval requirements, unbundling orders, interconnection orders, and so on. The basic presumption of the Act is that all telecommunications services are governed by tariffs that must be filed and approved by the Canadian Radio-television and Telecommunications Commission (the CRTC) prior to the offering of service, unless the CRTC has made a decision expressly to forbear from exercising its powers in certain circumstances (Government of Canada, 2006). In other words, telecommunications services cannot be offered to the public without prior approval by the state.[3]

The *ex ante* system of prior approval and supervision by the regulator has been criticized by commentators as being a blunt 'one size fits all' mechanism (Janish, 2005, p. 13). A new regulatory framework has been recommended by the Telecommunications Policy Review Panel (TPRP) Final Report, released in March 2006 (Tremblay et al. (2006). The TPRP approach would set out broad principles to prohibit anti-competitive conduct instead of detailed *ex ante* rules.

We propose three principles that ought to govern the balance between *ex ante* and *ex post* systems in the telecommunications world of the future. First, governments and regulators ought to focus on the prevention of harm and the serving of human needs in the order of their priority (Archibald, Jull and Roach, 2004). Recent scholarship has focused on the prevention of harm in regulation. In his book *The Character of Harms*, Malcolm Sparrow (2008) argues that regulators should focus on fixing the big problems that cause the most harm, by developing strategies that operate at the operational risk level (Sparrow, 2008). We see a role for *ex ante* regulation in preventing harm in the telecommunications sector (such as preserving the integrity of emergency 911 systems) and in protecting vulnerable groups within the meaning of human rights legislation. The application of harm theory to telecommunications policy puts the industry into the proper context when considering the role of regulation. In short, telecommunications may prevent harm from coming to people, but not to the extent of some other sectors (such as health, the environment and transport) that may warrant deeper regulatory oversight.

The CRTC has responded to the 911 issue by announcing that it will force the cellphone industry to upgrade Canada's 911 systems. Government officials said in January 2009 that they will impose a February 2010 deadline to install the necessary equipment to give 911 dispatchers the ability to locate cellular calls in an emergency.

In announcing the decision to the media, the CRTC uses language which is consistent with the thesis of this chapter: 'I thought that this would put the matter to rest', a CRTC official said in an interview. 'We are concerned about the safety and security of Canadians . . . that's sort of our guiding light if you wish' (Robertson, 2009).

The second principle that we propose is that regulation of competitive markets will encompass risk management and social regulation.[4] The financial crisis of 2008 underscores the need for regulation of risk management systems in economic markets.

Our third principle is that the regulatory state should offer multiple models that reflect the different needs and interests at stake within both the *ex ante* and the *ex post* world. Indeed, there is a growing recognition in legal circles that any sophisticated system ought to offer a full range of models based on the seriousness and economics of the case (Mullan, 2005). For example, within the *ex post* paradigm, there is a new principles-based model emerging, led by developments in the securities field. Ford (2008) describes this development and she identifies the principles-based and outcome-oriented regulation that has been advocated by the British Columbia Securities Commission as an example of the new governance. She describes the basic difference in the two systems as follows:

> The classic example of the difference between rules and principles or standards (to use another term) involves speed limits: a rule will say, 'Do not drive faster than 55 mph', whereas a principle will say, 'Do not drive faster than is reasonable and prudent in all the circumstances'. Put another way, a rule generally entails an advance determination of what conduct is permissible, leaving only factual issues to be determined by the frontline regulator or decision-maker. A principle may entail leaving both specification of what conduct is permissible and factual issues for the frontline regulator. (Ford, 2008, p. 8)

If one accepts that the *ex post* model might have two subsets within it, being rules-based and principles-based, then one must also accept that the *ex ante* paradigm also has two subsets within it, being rules-based and principles-based. The dichotomy of *ex ante* and *ex post* becomes a matrix of four models. An example of a rules-based *ex ante* solution in the highway traffic example would be the requirement that all trucks have governors installed that prevent them from travelling over 105 km per hour. An example of a principles-based *ex ante* solution would be the requirement that each driver pass a driver's test that is based upon criteria developed from a principled approach.

Table 6.1 illustrates the integration of the three principles that we are advocating.

Table 6.1 Ex ante *and* ex post *rules- and principles-based regulation matrix*

Paradigm	Timing	
	Ex Ante	*Ex Post*
Rules-based	Harm related: imposition of detailed technical rules for operation of 911 emergency services as part of licensing requirements.	Prohibition of contacting customers on a do-not-call list.
Principles-based	Imposition of licensing terms requiring reasonable, non-discriminatory access to services for disabled persons.	Principle that carriers must interconnect with each other for the exchange of traffic on reasonable terms.

6.2 PRINCIPLES OF REGULATION

6.2.1 First Principle: Governments and Regulators Ought to Focus on the Prevention of Harm and the Serving of Human Needs in the Order of their Priority

Types of harm

Sparrow (2008) argues that regulators should focus on fixing the big problems that cause the most harm, by developing strategies that operate at the operational risk level. He further observes (p. 10) that the financial sector laid an early claim to the phrase 'risk management'. Risk management is the process of weighing policy alternatives in light of the results of risk assessment, and selecting appropriate control options, including regulatory measures (Powell, 2000, p. 138). Sparrow's (2008, p. 11) major contribution is to inject the word 'harm' into the risk formula, which includes potential harms. He identifies five properties of harm that are common to many disciplines: invisible harms, harms involving conscious opponents, catastrophic harms, harms in equilibrium and performance-enhancing risks (p. 173). Each is worth briefly reviewing.

Invisible harms may be invisible by design, such as white-collar crimes. In other cases the harms are not visible because of a lack of reporting will or mechanisms. Sparrow (2008, Chapter 8) gives the example of the internal revenue service failing to measure the problems of non-filers, as by definition there was no record of those who did not file their tax returns.

One solution suggested by Sparrow would be to cross-reference outside databases. For example, a review of the Yellow Pages of businesses in a particular area could be cross-referenced to the number of businesses who actually filed.

In the telecommunications field, an example of invisible harm might be the exclusion of certain disadvantaged groups from service due to cost or geographic issues. As there is no record of non-subscribers, the harm to them is not easily measured but can only be speculated on.

Conscious opponents are described by Sparrow (2008, Chapter 9) as having a brain behind the harm. The opponents can range from fraud perpetrators to students cheating in exams. Sparrow observes that these types of harms confound our normal reliance on probabilities, as it is difficult to predict the intentional conduct of opponents. The methods of fighting this type of harm are traditionally found in the criminal justice system: use of surveillance, informants and confessions.

In the telecommunications field, as it has been historically highly regulated, it is likely that the incidence of harm caused by fraudulent behavior of service providers is relatively low. As the market becomes increasingly deregulated, this could potentially change. Although this is speculative, we predict that the level of intentional or fraudulent conduct will continue to stay relatively low as the market is deregulated, because the culture of most players in the industry has its origins in regulated compliance. This does not mean however that regulators should be complacent as the market becomes deregulated. Regulators must attempt to measure the possibility of non-compliance and be prepared to borrow techniques from the criminal justice area.

Catastrophic harms are rare events with very serious consequences, such as nuclear terrorism or earthquakes. In such events the telecommunications networks may be threatened and must have in place contingency plans and backup systems.

Harms in equilibrium relate to the imbalance that may be created in a competitive industry from varying levels of compliance. Sparrow (2008, Chapter 11) gives the example of an industrial sector largely out of compliance with environmental regulations. A company that desires to be compliant will be put at a disadvantage from a cost perspective relative to its peers who are non-compliant.

Performance-enhancing risks relate to the organization's performance goals. Sparrow (2008, Chapter 12) gives the example of aggressive, fraudulent or abusive billing practices which enable a corporation to maximize profits. Sparrow argues that departments responsible for compliance within an organization must have credibility if the organization is to resist performance-enhancing risks. The reduction of performance-enhancing risks relates directly to compensation. Compensation ought to be (and historically has not been) related to levels of compliance in addition to levels of economic gain.

Ranking of types of harm
One of the authors of this chapter has argued elsewhere that where legislatures codify risk assessment values that potentially impact on human life and health, the legislative branch should utilize the wealth of scientific expertise to the greatest extent possible (Archibald, Jull and Roach, 2004). The uneasy tension between politics, the public and experts' assessments has led to several problems in risk management (Slovic, 1999). A problem that emerges from this tension is identified by Justice Breyer (1993) as 'random agenda selection':

> Agency priorities and agendas may more closely reflect public rankings, politics, history, or even chance than the kind of priority list that environmental experts would deliberately create. To a degree, that is inevitable. But one cannot find any detailed federal governmental list that prioritizes health or safety risk problems so as to create a rational, overall agenda, an agenda that would seek to maximize attainable safety or to minimize health-related harms.

Breyer (1993) illustrates a very interesting comparison of the expert's assessment of various risks, against the public's assessment of risks, in Table 6.2.

The public policy implications of the above comparison are enormous.[5] The differences between expert and public assessment of risk are the subject of analysis by the Canadian Centre for Management Development (Hill, 1969). Serious issues such as air pollution, which the experts assert pose high risks to public health, are not viewed by the public as significant. People rank vehicle exhaust as fifteenth in priority, and yet the experts indicate that this problem should be ranked much higher.

Obviously, there is a serious knowledge gap between the public perception of risk and the experts' perspective. In setting regulatory standards, we invite the legislative branch to measure the potential gravity of impact upon our society in accordance with scientific methodology. In other words, governments should try to narrow the information gap between the experts and the public, by taking the lead in recognizing new scientific theories. Sparrow (2008, p. 13) makes a similar argument that: 'At this, the highest level, where one adopts the more paternalistic perspective of governments, the psychology of individual risk perception matters less, or matters for different reasons.' Science can identify a hierarchy of human needs within social psychology. Governments and regulators should be measured by their ability to serve these needs in the appropriate order.[6]

Table 6.2   Knowledge gap between the public perception of risk and the experts' perspective

Public	EPA experts
1. Hazardous waste sites	Medium-to-low
2. Exposure to worksite chemicals	High
3. Industrial pollution of waterways	Low
4. Nuclear accident radiation	Not ranked
5. Radioactive waste	Not ranked
6. Chemical leaks from underground storage tanks	Medium-to-low
7. Pesticides	High
8. Pollution from industrial accidents	Medium-to-low
9. Water pollution from farm runoff	Medium
10. Tap water contamination	High
11. Industrial air pollution	High
12. Ozone layer destruction	High
13. Coastal water contamination	Low
14. Sewage plant water pollution	Medium-to-low
15. Vehicle exhaust	High
16. Oil spills	Medium-to-low
17. Acid rain	High
18. Water pollution from urban runoff	Medium
19. Damaged wetlands	Low
20. Genetic alteration	Low
21. Non hazardous waste sites	Medium-to-low
22. Greenhouse effect	Low
23. Indoor air pollution	High
24. X-ray radiation	Not ranked
25. Indoor radon	High
26. Microwave oven radiation	Not ranked

Source: Breyer (1993), Table 4, p. 20, as reproduced with permission in Archibald, Jull and Roach (2004).

At this juncture, the reader may say, what does all of this have to do with telecommunications law? Several conclusions relevant to the practice of telecommunications regulation flow from the foregoing. First, telecommunications regulation ought to have as its primary focus the prevention of harm. The approach advocated by Sparrow (2008, p. 17) of preventing bad things as opposed to the construction of good things brings perspective to the progressive lessening of sector-specific economic regulation in telecommunications. Simply put, we should ask whether it is wise to put significant public resources into the detailed economic regulation of telecommunications when that money might be spent more wisely in other sectors that prevent harm to people. These sectors include health, the environment and security. (Of course, where issues of harm genuinely present themselves in telecommunications markets, they do merit regulatory attention.) Second, the latest scientific thinking can, and should, be invoked to focus harm-prevention efforts in accordance with human needs priorities. This means, for example, that a regulatory focus on matters such as the 911 service which have a dramatic impact on saving lives will have significant priority over measures focused, for example, on the pricing of communications services (where questions of human harm are less acutely engaged). Third, telecommunications regulatory practice organized around the prevention of harm, and duly informed by scientific thinking, will look fundamentally different from the practice of regulation today where public and private investments in regulation almost certainly reflect an inversion of this thinking.

We acknowledge that the pricing of telecommunications services can, in and of itself, engage human harm issues. For example, this would be the case if services determined to be essential to participation in society and the economy were priced at a level, or in a manner, that precluded customers from engaging in social and economic activity via the telecommunications network. There is a general recognition that local telephone service is an essential service and that it ought to be priced in a manner that makes it accessible to all users, including lower-income users. That being said, market circumstances in Canada have evolved to the point where the CRTC relies primarily on competitive markets to constrain the pricing of local telephone service. In sum, we acknowledge that supplier pricing behavior for the subset of telecommunications services determined to be essential services can raise human harm issues. This is why essential services have been historically subject to economic regulation where markets forces, alone, were not considered sufficient to constrain pricing.

6.2.2 Second Principle: Regulation of Competitive Markets will Encompass Risk Management and Social Regulation

The second principle that we propose is that the regulation of competitive markets will encompass risk management and social regulation. The 2008 financial crisis emphasized the need for regulation of risk management systems in competitive markets. A leading text used in business schools to teach risk management is *Seeing Tomorrow* by Ron Dembo and Andrew Freeman (1998). Ten years before the 2008 crisis, Dembo and Freeman

(p. 28) warned that: 'entire businesses have been brought to their knees by single-scenario forecasting'.

Regulation cannot of course impose prudent investment strategies, but what it can do is require that a risk assessment structure is in place and that investors have maximum knowledge of the risks that they may face. The US Securities and Exchange Commission has recently encouraged companies to take a more risk-based approach to complying with legislation such as the Sarbanes–Oxley corporate reform package (*Globe and Mail*, 2007). Following the financial crisis of 2008 there is a need for clarification about proper risk management techniques versus risk assessment which merely measures the level of risk. Risk management is the process of weighing policy alternatives in light of the results of risk assessment, and selecting appropriate control options, including regulatory measures. Risk management identifies the priorities of an organization and in particular when audits of these priorities may be required (Archibald, Jull and Roach, 2004, Chapter 7).

Applying the second principle to the reform of telecommunications requires regulation of risk management in competitive markets. The tariff scheme in the present Telecommunications Act requires prior approval, which increases the risk of discouraging legitimate competition. The TPRP Final Report (Tremblay et al., 2006) cites Posner (1999) for the proposition that economic regulation leads to activity to circumvent those constraints, which is a waste from the point of view of the economy.[7]

Increased reliance on market forces brings the telecommunications sector in line with other competitive industries. The TPRP recognizes that new principles and rules must be drafted that will then be enforced: 'Consistent with the deregulatory approach adopted by the Panel, the new regulatory framework should set out broad principles to prohibit anticompetitive conduct instead of detailed *ex ante* rules' (Tremblay et al., 2006, p. 22).

Returning to the TPRP recommendation, what will the broad principles that will prohibit anticompetitive conduct look like? The starting point is the existing Competition Act:

> To address anti-competitive conduct in the telecommunications market, the provisions of the current Competition Act cannot be adopted word for word but can serve as a framework. Telecommunications is a network industry, with large sunk costs and significant economies of density and scope as well as positive externalities. In such an industry, network effects are important, which naturally allow some players to have very large market shares in equilibrium. As well, the definition of the proper market for further analysis can be particularly difficult in telecommunications markets. (Tremblay et al., 2006, p. 24)

The lessening or removal of sector-specific economic regulation does not require the abandonment of social regulation. Historically, globalization forced deregulation in some industrial sectors, but at the same time, expanded regulation occurred in 'framework sectors' such as the environment. The Canadian government has recognized this distinction with its direction that the CRTC, 'when relying on regulation, use measures that are efficient and proportionate to their purpose and that interfere with the operation of competitive market forces to the minimum extent necessary to meet the policy objectives' (Governor in Council, 2006).

The TPRP recommended a new approach to control anticompetitive conduct in telecommunications markets on the basis of complaints made on an *ex post* basis, rather than by prescribing detailed *ex ante* restrictions governing the provision of services. We argue in the next section that a two-model dichotomy is not sufficient, but rather that there is a matrix of four models to choose from.

6.2.3   Third Principle: The Regulatory State should Offer Multiple Models that Reflect the Different Needs and Interests at Stake within both the *Ex Ante* and the *Ex Post* World

Definition of terms: *ex ante* versus *ex post*
There is some confusion about the distinction between *ex ante* and *ex post* systems. We wish to define our terms. The central concept that underlies an *ex ante* system is prior approval of an activity. For example, the CRTC may require that a tariff be approved, or a license be obtained, prior to the provision of telecommunications services to the public. Rates, terms and conditions for the provision of service are imposed *ex ante*, but any breach of these may be addressed *ex post*. An *ex post* approach is distinguished from an *ex ante* approach by the fact that the former imposes remedies only after a legal finding of wrongdoing, whereas the latter imposes remedies, preventatively, in advance of any wrongdoing.

These concepts may seem foreign, as the practical reality is that there are very few telecommunications-related enforcement actions brought *ex post* in Canada, despite the fact that there are a range of remedies presently available to the CRTC. We suspect that the same situation applies globally. As a result of low enforcement, the perception may be that the existing framework of prior approval is toothless. This paucity of enforcement activity may simply reflect high levels of compliance.

Other areas of the law also use a combination of *ex ante* prior approval by a regulator combined with *ex post* enforcement. For example, in the area of environmental conservation, it is a requirement of Canadian landowners that they seek permits for any work adjacent to floodplain areas.

Unlike telecommunications, the incidence of non-compliance is higher. Regrettably, some landowners choose to ignore or violate the permit process and undertake development without permits. In such a case, the landowner can be prosecuted for committing the regulatory (strict liability) offence of development without a permit. In this area, it makes sense that a specialized regulator may give *ex ante* approval for development in sensitive areas, and that a failure to seek such approval will be enforced with a regulatory offence that deals with *ex post* conduct.

In the area of telecommunications, most are familiar with the type of *ex ante* orders made through various mechanisms including tariff approval and licensing processes. Less familiar are the *ex post* remedies presently available to enforce regulatory non-compliance. There are remedies provided by the Telecommunications Act that may be enforced by courts, and these broadly include regulatory offence proceedings and damages awards, based upon a statutory breach of the Act.

The TPRP recommended a new approach to control anticompetitive conduct in telecommunications markets on the basis of complaints made on an *ex post* basis, rather than by prescribing detailed *ex ante* restrictions governing the provision of services. By definition, *ex post* enforcement occurs only subsequent to a violation of rules, and is often triggered by a complaint. The *ex post* model is used in most legal systems, ranging from regulatory to criminal law. A pure *ex post* system has its origins in the common law system of privately enforced rights. Here, the focus is on redressing the harm caused by a breach of rules or standards.

Where health and safety of consumers is at risk, we can envision some justified need for *ex ante* approval by a regulator

Ex ante control has the advantage of using a rational approach towards an issue in advance (Yael Aridor Bar-Ilan, 2007). The CRTC has recognized this by recently reissuing its 911 requirements for Voice over Internet Protocol (VoIP) service providers. In its 911 Circular, the CRTC articulates detailed rules that must be followed by VoIP providers when handling 911 calls. This would be an example of a rules-based *ex ante* system, as it would be a requirement for all telecom providers. In the case of the 911 service, *ex ante* rules concerning the service could be coupled with additional measures to ensure compliance (periodic audits or inspections, for example).

There is precedent for the use of *ex ante* prior approval to protect human life. For example, in Canada one must obtain prior approval from a regulator to build in areas adjacent to floodplains, given the risks to human life and safety that may be created by structures that are in the path of a flood. As a general principle, our reference to health and safety is not restricted to life-threatening situations. We use the term in the tradition of prevention of harm to human health, which is a wide-ranging concept and is consistent with Sparrow's concepts. In particular, prevention of catastrophic harms justifies an *ex ante* approach.

Where vulnerable minority groups[8] within the meaning of human rights legislation may not have appropriate access to telecommunications, an *ex ante* approach fills the gap

Disabled persons, for example, may not have the resources to enforce their rights in an *ex post* world, depending upon the level of regulatory enforcement. Moreover, access to telecommunications for disabled persons may present special needs that are more appropriately dealt with *ex ante*. Sparrow might characterize the risk to disabled people as an invisible harm, as they are less likely to have the resources to lobby for coverage.

The CRTC Discussion Paper (Government of Canada, 2006) submitted to the Telecommunications Policy Review (TPR) noted that competition law does not contain mechanisms to ensure access to telecommunications networks by disabled persons. We recognize a limited role for *ex ante* regulation to protect groups such as the disabled, in conjunction with *ex post* remedies (discussed later in this chapter).

The choice between rules-based and principles-based systems will be assisted by analyzing the amount of flexibility that is required

Within both the *ex ante* and *ex post* paradigm, there may be both rules-based and principles-based models. We believe that this distinction may have been lost in the fray of the fight between *ex ante* and *ex post* approaches. Every regulatory framework can be placed along a continuum between rules-based and principles-based. Cunningham (2006) made this observation in his study of Principles and rules in public and professional securities law enforcement: a comparative US–Canada inquiry:

> Rules and principles are both necessary and desirable features of an overall securities regulation system. Whether laws are better cast in one form or the other depends on trade-offs involving certainty versus contextual judgment on the one hand and the relative novelty versus norm recognition on the other. Aspirations to create securities regulation regimes denominated as principles-based or rules-based likely are conceptual fantasies, given the nature, complexity and capaciousness of these labels.

As noted above, a rules-based *ex ante* model would be the requirement that all trucks must have governors on them that prevent the truck speed from exceeding 105 kph (Ministry of Transportation, 2008). Such a rule is proactive, and protects the lives of drivers on the road. The rule is overinclusive, in that it includes truck drivers who have impeccable driving records and who are safe drivers. The rule is necessary, however, because recent tragedies have demonstrated that there are some truck drivers who, for various reasons, are not obeying speed laws. Given the size of the trailers carried by trucks, the results of excessive speed can be deadly. The requirement of governors on all trucks obviously lacks flexibility. By contrast, a principles-based license is granted by a regulator (or examiner in the case of transport) executing discretion based on principles of safety.

Sparrow (2008, Chapter 6) devotes a chapter of his book to the puzzles of measurement. This novel analysis of measurement of harm contains several insights for the *ex post* versus *ex ante* debate. In an *ex ante* world, the focus is on measurement of process, being the process of regulatory approval. In an *ex post* world, the focus shifts to measurement of harm. This raises an important issue about resources. A shift from *ex ante* regulatory approvals will free up resources to be devoted to the prevention of harm in an *ex post* world. It is essential that the resources be devoted appropriately to ensure the robust nature of the *ex post* system.

In telecommunications, as noted above, there is a strong case for *ex ante* regulation to ensure effective access to telecommunications services by disabled persons. In an area of rapidly evolving technology – cellular and PDA (personal digital assistant) handset design, for example – it may not make sense to adopt an exclusively rules-based, *ex ante* approach because overly prescriptive rules about device design[9] will rapidly fall out of date as technology evolves. In this case, a principles-based approach may be preferable. This approach would impose principles-based accessibility requirements on an *ex ante* basis. An example of this would be an obligation that communications devices become reasonably accessible to various classes of disabled persons. This principles-based approach to accessibility reflects the approach of Canadian human rights legislation. Where a human needs analysis indicates a requirement for a more precise, consistent and industry-wide implementation of an accessibility requirement (as might be the case with the Braille inscriptions on telephone keys), a rules-based, *ex ante* approach would be suitable. This approach could, of course, coexist alongside principles-based *ex ante* requirements for accessibility.

A rules-based system, whether it is *ex ante* or *ex post*, is better suited to regulated actions that are simple, stable and do not involve huge economic interests.[10] Rules may suffer from being either overinclusive or underinclusive, due to the static nature of the legislative process. Economists predict that one will find more overinclusive rules in areas where the sanction is simple damages or where transaction costs are low (Ehrlich and Posner, 1974).

Industries like telecommunications are undergoing rapid and fundamental technological and business transformation with the result that they are not simple or stable and the economic stakes are very high. In such contexts, a principles-based approach may be more appropriate. Cass Sunstein (1995) argues that rules can be overtaken by changing circumstances in fields such as telecommunications: 'In the face of rapidly changing technology, current rules for regulation of telecommunications will become ill-suited to future markets. For this reason it may be best to avoid rules altogether or at least create only a few simple rules that allow room for private adaptation' (Sustein, 1995, pp. 993–4).

As is clear from our taxonomy, we would not go so far as rejecting all rules in the telecommunications area. Braithwaite's (2002, p. 54) study of rules versus principles makes this sector-specific observation: 'What a telephone means today may be something quite different tomorrow. In terms of the conclusion of this article, so long as the principles that underpin the redefining are clear, redefinable rules can regulate a transitional technology with more certainty than fixed rules'.

Some have argued that principles and standards are more conducive to the development of an altruistic state (Goebel, 1993, p. 68). The debate about our ability to control the world *ex ante* is a debate that has continued throughout human history. Aristotle, in his *Nicomachean Ethics*, recognized that all universal laws made up in advance reflect human capacities and accordingly are going to be defective in addressing particular decisions (Bar-Ilan, 2007, p. 936). For example, where economic regulation is concerned, it is extremely difficult, if not impossible, to predict market outcomes before the fact.

In summary, there is a continuum from *ex ante* to *ex post*, with both models using a combination of rules and principles where appropriate.

Where economic regulation is concerned, it is difficult, if not impossible, to predict market outcomes before the fact. In those sectors where rules are regulating economic decision making, a consistent application of a prescriptive *ex ante* system will have major disadvantages. Instead, it is likely that a menu of regulatory models will be more appropriate and that the choice of model will vary according to the level of competition in the relevant market

A more flexible regulatory approach is justified in recognition of the dynamic and complex nature of the market, the waning of market power, and the ability of entrants to act as a check on the behavior of the incumbent. An *ex post* regime maximizes flexibility without abandoning regulatory control.

A rules-based, *ex ante* approach will be appropriate in the most extreme

cases of market failure, for example, in those rare instances where a carrier
has control over an essential facility. In such an instance, establishment
of non-discriminatory access to that facility will require the imposition of
ex ante rules governing the rates, terms and conditions. At the other end
of the regulatory spectrum are classes of services that are competitive, if
not forborne.[11] For these services, greater reliance can be placed on the
market as a disciplining force and a more flexible regulatory approach will
be justified. Indeed, the dynamic nature of these markets strongly counsels
against a heavy-handed rules-based, *ex ante* approach because the costs
imposed by regulation in such circumstances outweigh the benefits to customers. In this latter case, a principles-based requirement that rates be just
and reasonable in the circumstances, coupled with *ex post* enforcement,
may be sufficient to protect the interests of customers.

A major theme of this chapter is that the *ex post* paradigm is not mono-
lithic, but in fact contains a spectrum of different models that range from
rules-based to principles-based. There is a tendency in some quarters to
view the *ex post* world as a version of the old Wild West, where lawlessness
prevailed. The opposite is true, as the *ex post* paradigm governs criminal
and regulatory law which has historically protected our society. *Ex post*
systems are also preventative in nature and do not only apply after damage
is done. The motorist who drives by a person who has been stopped by
the police for speeding will be subject to general deterrence and may slow
down accordingly. This is an example of the proactive power of deterrence
within an *ex post* model. Sparrow (2008, Figure 6.1, p. 137) makes this
point by developing the concept of the unfolding chronology of harm.
Different types of preventative steps may be separated by short periods of
time in this chronology and may be drawn from both *ex ante* requirements
or *ex post* rules and principles.

**In those areas of the economy where technology is moving very quickly, an
ex post, principles-based system is likely superior to an *ex post* rules-based
system, as it maximizes flexibility**

Ex post rules, as distinct from principles, may suffer from inflexibility,
related to the slow pace of legislative reform. Accordingly, a principles-
based system may be superior. Interconnection in a multi-network, next
generation environment provides a good example of this. The fast-changing
and heterogeneous nature of the environment, with many providers, many
platforms (wireless, cable, copper) and many different technologies within
those individual platforms, demonstrates the limits of both *ex ante* and
rules-based *ex-post* systems. Neither can fully anticipate, and adequately
codify, the range of interconnection arrangements that are likely to emerge.
In this type of situation, it makes more sense for a specialized tribunal to

articulate a principle (carriers must interconnect with each other for the exchange of traffic on reasonable terms). This specialized tribunal would then have an *ex post* discretion in the interpretation of reasonable standards (relative to the implementation of the principle) after a carrier had made out a case that the principle was being breached.[12]

A rules-based *ex post* system is suited to certain types of 'social' regulation as contrasted to 'economic' regulation

The recent legislation relating to a 'do not call list' reflects the *ex post* rules-based approach. If a telecommunications provider fails to comply with a direction to not call, this may be the result of inadvertence or it may be the result of conscious harm, to use Sparrow's term. Moreover, unless all telecommunications providers are treated equally, there is a risk of equilibrium harm.

A rules-based system may be very precise, depending upon how the statute is drafted. An archetype of precise standards is found in the occupational health and safety legislation. For example, the regulations require that a guardrail system 'shall be located at least 0.9 meters but not more than 1.1 meters above the surface on which the system is installed'.[13] The occupational health and safety regulations are not so much prohibitions as detailed prescriptions as to how one ought to conduct business. These rules are an example of the potential for rules to educate, as well as making prosecutions easier to win (Baldwin, 1990).

There are some disadvantages to precision. Rules may come close to an *ex ante* model of control, if they start to resemble detailed licenses. One could argue that occupational worker protection law should be *ex ante*, and the only reason that it is not is a historical anachronism. Conversely, telecommunications has largely been *ex ante* also as a result of historical antecedents. Back to the point of the chapter, we argue that history ought not to unreasonably guide or bind us, but rather help us choose the best model. Perhaps worker safety regulations ought to consider an *ex ante* model requiring the filing of business proposals before construction, which set out a plan of safety, given the seriousness of the issues at stake. In some cases, the technological change may result in standards being set too low (an example would be an environmental scenario where better methods are available to remove or prevent contamination). In other cases, technological change may make the regulations redundant, or worse, counterproductive.

In summary, as applied to telecommunications, there is a fear in some quarters that deregulation in favour of an *ex post* system will lead to the equivalent of a lawless Wild West. Precise rules, such as occupational regulations, demonstrate that this need not be the case. In the areas of social regulation for telecommunications, such as health and safety regulations for wireless radio equipment (transmitters, receivers, handsets), there is a justified basis for detailed *ex ante* rules coupled with *ex post* monitoring and enforcement.

The type of rule, as explored above, should not be confused with the appropriate remedy for the breach of the rule
There is a danger in the *ex ante* versus *ex post* debate of simplifying the remedy issues. For example, some have argued that a disadvantage of the *ex post* framework is that it is slow to react to damage that has occurred, which may cause irreparable harm. This criticism fails to consider injunctive relief that is available in the *ex post* model to prevent irreparable harm. Quasi-judicial orders are exemplified by the existing power of the CRTC to make mandatory and restraining orders under section 51 of the Act. In an *ex post* world, injunctions are a well-known remedy which have developed their own body of precedent (Sharpe, 1998).

Telecommunications regulators will have a menu of potential remedies available to them, including administrative monetary penalties, regulatory fines, and criminal sanctions. The full spectrum of remedies should be considered and suitably applied to address actual or potential breaches of regulations. This leads to our last point.

As a general approach to remedies within the *ex post* world, we advocate an approach towards enforcement that has its origins in restorative justice
Braithwaite (1985), a pioneer in the field of restorative justice, developed the concept of a regulatory pyramid in his book published in 1985. Ayres and Braithwaite (1992) elaborated further on the enforcement pyramid. The base of the pyramid was concerned with persuasion of the regulated by the regulators. If this approach proved ineffective, the regulatory effect could be incrementally escalated through warnings, civil penalty, criminal penalty, license suspension and finally licensing revocation. Sparrow makes a similar point that administrative agencies often jump to the conclusion that they need the biggest weapon, when in fact a more effective solution would be increased vigilance in surveying the problem.

Warning letters sent by the regulator may serve to reduce the amount of damage as soon as a complaint is made, and fit well into the regulatory pyramid of proportionate responses. There is precedent for this type of approach in the United Kingdom Communications Act 2003. Under this regime, the Office of Communications (Ofcom) may send a warning letter which specifies a period during which the person notified may make representations, bring itself into compliance or remedy the consequences of the contravention. Warning letters may also avoid the vagueness problem that has been identified with a principle-based system, as they alert the regulated that they are potentially in violation of the principles (Halpern and Puri, 2008, p. 212).

Following the warning letter, if appropriate compliance steps have not been taken, the next step up in the pyramid is an enforcement notification. This notification may impose a requirement to take specific steps to comply, and may be enforced by injunctive relief. Finally, penalties may be imposed for contravention of the enforcement notification. A pyramid approach is built into the penalty section by explicit reference to the prior failure to comply with lower remedies such as the warning letters. The restorative justice approach should not be construed as soft on enforcement. To the contrary, the restorative justice enforcement pyramid works because, at the highest point, there is enforcement of the most serious criminal penalties which apply after lower-level approaches have failed. Part of the reason for the effectiveness and success of cooperative measures at the base of the pyramid is the threat of escalation.

6.3 CONCLUSION

The intention of this chapter has been to provide a framework that can form the basis for new approaches to the regulation of telecommunication. Three points are fundamental in this regard and have implications for the regulation of telecommunications.

First, governments and regulators ought to focus on the prevention of harms and the serving of human needs, in the order of their priority, when they make decisions about the allocation of scare public resources.

Second, it is important to emphasize that regulation of competitive markets will encompass risk management and social regulation. Robust regulation that ensures proper risk management and transparency is essential in a competitive economy. In addition, the new rules will target anticompetitive behavior in line with other competitive industries that fall within the Competition Act.

Third, the regulatory state should offer multiple models that reflect the different needs and interests at stake within both the *ex ante* and the *ex post* paradigms. There is growing recognition in legal circles that any sophisticated system ought to offer a full range of models based on the seriousness and economics of the case. In the field of telecommunications regulation, this insight requires: (1) that a menu of regulatory approaches be acknowledged and available to regulators; (2) that principles be articulated for making choices within these approaches; and (3) that these approaches

146   *Regulation and the evolution of the global telecoms industry*

be applied with an alert sensitivity to subject matter, the human needs at stake and the economics of a given case.

Each sector ultimately is enforced *ex post* if there is a breach of the rules or principles. The choice between systems relates to types of rules and principles on a continuum. Prior approval is suitable in cases where serious harm may be prevented, whereas principles are generally more suitable for economic regulation in the market circumstances now prevailing in Canadian telecommunications.

NOTES

1. An earlier version of this chapter was published in the *Canadian Business Law Journal*.
2. Robertson, G. (2008).
3. Since the passage of the Telecommunications Act in 1993, the CRTC has issued a series of forbearance rulings with the result that most telecommunications services can now be offered without prior approval. Economic regulation, and the accompanying requirement for prior approval of rates, is now largely confined to local telephone service in remote and rural areas and wholesale services offered to competitors. However, it should be emphasized that even where the CRTC has forborne, such forbearance is typically partial and conditional, in one way or another, with the result that the CRTC still maintains some form of regulatory oversight over virtually all telecommunications services in Canada, albeit that the CRTC does not impose price controls on forborne services. Consequently, even after substantial forbearance from economic regulation (or because of it), it is necessary and important to consider new and better ways to approach regulation in the telecommunications sector.
4. According to Posner (2003): 'The movement, which may be said to have begun in the 1970's in the airline industry, though there were earlier anticipations, coincided with increased regulation of health, safety, and the labor markets, so that to speak of "deregulation" in the large is misleading; regulation has changed, rather than diminished.'
5. Public risk perceptions may be a poor guide for policy-making (Cross, 1998) and public participation may lead to overemphasis on some issues and underemphasis on others (Green, 1997).
6. There is a school of political science that attempts to build a political theory on the basis of human need rankings. Bay (1981) argues that the only legitimate justification of government is its ability to serve human needs in the order of their priority as established though psychological research. See also Manzer (1974).
7. 'The effort to constrain, I argue, is more likely to produce distortions than to bring about a reasonable simulacrum of competitive pricing and output. This is primarily because of information and incentive problems of regulators and because of efforts by the regulated firms to neutralize regulation or to bend it to their advantage' (Posner, 1999, p. vii).
8. We restrict the definition of minority to that used in human rights legislation, which recognizes historically disadvantaged minority groups who may lack power within a democratic majority.
9. Such *ex ante* rules might focus, *inter alia*, on key size, keyboard layout, and LCD screen size and design, but would fall out of date very rapidly as technology evolved and accordingly would have every limited benefits.
10. Braithwaite, J. (1985).
11. Residential telephone service is one such example. The service remains rate-regulated in much of North America notwithstanding the presence of interplatform competition among

cable, wireless, broadband and incumbent telephone carriers. Where local telephone service is not forborne from regulation, it is nonetheless often subject to both inter- and intra-modal competition and this fact should inform choices about regulatory models.
12. The FCC has exhibited a basic humility about the ability of prescriptive rules to anticipate, codify and respond to network neutrality issues. Instead of developing a framework of prescriptive rules, the FCC has adopted a set of basic principles to address network neutrality issues.
13. Occupational Health and Safety Act, O. Reg. 213/91, s. 26.3(4), amended to O. Reg. 527/00.

REFERENCES

Archibald T., K. Jull and K. Roach (2004 – updated annually). *Regulatory and Corporate Liability: From Due Diligence to Risk Management*. Ontario: Canada Law Book, a Division of the Cartwright Group, Chapter 1, pp. 1–21.

Ayres, I., and J. Braithwaite (1992). *Responsive Regulation: Transcending the Deregulation Debate*. New York: Oxford University Press.

Baldwin, R. (1990). Why rules don't work. *Modern Law Review* **53**, 321.

Bar-Ilan, Y.A. (2007). Justice: when do we decide? *Connecticut Law Review* **39**, 923–32.

Bay, C. (1981). *Strategies of Political Emancipation*. Notre Dame, IN: University of Notre Dame Press.

Braithwaite, J. (1985). *To Punish or Persuade: Enforcement of Coal Mining Safety*. Albany, NY: State University of New York Press.

Braithwaite, J. (2002). Rules and Principles: A Theory of Legal Certainty. *Australian Journal of Legal Philosophy*, **27**, 47–82.

Breyer, S. (1993). *Breaking the Vicious Circle: Toward Effective Risk Regulation*. Cambridge, MA: Harvard University Press.

Cross, F.B. (1998). Facts and values in risk assessment. Reliability Engineering & System Safety.

Cunningham, L. (2006). *Principles and rules in public and professional securities law enforcement: a comparative US–Canada inquiry*. Research study commissioned by the Task Force to Modernize Securities Legislation in Canada.

Dembo, R.S. and A. Freeman (1998). *Seeing Tomorrow: Weighing Financial Risk in Everyday Life*. Toronto: McClelland & Stewart.

Ehrlich, I. and R. Posner (1974). 'An economic analysis of legal rulemaking'. *Journal of Legal Studies*, **3** (1), 257–72.

Ford, C. (2008). New governance, compliance, and principles-based securities regulation. *American Business Law Journal*, **45** (1), 1–60.

Globe and Mail (2007). 'Sarbanes, Oxley defend law's strict measures', 21 July, p. B5.

Goebel, J. (1993). Rules and standards: a critique of two critical theorists. *Duquesne Law Review*, **31**, 51–68.

Government of Canada (2006). Canadian Radio-television and Telecommunications Commission, Discussion paper, 15 August, submitted to the Telecommunications Policy Review.

Governor in Council (2006). Order issuing a direction to the CRTC on implementing the Canadian telecommunications policy objectives. P.C. 2006-1534, 14 December.

Green, A.J. (1997). Public participation and environmental policy outcomes. *Canadian Public Policy Analysis*, **22** (4), 234–435.

Halpern, P. and P. Puri (2008). Reflections on the recommendations of the Task Force to modernize Securities Legislation in Canada: a retail investor perspective. *Canadian Business Law Journal*, **46**, 199–212.

Hill, S. (1969). A foundation for developing risk management learning strategies in the public service. CCMD Roundtable on Risk Management (Canada). Retrieved from http://www.ccmd-ccg.gc.ca/research/publications/pdfs/risk_mgnt_rt_e.pdf.

Janisch, H. (2005). Response of Hudson Janisch to the CRTC's Discussion Paper, Annex C, Reply Comments of TELUS Communications, *Telecommunications Policy Review*, 15 August.

Manzer, R. (1974). *Canada: A Socio-Political Report.* Toronto: McGraw-Hill Ryerson.

Ministry of Transportation (2008), Ontario to limit truck speed. Ontario Government, 16 June. Retrieved from http://ogov.newswire.ca/ontario/GPOE/2008/06/16/c4593.html?lmatch=&lang=_e.html.

Mullan, D. (2005). Tribunals imitating courts: foolish flattery or sound policy? *Dalhousie Law Journal*, **28**, 1.

Posner, R. (1999). *Natural Monopoly and its Regulation.* Washington, DC: Cato Institute, p. vii.

Posner, R. (2003). *The Economic Analysis of Law*, 6th edn. New York: Aspen Publishers.

Powell, D. (2000). Risk-based regulatory responses in global food trade: a case study of Guatemalan raspberry imports into the United States and Canada, 1996–1998. In G. Bruce Doern and T. Reed (eds), *Risky Business: Canada's Changing Science-Based Policy and Regulatory Regime* (p. 134). Toronto: University of Toronto Press.

Robertson, G. (2008). Canada's 911 emergency. *Globe and Mail*, 20 December, p. A1.

Robertson, G. (2009). Cellphones ordered to fix 911 systems to save lives. *Globe and Mail*, 7 January, p. A1.

Sharpe, R.J. (1998 – updated annually). *Injunctions and Specific Performance*, 2nd edn. Ontario: Canada Law Book, a Division of the Cartwright Group Ltd.

Slovic, P. (1999). Trust, emotion, sex, politics and science: surveying the risk-assessment battlefield. *Risk Analysis*, **19**(4), 689.

Sparrow, M. (2008). *The Character of Harms: Operational Challenges in Control.* New York: Cambridge University Press.

Sunstein, S. (1995). Problems with rules. *California Law Review*, **83**(4), 953.

Tremblay, A., G. Sinclair and H. Intven (2006). The Telecommunications Policy Review Panel Final Report. Retrieved from http://www.telecomreview.ca/eic/site/tprp-gecrt.nsf/vwapj/report_e.pdf/$file/report_e.pdf.

7. The transformation of telecoms industry structure: an event study

Olaf Rieck

7.1 INTRODUCTION

The telecommunications industry is currently going through yet another phase of transformation. This transformation does not merely change the way a set of given services is offered. Rather, we observe that technological changes and service innovations are fundamentally altering peoples' communication patterns. Examples include the shift of the telephone network from being a physical network to being a (potentially converged) Internet application, the shift from a focus on connectivity to a focus on content, and the shift from dyadic communication relationships to social networking.

In this emerging new telecommunications paradigm, traditional business models are increasingly replaced by alternative models. Players that traditionally controlled the telecoms value chain are rapidly losing ground against new entrants, some of which entered from 'outside' the traditional telecommunications service industry or did not even exist at the time when the 1996 Telecommunications Act was drafted. For example, Google emerged as world market leader for online searches and has thereby established itself as a significant player in the telecoms value chain. Google is also – directly or through investments – involved in the roll-out of Internet access infrastructure and the provision of content. Not surprisingly, in September 2007 the combined market capitalization of Google and its close competitor, Yahoo!, surpassed the market capitalization of AT&T.

While there exist a number of studies that analyze telecommunications carriers' performance in their traditional line of business, only a few studies (Rieck and Doan, 2007) have attempted to assess quantitatively the new strategic directions that firms take in this changing environment. This chapter studies activities (conducted by telecom carriers or by other related firms) that aim at vertically integrating different segments of the telecom value chain. It then looks at the implication of these activities on telecom carrier performance. I am interested in finding out to what extent telecom carriers are threatened by other firms' initiatives targeted at 'invading' their turf, and how effective they are in countering such threats.

This chapter is organized as follows. Section 7.2 will provide a brief introduction into the research background and method, and the data collection. Section 7.3 will discuss the empirical results. Section 7.4 will summarize and conclude.

7.2   RESEARCH BACKGROUND AND METHOD

7.2.1   The Telecommunications Value Chain

Since this chapter is first and foremost concerned with changes in the telecommunications value chain, it is important to begin by providing a simplified model of this value chain. Value chain models can generally not be assessed in terms of whether they are right or wrong, but rather in terms of whether they are useful in structuring a given problem or illustrating a point. For this chapter I we will use a modified version of the layer model discussed by Fransman (2002). This model breaks the industry into five layers, which are depicted in Figure 7.1.

Firms in Layer 1 develop and manufacture telecommunications equipment, including networking and customer premises equipment. Examples for such companies are CISCO, Nortel, Nokia, Motorola, Sony-Ericsson and Apple (due to their widely celebrated iPhone). Layer 2 includes some of the key elements of the traditional telecommunications operator business, namely the ownership of network facilities. These include both cables and licenses to spectrum. Companies that dominate this layer are the likes

Layer 5	Content Provision
Layer 4	Content Integration
Layer 3	Network Operation
Layer 2	Facilities Provision
Layer 1	Equipment Manufacturing

Figure 7.1   Telecommunications value chain

of Verizon, AT&T, Deutsch Telekom and SingTel, but also cable operators, or the mobile carriers such as Vodafone, Orange and M1. Layer 3
comprises of all activities related to the operation of networks. These may
or may not involve the companies owning the underlying infrastructure.
The separation of facility ownership from service operation is common
in fixed line telecommunications, but may also become a more common
model in the mobile communications value chain. After all, the regulatory foundations were laid as recently as March 2008 – the conditions for
the Federal Communications Commission (FCC)'s 700 MHz frequency
band auctions include an open access rule, which allows facilitates market
entry of purely 'service-based' mobile operators. Layer 4 is concerned
with content integration (Fransman refers to this layer as 'Navigation
and Middleware') and is dominated by Google, followed by Yahoo! and
Microsoft. Finally, the top layer encompasses all activities involving the
provision of content and applications. This can be, for example, the production of movies or news, the provision of online brokerage services,
online travel brokers, or the development of games. Notable companies in
this layer are Time-Warner, BMG, NBC and Travelocity.

It is important to note that, while most companies originated predominantly from one of these layers, many of them have tried to integrate
vertically into activities in different upstream or downstream layers of
the telecommunications value chain. Again, to give some example, Time
Warner acquired AOL, and Deutsche Telekom bought (and later sold) the
online newspaper *Bild.de*.

Attempts to extend market power in the telecom value chain are not
necessarily limited to the acquisition of ownership stakes. Rather, they
often take the form of organic entry. Some of the most discussed examples
of the past years are Apple's launch of the iPhone, which threatens various
traditional players in the value chain. Besides effectively becoming a new
player in the hand phone equipment market (Layer 1), Apple has managed
to (technically) tie its handset with its online music and video store iTunes
(Layer 4), and has even attempted to negotiate cuts in the mobile operators' revenues (Layer 2) in exchange for letting them exclusively sell their
iPhones. Apple has also secured a number of content deals with content
providers, thereby extending its market power in the content integration
layer (Layer 5).

Nokia and Sony-Ericsson have followed Apple's lead by launching their
own content platforms. At the same time, Google has been working on
invading the handset market, first by working on a new operating system
for Smartphones (Android), and later by launching its own Smartphone
('G1'). Google has also exerted its influence extending beyond content
integration by first lobbying for an open access rule in the US 700 MHz

spectrum auctions (such as to open the spectrum to resale if licenses were auctioned off at a price of at least US\$4.6 billion) and then bidding the price up this threshold such as to trigger the open access rule.

Traditional telecommunications carriers, while under threat from all sides to be reduced to 'bit-pipes', have tried to counter the threats by also engaging in activities in vertically related markets. These include initiatives like joining the open handset alliance, the development of mobile portals, or striking deals with content providers such as to strengthen their position in the content integration layer.

In light of all of these activities, the questions arising are precisely how the structure of telecommunications markets is likely to evolve, which companies will be able to extend their reach and power, and whether there are companies whose role may be reduced in future. In my view, the best places to look for an answer to these questions are the (forward-looking) financial markets, which do react to the changes described above and reflect their evaluation in the market stock prices of the involved companies. I hence conducted a set of event studies to assess the impact of various strategic initiatives ('events') on the valuation of telecom operators.

In order to get a comprehensive view of changes in the telecom value chain, I studied a broad range of events where telecom industry players entered vertically related segments of the telecom value chain. I classified these events into 'vertical integration through M&A' (mergers and acquisitions) and 'vertical integration through organic entry', and conducted two separate studies for these two classes of events. I felt that the characteristics of M&A and organic entry events are too different to integrate them in one joint study.

M&A have been frequent occurrences in the telecommunications industry ever since the break-up of the telecom monopolies. I was therefore able to collect data on a relatively large number of M&A events that constitute some degree of vertical integration and of M&A events that occurred within one layer of the value chain (the latter is typically related to geographical expansion). Given the amount of information I had, I found it useful to compare and benchmark the performance of vertical M&A against that of intra-layer M&A.

Besides, a significant incidence of organic entry is still relatively rare. Organic entry tends to be somewhat less well defined than M&A and harder to pinpoint to a particular point in time. I therefore needed to restrict my study to a few pivotal events in this category, which makes the treatment of organic entry rather exploratory in nature. Even rarer is organic entry of telecom carriers into markets that are in the same layer of the telecom value chain. As a result, I was not able to compare and benchmark vertical organic entry against intra-layer organic entry. I therefore proceeded by treating M&A and organic entry in two separate event studies, as detailed below.

7.2.2 The Event Study Method

There have been numerous event studies undertaken in a variety of areas, foremost in the fields of accounting, finance and strategic management. For instance, Subramani and Walden (2001) study the returns to shareholders in firms engaging in e-commerce. Johnson et al. (2005) examine the impact of ratings of board directors by the business press on stockholders' wealth. Another area in which event studies have been widely used is in the evaluation of M&A. For instance, Wilcox et al. (2001) analyze M&A events in the telecommunications industry by testing the impact of near diversification, far diversification, and the size of the firms on the shareholder value. Uhlenbruck et al. (2006) focus on acquisitions of Internet firms and the potential for the transfer of scarce resources in a resource-based view. Finally, Rieck and Doan (2007) study the value effects of different types of M&As in the telecommunications industry. However, event studies also lend themselves well to evaluating the impact of other kind of corporate activities, as long as they can be traced to some kind of singular event such as a corporate announcement or a service launch.

The event study method is based on the assumption that capital markets are efficient. Based on this assumption, it allows estimation of the impact of new information on anticipated future profits of the firms. If information communicated to the market contains any useful and surprising content, an abnormal return – which is defined as the difference between the actual post-event return and the return expected in the absence of the event – will occur (MacKinlay, 1997).

The assumption of efficient capital markets has also given rise to criticism of the event study method, particularly in times of stock market bubbles and busts. In response to these valid concerns I would like to note two points. First, the alternative to event studies in the realm of quantitative performance analysis is the use of accountancy data. Confidence in accountancy data has been periodically shaken (recall the Enron scandal), as has the confidence in the rationality of stock markets. McWilliams and Siegel (1997) compare the event study methods which use accounting measures and state that: 'The event study method has become popular because it obviates the need to analyze accounting-based measures of profit, which have been criticized because they are often not very good indicators of the true performance of firms'. Second, the event study method looks at stock returns net of the general market trend in an industry. Hence, the presence of a general euphoria or panic will not affect the results of this method. In the event study method, what matters is the stock performance relative to the industry performance. Overall it is expected that event studies will continue to be a valuable and widely used tool in economics and finance.

According to MacKinlay (1997) and Rieck and Doan (2007) an event study can be roughly categorized into the following five steps:

- identifying of the events of interest and defining the event window size;
- selection of the sample set of firms to include in the analysis;
- prediction of a 'normal' return during the event window in the absence of the event;
- calculation of the abnormal return within the event window, where the abnormal return is defined as the difference between the actual and predicted returns;
- testing whether the abnormal return is statistically different from zero.

The events of interest in this study are of two kinds:

1. M&A announcements of major telecommunication companies listed in the US or European stock exchanges. These may or may not involve some degree of vertical integration in the telecom value chain. Looking at both types of M&A (that is, with and without vertical integration) allows us to contrast empirically these two types of strategies.
2. Any events that constitute an instance of organic vertical integration in the telecommunications value chain. These include any new ventures that do not fall into the category of M&A activities. Note that here we are looking not only at the telecommunications carriers' own strategic activities to integrate within the value chain. Rather, in the context of this study an event is defined as any activity that aims at extending a firm's position in the telecom value chain. However, also note that for all of these events I study only their value effects on the telecommunications carriers.

Defining an event window means fixing the time period before and after a focal event, within which any abnormal return is attributed to this focal event. This study uses three symmetric event windows: three-day (one day prior to the event day and one day after the event day), five-day (-2; +2) and 11-day (-10; +10) event windows. These window lengths are appropriate to capture any news that might have leaked shortly before the official

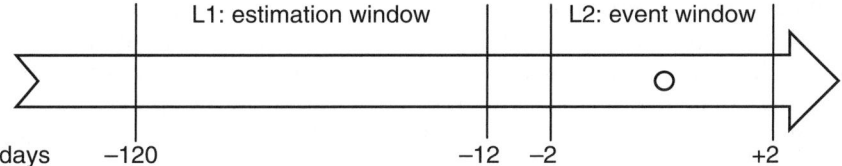

Figure 7.2   Estimation and event window on a timeline

announcement was made and also to consider any short-term stock price reactions linked to the event after the announcement.

The estimation window is the control period which precedes the event period. In this study, the estimation window (denoted as L_1) for all events ends 12 days before the event and extends back to 120 days prior to the event date. Estimation periods generally end before the event of interest so that the returns during the event window will not influence the model parameters. Figure 7.2 illustrates the situation for the case of a 5-day event window.

The companies examined in this study are telecommunications carriers that are listed on (at least) one of the major US or European stock exchanges. For this study of M&A events, the population consists of all M&A announcements released by these telecommunications companies. For the study of other (non-M&A) events, the population consists of a set of initiatives that are either announced by the telecom carriers themselves, or by other companies, but directly affect the telecommunications companies one way or the other. In practice, the population list of M&A announcements is generated by searching for specific terms in the titles of media and news releases.

If the impact of M&A announcements on stock returns is to be examined, a measure of what shall be the 'normal' return for the given stock is required; 120 days of historical stock data will be used for each event. These 120 days are enough to calculate valid estimators needed for the event study model (MacKinlay, 1997).

The event study methodology relates the historical stock data in the estimation window to the market index. In the case of this study, the Dow Jones Titans Telecommunications Sector Index was used as the market index, as detailed below. To predict each firm's market model, daily returns were used to estimate a regression equation over the estimation period. The underlying securities are assumed to be independently and jointly normally distributed and shall be identically distributed through time (MacKinlay, 1997). For any company i, the market model is specified as:

$$R_{i\tau} = \alpha_i + \beta_i R_{m\tau} + \varepsilon_{i\tau} \qquad (7.1)$$

where $R_{i\tau}$ is the return of security i and $R_{m\tau}$ is the rate of return of the market portfolio in period τ. $\varepsilon_{i\tau}$ is the zero-mean disturbance term. α_i and β_i are firm specific parameters of the market model. The market model assumes that in the absence of the event, the relationship between the returns of firm i and returns of the market index remains unchanged and the expected value of the disturbance term $\varepsilon_{i\tau}$ is zero. Using this approach the resulting regression coefficients and the firm's actual daily returns were then used to compute abnormal returns for each firm over each day of the event window period. The sample abnormal return $AR_{i\tau}$ on the event day τ is calculated for the ith firm by subtracting the predicted return of the market model from its observed return:

$$AR_{i\tau} = R_{i\tau} - (\hat{\alpha}_i + \hat{\beta}_i R_{m\tau}) \qquad (7.2)$$

where the coefficients $\hat{\alpha}_i$ and $\hat{\beta}_i$ are ordinary least squares estimates of α_i and β_i. The Cumulative Abnormal Return (CAR) for firm i over the event period $\tau1$ to $\tau2$ is then calculated as follows:

$$CAR_\tau(\tau1, \tau2) = \sum_{\tau=\tau1}^{\tau2} AR_{i\tau} \qquad (7.3)$$

where $(\tau1, \tau2)$ is the event window interval; and all other terms as previously defined. The abnormal returns represent the extent to which actual realized returns on any of the event days deviate from the expected returns based on the estimated firm-specific market model. In this sense, the abnormal returns can be seen as prediction errors ($\varepsilon_{i\tau}$). MacKinlay (1997) then goes on developing the methods to test whether or not the computed CAR is significantly different from zero. The respective formulas are omitted at this point, but can be found in many of the papers quoted above.

7.2.3   Data

The two data inputs required for the M&A event-study model are the events themselves (in this case M&A announcements and other non-M&A events), and historical stock price data (security prices and the reference index). They were both gathered from ABI/Inform and Yahoo.com.

In order to explore the effects of the selected events on stock prices, this research limits its scope to companies that are either listed on one of the major European stock exchanges (London, Paris, Frankfurt, Madrid, Amsterdam) or on a US stock exchange (NYSE, NASDAQ). The reason is that most of the telecommunication operators are listed on at least one of these stock exchanges. It is generally known that these exchanges inspire high public confidence due to the high standards and listing requirements.

Thus, using only major exchanges also increases the price stability of the securities – an important prerequisite for the event-study method.

Diversification through mergers and acquisitions

For my M&A analysis I collected a sample of M&A completed by publicly traded acquirers between 1998 and 2006. Telecommunication operators were selected from the Thomson One Banker database indexes by searching for companies within the following industries: companies with primary SIC Code 4813 (Telephone Communications except Radiotelephone), 4812 (Radiotelephone Communications) or 4842 (Cable and Other Pay Television Services). The results were limited to companies that were (at the time of the event) listed on a major US or European stock exchange.

Next, with a list of 56 potential acquirers, the M&A events which are associated with every one of these carriers were separately retrieved from the Highbeam™ Research database.[1] The following search on Highbeam™ using a set of relevant search terms was performed and the earliest dates announcing the event were recorded:

- Search in article title only:
  {*company name from list*} AND *buy* OR *acquires* OR *bid* OR *merge* OR *takeover* OR *merger* OR *acquisition* OR *buys* OR *merges* OR *merging* OR *acquiring*
- Search in the following sources:
  {*news sources from the following list: Business Wire, Associated Press, PR Newswire, Reuters, Wall Street Journal, Washington Post*}
- Dates:
  1 January 1998 and 31 March 2008

Events that were identified using these criteria were consolidated into a master list with duplicates removed. The preliminary sample frame had 512 M&A events. Out of this, an acquisition that resulted in a controlling stake for the acquirer, that is, a stake greater than 50 percent, was chosen. This means that acquisitions giving the acquirer lower than 50 percent in stakes are by definition not considered as acquisitions. It is true that that in practice acquirers may de facto control the target even though they hold less than 50 percent of the shares. Moreover, it is also true that owning only a minor stake, say 10 percent of the shares, could hardly result in a significant degree of control of the target. Determining whether or not an acquisition resulted in a controlling stake is difficult without using a great deal of insider knowledge and subjectivity. I therefore chose to apply the 50 percent rule outlined above, acknowledging the practical limitations of our approach.

Moreover, only M&A announcements containing accurate and detailed information about date of announcement, partner and transaction value were included. Further, only those events with at least 120 days of historic stock data[2] available were selected. Last, to avoid possible confounding effects within the event window, a number of M&A events were selectively omitted. Excluded events are those that coincided with other major firm-specific events that might affect the stock price, such as earnings alliance announcements, earnings, large investment decisions or new product introduction (McWilliams and Siegel, 1997). Confounding events were identified using the Highbeam™ Research function. It displays all company news within a four-day range of the specific date. After meeting all these criteria, the final sample contained 88 M&A announcements of 37 telecommunication firms. Once the M&A announcements were isolated and the event window was defined, the 'normal' (expected) returns for that window needed to be estimated. This was done by using historical stock price data (adjusted closing prices) for all acquiring companies listed in the master list. For each event 132 days of historic stock data before the event date and ten days of stock data after the event date were downloaded through either the Center of Research on Security Prices database (CRSP) or the Yahoo! Finance database.

To measure the direction of diversification of the M&As, this study employs the SIC[3] classification as conducted by Berry (1971) and Ferris and Park (2001). Many industrial organization studies have used objective measures based on standard industrial classification (SIC) counts to capture the aspect of diversification (Ramanujam and Varadarajan, 1989). The first number assigns a product to a very broad category. Each subsequent number distinguishes the product at a progressively finer level. The SIC classification has been widely used among economists to determine in which industry segments the company is operating. As all acquirers in this study are listed telecommunications companies, they all operate with the two-digit 48xx SIC code. The acquirer's strategy can be determined by comparing the SIC codes of the acquirer and the target. Telecommunications M&A occurring solely within the 48xx SIC code (that is, both acquiring and target firms) are termed non-conglomerate mergers. M&A where the target has a SIC code other than 48xx are classified as conglomerate mergers (Ferris and Park, 2001; Ramanujam and Varadarajan, 1989). Conglomerate mergers include all mergers across different parts of the value chain.

Organic diversification events

The data collection for relevant organic diversification events was conducted in a similar fashion as for the M&A events; however, it was somewhat less systematic and more exploratory. The reason is that in the case of M&A it is intuitively clear what constitutes a 'significant' event (that is, the acquirer purchases a significant number of that results in a voting majority). If a company organically enters a part of the telecom value chain, it is not immediately clear what criteria to apply in order to select only significant events, that is, events that are significant enough to make a noticeable impact on the industry structure. In this chapter, I will report the impact of a few such events, which I consider intuitively as significant. These events are:

1. Apple's launch of the iPhone (and the signing of exclusive contract deals).
2. Nokia's announcement of OVI.
3. Sony-Ericsson's announcement of a major content deal with three of the four major music labels.
4. Google's announcement of Android.
5. Surpassing of the US$4.6 billion mark in the 700MHz license auctions, which triggered the open access rules.
6. Cingular Wireless, Verizon and Sprint announcing their music download portals.

Since I expected the respective announcements to be somewhat less of an immediate surprise, I chose a 21-day event window. I studied the impact of the six events on the affected telecommunications carriers' evaluation. In other words, I studied the impact of the iPhone announcement in the UK on the Telefonica (O2) valuation, as well as on the evaluation on its competitors. I did not however consider the impact on the valuation of Apple Inc. Similarly, I studied the impact of the OVI announcement on the affected telecom operators (note that most large operators are trying to offer similar services, so OVI is a clear threat to their business expansion into the content integration market), but I did not study the impact of the OVI announcement on the Nokia share price. Beyond that, I followed the same data collection procedure as outlined in the case of the M&As.

7.3   EMPIRICAL RESULTS

Table 7.1 presents a summary of the results from the event study of M&A and shows the statistical values of firm diversification strategy, that is, conglomerate and non-conglomerate mergers.

The first two columns of Table 7.1 present coefficients and significances

Table 7.1   Event study results: diversification through M&As

Event types	N	Z-value for mean CAR	
		(-1,+1)	(-2,+2)
Complete sample	88	2.234*	1.871*
		(0.85%)	(0.86%)
Firm diversification			
Non-conglomerate merger	58	1.478#	1.427#
		(0.63%)	(0.61%)
Conglomerate merger	30	1.764*	1.219
		(1.29%)	(1.19%)

Note: The symbols #,*,** and *** denote statistical significance at the 10%, 5%, 1% and 0.1% levels, respectively, using a 1-tail test.

for the 'additive event-window abnormal returns'[4] in the respective event window (also referred to as 'mean cumulative abnormal return' – CAR[5]). The abnormal returns are calculated using the market model estimated from 132 to 12 trading days prior to the event announcements. The Mean CARs (given in parentheses) represent the cumulative market model-adjusted abnormal returns over the relevant event window. The Z-statistics for the (-1,+1) and (-2,+2) event windows are based on the standardized abnormal return method according to Patell (1976).

The first row of Table 7.1 reports the results for the complete sample. There is significant support for the hypothesis that M&A activities will, in general, have a positive impact on telecommunication firms participating in these activities. The mean CAR for both windows is approximately +0.85 percent and significant at the 5 percent level. If we break the set of M&A events up into conglomerate mergers (that is, mergers within a segment of the value chain), non-conglomerate mergers (that is, diversification across segments of the value chain), we get the following results: non-conglomerate mergers will generate positive abnormal returns. CARs of +0.63 percent and +0.61 percent are reported for both windows, respectively, both which are weakly significant at the 10 percent level.

The CAR for the conglomerate mergers is twice as high (+1.29 percent) and is significantly positive at the 5 percent level for the (-1,+1) window. However, no statistically significant evidence was found in the (-2,+2) window. The t-score for the paired t-test that measures the difference for both CAR means (Δ = *NonConglom.CAR - Conglom.CAR*) is negative, but not significant. Hence, there is no significant evidence

Table 7.2   Effects of organic diversification on carrier valuation

Sample type	N	Z-value for mean CAR
		(-10,+10)
Complete sample	35	-0.97***
Subclasses of events – effects on telco carriers		
Outsiders entering content integration markets	19	-0.47*
Telcos' own entry into content integration market	3	0.05
Google launching Android	7	-0.26*
Triggering of open access rule in spectrum auctions	3	-0.43***
Telcos' exclusive iPhone deals	5	0.14

Note: The symbols #, *, ** and *** denote statistical significance at the 10%, 5%, 1% and 0.1% levels, respectively, using a 2-tail test.

that conglomerate mergers show different abnormal returns than non-conglomerate mergers.

Table 7.2 shows the CARs related to a number of events that alter the market structure in the value chain, but are organic in nature (that is, not related to M&As). While these results are fairly exploratory in nature, there are a number of interesting trends to be discussed. First, overall telecommunications carriers appear to be highly threatened by the majority of changes that currently occur in the value chain. This is indicated by a highly significant negative CAR of the complete sample of events. When looking at some individual components, it appears that it is particularly the 'outsiders', such as Google, Nokia and Ericsson, with their attempts to reach out into different parts of the value chain that impose a threat to the telecom carriers by lowering their forward-looking valuation.

So far the carriers themselves could do little to convince markets that they are capable of launching promising activities. Neither the launch of carriers' own music download portals, nor the exclusive iPhone deals were greeted with a great deal of enthusiasm. In both cases, the changes in CAR were insignificant.

7.4 DISCUSSION AND CONCLUSION

The results of the M&A event study are consistent with previous event studies in showing that M&A in the telecommunication industry generally result in significant gains in the market values of the acquirer. Therefore, it can be concluded that the market is generally optimistic with regards to the potential of telecom carriers to add value in this industry. The highly competitive marketplace in the telecom sector means that high returns are no longer guaranteed for big telecom firms. Telecommunications networks have typically high fixed costs but comparably low marginal costs. As a result, the potential for economies of scale and scope remains enormous in this industry. All rival operators are racing to grow fast to reap those benefits. Investors may have realized that long-term growth depends on capital being diverted to productive purposes. However, a reason for cases where firms do not show positive gains after an M&A announcement can be that it is not always easy for a company to achieve synergies and to reap scale and scope. High integration costs and differences in corporate culture are reasons why M&A fail to add value. This shows investors' skepticism about the likelihood that the acquirer will be able to realize these synergies required to justify the premium paid (Selden and Colvin, 2003).

Besides engaging in synergistic M&As, telecom operators have also been engaging in conglomerate mergers, diversifying into different parts of the value chain. This type of M&As were also found to be wealth-creating. Furthermore, wherever I looked at incidences of organic carrier diversification, the announcements of these activities did not show any significant impact on firm valuation. It appears that forward-looking markets did not put a great deal of confidence into the own innovative strengths of the carriers in my sample when it came to business ventures outside their traditional core business. Unfortunately for the carriers, in times of reducing margins in the core business and commoditization of the traditional network operation business it is precisely those innovative activities in vertically related markets that could help them regain their competitive strength. The Japanese and Korean operators provide a good illustration of this point.

At the same time the telecom operators I studied appeared to be under threat by 'outsiders', whenever these outsiders engaged in those kind of activities that shift the market power in the value chain. Financial markets, therefore, appear to support the nation that telecom carriers may gradually become reduced to 'bit-pipes'. The carriers may in future need to accept that they no longer act from a position of control or strength, but rather from a defensive position. From a regulatory perspective, telecommunication firms' market power is increasingly threatened by 'outsiders' who penetrate their traditional markets. Thus, telecom regulation may need to move further away from the focus on 'carrier regulation' and towards a holistic oversight of competition in the entire telecom value chain.

This chapter has analyzed the financial markets' reaction to corporate activities that aim at changing the market structure within the telecom value chain. While I do appreciate that financial markets often appear to behave in a less than rational fashion, I also believe that this reaction is arguably one of the best gauges of future developments at our disposal. I would also like to remind the reader that parts of this study are exploratory in nature. Particularly, my analysis of organic entry is based on a rather small sample size. As time passes, more companies will try to extend their position in the telecom value chain, which will allow researchers to conduct studies based on larger sample sizes. Nevertheless, with this study I hope to have contributed to an initial understanding of current fundamental changes in the telecommunications industry. It can be a basis for further research in this field and might also be helpful for research on different industries that share a similar structure and conditions to telecommunications.

NOTES

1. HighBeam™ Research is an online research engine which sorts free, paid and proprietary online articles and databases. It is a tool for serious business, education and personal research; http://www.highbeam.com.
2. They were retrieved through Yahoo! Finance and the CRSP database.
3. A standard industrial classification (SIC) code categorizes US business establishments based upon the type of business activity performed at their location. All fields of economic activity are included in this system including both manufacturing and non-manufacturing operations. The system is governed by the Office of Statistical Standards.
4. They represent the sum of the mean cumulative abnormal returns over all event window days.
5. Diagrams for the mean cumulative abnormal returns for the complete sample were generated for the five-, ten- and 20-day event windows.

REFERENCES

Berry, C. (1971). Corporate growth and diversification. *Journal of Law and Economics*, **14**(2), 371–83.

Ferris, S. and K. Park (2001). How different is the long-run performance of mergers in the telecommunications industry? SSRN Working Paper.

Fransman, M. (2002). Mapping the evolving telecoms industry: the uses and shortcomings of the layer model. *Telecommunications Policy*, **26**(9), 473–83.

Johnson, J., A. Ellstrand, D. Dalton and C. Dalton (2005). The influence of the financial press on stockholder wealth. *Strategic Management Journal*, **26**(5), 461–71.

MacKinlay, A.C. (1997). Event studies in economics and finance. *Journal of Economic Literature*, **35**(1), 13–39.

McWilliams, A. and D. Siegel (1997). Event studies in management research: theoretical and empirical issues. *Academy of Management Journal*, **40**(3), 626–57.

Park, M., D. Yang, C. Nam and Y. Ha (2002). Mergers and acquisitions in the telecommunications industry: myths and reality. *ETRI Journal*, **24**(1), 56–68.

Patell, J. (1976). Corporate forecasts of earnings per share and stock price behavior. *Journal of Accounting Research*, **14**, 246–76.

Ramanujam, V. and P. Varadarajan (1989). Report on corporate diversification: a synthesis. *Strategic Management Journal*, **10**, 523–51.

Rieck, O. and C.T. Doan (2007). Shareholder wealth effects of mergers & acquisitions in the telecoms industry. Proceedings of the ITS 18th European Regional Conference, 4–6 September, Istanbul, International Telecommunications Society.

Selden, L. and G. Colvin (2003). M&A needn't be a loser's game. *Harvard Business Review*, **81**(3), 70–73.

Subramani, M. and E. Walden (2001). The impact of e-commerce announcements on the market value of firms. *Information Systems Research*, **12**(2), 135–54.

Uhlenbruck, K., M. Hitt and M. Semadeni (2006). Market value effects of acquisitions involving internet firms: a resource-based analysis. *Strategic Management Journal*, **27**, 899–913.

Warf, B. (2003). Mergers and acquisitions in the telecommunications industry. *Growth and Change*, **34**(3), 321–44.

Wilcox, H. Dixon, K. Chang and V. Grover (2001). Valuation of mergers and acquisitions in the telecommunications industry: a study on diversification and firm size. *Information and Management*, **38**, 459–71.

PART III

Access regulation and performance in the deployment of NGNs: international experiences

8. From the pursuit of efficiency to the pursuit of competition in New Zealand's evolving telecommunications market[1]

Bronwyn Howell

8.1 INTRODUCTION

From an economic perspective, efficiency is the defining performance benchmark for any industry or sector – not least telecommunications. Consequently, the primary normative objective of law-and policy-making is the promotion of economic efficiency (in both its static and dynamic forms) via the elimination of market inefficiencies (Schmalansee, 1981; Kahn, 1970, 1975).

A minority of economists, and many consumer advocates, propose the use of law- and policy-making powers principally as a means of achieving distributional objectives, independent of their effects upon total efficiency (for example, Feldstein, 1972a, 1972b). However, in practice it is extremely difficult to achieve desired distributional outcomes through laws and policies (Schmalansee, 1981), and attempting to do so may well be counterproductive (Kahn, 1975; Peltzman, 1976). Furthermore, because distributional objectives are highly subjective, it is very difficult to adjudge the 'successes' of any distribution-motivated intervention. By contrast, efficiency is an objective measure that provides a useful benchmark for the economic assessment of law- and policy-making performance. Even if redistribution is a primary consideration, the Kaldor–Hicks criterion requires efficiency gains sufficient that the winners from a law or policy change could compensate the losers and still be better off relative to the status quo (Connolly and Munro, 1999).

Competition is one important means of increasing both static and dynamic efficiency; another is regulatory intervention. Although intervention undoubtedly has many potential pitfalls, its use may be justified in industries where pursuit of competition (in the form of many market participants) is unlikely to generate greatest total efficiency in the long run; for example, natural monopoly, where high fixed and sunk infrastructure costs and preservation of long-term incentives to invest in new, higher-quality technologies that maximize dynamic efficiency gains may require protection of a single operator from competitive entry (Alleman and Rappoport, 2005). In such industries the challenge for governments is to determine how to allocate responsibility for industry regulation and governance between two competing forms of intervention, competition law and industry-specific regulation, such that the pursuit of greatest total welfare ensues in subsequent industry activity.

Competition law can govern interactions to promote increased efficiency in most industries, but has limitations. Courts are reactive, responding only to those cases and those points of law brought before them, and thus influence only a small range of issues that are not necessarily the most important from a broader efficiency perspective. Moreover, even when cases are brought, they are often adjudicated by generalist antitrust judges who in many cases may lack the industry-specific economic knowledge upon which efficiency decisions may turn.

Where the underlying economic conditions are sufficiently different – as in network industries such as telecommunications – industry-specific regulation offers advantages. Industry-specific regulators in most cases have the requisite economic knowledge to give due weight to efficiency considerations and to balance the tensions between short-term static and long-term dynamic efficiencies. Rather than being reactive, they can be proactive – a power that, when applied appropriately, can lead to increases in efficiency. However, the weakness of industry-specific regulation is the risk of capture, leading to the pursuit of other objectives (such as distributions overly favorable to one group, or competition as an end in itself, rather than a means) over the pursuit of efficiency.

This leads to the motivating question for this chapter: 'Can an industry-specific regulatory regime be designed to maximize the relative benefits of industry-specific regulation by binding the regulatory institution to pursue an economic efficiency objective, whilst simultaneously avoiding the risks of regulatory capture?'

Examination of New Zealand's telecommunications sector over the past 20 years provides some interesting insights on this question. Although the New Zealand government initially prioritized pursuit of economic efficiency in its 'light-handed' regime (particularly the preservation of long-run incentives to invest in new networks and technologies), and attempted to safeguard its pursuit in subsequent telecommunications sector legislation, the objective was nonetheless vulnerable to capture by groups with strong vested interests. However, as the regulator was bound to prioritize efficiency, politicians rather than the regulator were the subject of capture exercises, as they held the power to alter the terms of the regulatory objective. Consequently, successive governing parties would alter the legislation enacted by their predecessors in order to pursue different sector objectives.

Thus, even legislation cannot ensure that efficiency remains a consistent goal of sector regulation, as long as regulators remain the agents of political principals. This suggests that, imperfect though it may be, competition law overseen by an independent judiciary offers the best chance of enshrining the pursuit of efficiency in industry governance, even in industries that are normally the focus of industry-specific regulation.

The remaining sections of this chapter describe the development of the regulatory framework for New Zealand's telecommunications industry, noting the shift in emphasis from a light-handed regulatory style with an efficiency focus ranking long-run dynamic efficiency considerations highest, to a more interventionist industry-specific regime with a focus on achieving certain distributional outcomes and fostering competition as the pre-eminent objective (where greater weight was given to the pursuit of static efficiency gains over total efficiency in the long run). The chapter concludes by summarizing the pattern of change in New Zealand, and drawing potential lessons for regulatory design.

8.2   INCEPTION OF NEW ZEALAND'S LIGHT-HANDED, EFFICIENCY-FOCUSED REGIME

New Zealand led the world in light-handed regulation when, from 1984, the government embarked upon a comprehensive restructuring of the country's economy. The government's clearly articulated objective was to increase economic efficiency and create, as Evans et al. (1996, p. 1863) cited: 'wherever possible, a competitive environment in which markets can operate relatively free from subsequent intervention by Government'. As part of this process, the government enacted the Commerce Act (1986), which was designed to promote competition for the long-term benefit of consumers, and the Telecommunications Act (1987), which avoided industry-specific regulation in favor of generic competition law under the provisions of the Commerce Act.

Efficiency considerations were paramount in the choice of institutional mechanisms for industry governance in New Zealand. Government advisors explicitly identified that regulatory weaknesses such as high costs, inflexibility and bureaucratic capture could be avoided under a competition law regime governing all commercial activity (Ministry of

Commerce and The Treasury, 1995; Blanchard, 1995). However, there were challenges in the enactment of effective competition law, not only as a result of the limitations attendant with non-specialist judges deciding on competition cases in a range of specialized industries, but also due to New Zealand's specific economic circumstances: small population, low population density, geographical isolation, challenging terrain, thin capital markets and historically highly concentrated industries. Almost all significant industries in New Zealand have three-firm concentration ratios in excess of 90 percent (Arnold et al., 2003), making oligopolistic rather than perfect competition the most likely pattern of interaction observed in all industries, not just those with high fixed and sunk costs.

On balance, the government favored establishment of a single competition law-based institution to capture institutional scale and scope economies as well as to ensure consistency in application across industries. The presumption was that holding a dominant position was acceptable (indeed it was unavoidable in almost all industries), but that exertion of that dominant position in such a manner that it harmed the long-term interests of consumers (as determined under competition law) was illegal. These arrangements preserved investment incentives for firms in industries with natural monopoly characteristics by permitting dominance (important, given high investment requirements in these industries, historically thin capital markets and a small market in the first place), but relied upon competition law to ensure that the behavior of firms with dominant positions did not adversely affect the achievement of static and dynamic efficiency objectives. However, Part IV of the Commerce Act explicitly provided for the government to impose price controls in industries where market power existed, should this be deemed necessary (for example, in the presence of repeated, proven abuse of a dominant position).

Although 'lightly' regulated under competition law, the telecommunications sector was far from unregulated. When the state-owned monopoly incumbent Telecom Corporation of New Zealand Limited (Telecom) was privatized in 1990, contractual obligations known as the 'Kiwi Share' (subsequently the Telecommunications Service Obligation or TSO) imposed rural–urban universal service and free local calling obligations upon Telecom, and a price cap on residential line rentals. These terms could be broken only with the express permission of the Minister of Communications, and even then only where it could be demonstrated that Telecom was under financial duress. The paramount principle was that contractual agreement, enforceable under contract law, rather than overt regulation, subject to high costs and risks of political capture, offered the most cost-effective means of advancing the pursuit of increased efficiency.

The New Zealand arrangements thus involved the government actively suspending its 'primary tendencies' to bow to vested distributional interests in order to create a set of governance arrangements that would insulate the sector from politically motivated actions (which would have diluted the pursuit of efficiency). Successful implementation would require the exercise of significant ministerial restraint in order to allow confidence in the court- and contract-based regulatory instruments to build up.

8.3 EXPERIENCES UNDER COMPETITION LAW

During the period of light-handed regulation, only two Commerce Act actions alleging exertion of a dominant position were brought against Telecom. In both cases, Telecom was found not to have acted anticompetitively. In one case, entrant Clear Communications alleged that Telecom had acted anticompetitively by including a component to recover the costs of the universal service obligation in its interconnection prices to competitors (the *Clear* case). After three years and three hearings (in the High Court, Court of Appeal and Privy Council), Telecom was adjudged to have acted legitimately in adopting Baumol-Willig Efficient Component Pricing (ECP). The other case was brought by the Commerce Commission in 1999, alleging that Telecom's charging for residential dial-up Internet calls was anticompetitive. This case was finally adjudicated in 2008, and again found that Telecom had acted as any competitive firm would in the circumstances (Howell, 2007).

Despite multiple claims of 'failure', performance data indicate that New Zealand's light-handed regime performed, on a range of static and dynamic efficiency grounds, at least as well as other industry-specific regimes. The New Zealand residential telephony price index fell by more than the Organisation for Economic Co-operation and Development (OECD) average during the 1990s (reflecting substantial static efficiency gains passed on to consumers), and free residential dial-up Internet access helped the country become a world leader in Internet connection and use – indicating that dynamic gains from consumer adoption of new applications were substantial and in excess of those achieved in other jurisdictions (Howell, 2007). Furthermore, dynamic efficiency gains from investment in new generation technologies did not appear to be impaired: Telecom was one of the OECD's earliest digital subscriber line (DSL) adopters (January 1999), using a high-speed (2 Mbps) service that was made widely available (85 percent of the population had access by 2003) and that was priced at a very low level, taking speed into account. Furthermore, by 2001, new networks bypassing Telecom's infrastructure had been deployed in several business districts and three substantial residential markets, including the capital city Wellington (Howell, 2003).

Nationally, however, Telecom still held a very large market share, and was clearly charging prices in excess of marginal cost in some markets (even if only to recoup universal service obligation costs). In response, many market participants – principally competitors to Telecom – and political opponents of privatization strongly advocated the introduction of industry-specific regulation.

8.3.1   Arguments for a Move Away from Competition Law Regime

The justification most commonly cited for a proposed shift to industry-specific regulation was the extent to which Telecom's prices exceeded marginal cost – an issue highlighted in the *Clear* case. Although the final Privy Council ruling was in Telecom's favor, the Court of Appeal had found that any price including a component of monopoly rent (that is, above marginal cost) was anticompetitive.

The essence of the Court of Appeal decision in the *Clear* case – and of most arguments for the introduction of industry-specific regulation – was a distributional rather than an efficiency concern. The implication of the Court's judgment was that Telecom alone should bear the costs of the social obligations. If the decision had been enforced, there would have been some static efficiency gains available to competitors (and potentially consumers) in low-cost (urban) areas, but with welfare losses in other higher-cost (rural) areas and longer-term dynamic efficiency consequences from likely inefficient entry that would be incentivized.

Under the Court of Appeal decision, competitors' costs would have been lower than Telecom's prices in areas where Telecom recovered the costs of the social obligations (urban areas), making selective entry into these low-cost areas profitable for competitors but making Telecom's business financially unsustainable. Profits that Telecom was using to subsidize otherwise unprofitable connections would instead be transferred to competitors, and Telecom would be forced to raise prices for all its remaining customers to remain financially viable. In low-cost areas, such price rises would accelerate consumer substitution to lower-price entrants.

Ultimately, Telecom alone would be left servicing all high-cost areas at higher prices than previously, lowering consumer welfare in those regions; and entrants would be servicing the low-cost areas, with higher consumer welfare only if the prices charged were lower than those originally charged by Telecom. Furthermore, the resulting two-price equilibrium would be at odds with the distributional intention of the universal service obligation that was the source of the cost–price dichotomy in the first place.

Moreover, dynamic efficiency would be potentially lowered as entrants could profitably enter in low-cost areas even if their costs exceeded Telecom's, as long as their prices fell below Telecom's (which included social cost recovery not imposed on entrants). Although such entry would lead to decreases in efficiency overall, it was favored with Telecom's competitors and opponents of privatization largely because of the distributional consequences. Under the guise of promoting competition (competitive entry) these stakeholders argued that a disproportionate share of the costs of social obligations fall on the private owners of the formerly government-owned network, rather than being spread across the industry, and thereby across all consumers of telecommunications services, irrespective of their provider, as part of Telecom's interconnection charge or as a tax (Armstrong, 2001).

By contrast, the Privy Council (prevailing) decision did not imply distributional change, and thus preserved existing network investment incentives and Telecom's financial sustainability. The Council ruled that Telecom had priced exactly as a competitive firm could be expected to price its products, based upon its opportunity cost. However, in making this ruling, the Council enshrined in precedent a pricing rule that could, under certain circumstances, result in less than efficient entry and hence potentially lower product variety than optimal in downstream markets (short-term dynamic efficiency; Economides and White, 1995). The decision was interpreted by some commentators at the time as an example of how court-based pursuit of 'competitive' outcomes might compromise pursuit of the efficiency objectives governing the design of the light-handed regime (Blanchard, 1994a, 1994b, 1995). That is, pursuing competition (the means) potentially compromised efficiency (the end). In practice, however, the decision likely averted potentially more damaging long-term dynamic efficiency consequences, and preserved the distributional objectives for consumers inherent in universal service obligations.

To overcome the perceived shortcomings with court-based decision-making in the telecommunications industry, some proposed not the adoption of full industry-specific regulation, but rather a light-handed arbitration alternative. The envisioned alternative would sit under the Commerce Act and the courts, enabling swifter resolution of disputes and with a wider mandate to consider efficiency issues not directly part of court pleadings (Blanchard, 1995).

8.3.2 Political Review of Competition Law Regime

In 1995, the New Zealand Treasury and the Ministry of Commerce led an inquiry into the proposed alternatives to competition law (Ministry of Commerce and The Treasury, 1995). Based on the review findings, the Minister of Communications demonstrated ministerial restraint when expressing confidence in the existing system and stating that no changes would be made. Thus, the pursuit of total economic efficiency as the prevailing sector objective and the use of 'light-handed' regulation to deliver it was, for the time being, reinforced at the political level (ibid., para. 209), despite the recognized (but assessed to be small) risk that this might have on downstream competition.

However, political change (in the 1999 election, which brought a Labor Party-led coalition government into power) saw the competition law regime come under review again in 2000. The Labor Party had campaigned on a promise to reform the Commerce Act to tighten controls on firms with a dominant position, and to launch an inquiry into the conduct of both the telecommunications and electricity industries. The promised Commission of Inquiry was established soon after the election, with a brief to address (amongst other issues): alternative means of establishing interconnection terms and conditions; pricing principles and other terms and conditions for interconnection; processes applying to interconnection negotiations, including dispute resolution and enforcement mechanisms; local loop unbundling; resale of telecommunications services; information disclosure; the 'Kiwi Share obligations'; the numbering regime; number portability and the development of an information economy; and the effect of regulatory regimes on incentives to invest in new technologies.

The Commission of Inquiry appeared to interpret its brief as expressed in the government policy statement in efficiency-related terms. 'Cost-efficient' was presumed to mean that services are produced 'at the lowest cost and delivered to consumers at the lowest sustainable price' (MED, 2000, p. 11) (that is, perfect productive and allocative efficiency), 'timely' to mean 'the absence of barriers that would impede the implementation and uptake of innovative services' (dynamic efficiency) and 'ongoing' to mean that 'regulation should be forward-looking, robust, durable and consistent over time, and not sacrifice long term [*sic*] gains for short-term considerations' (the trade-off between dynamic and static efficiency). In addition, the Commission took 'fair and equitable' to mean 'the way in which services are provided, the conduct of the industry players and their interactions', suggesting weight would be given to competitor equity as well as consumer welfare (and efficiency).

However, because the government's policy statement imposed multiple and conflicting objectives, the Commission of Inquiry was clearly unable to deliver on all of them. Both static and dynamic efficiency objectives were to be addressed simultaneously, with no guidance given as to which should take priority. Moreover, the policy statement required both efficiency and distributional objectives also to be simultaneously satisfied. Without clear prioritization criteria, the policy statement left considerable scope for the Commission of Inquiry itself to be captured by vested interests.

Unsurprisingly given these conflicting instructions and the strongly polarized views held by industry participants and political agents, the Commission of Inquiry's findings were focused on efficiency in some instances, and not in others. For example, its report considered dynamic efficiency in rejecting local loop unbundling. Echoing the Court of Appeal decision, the Commission also considered efficiency in concluding that Telecom's prices still included elements of prices above cost (and thus was not perfectly allocatively efficient). However, in proposing a solution to the perceived failure of light-handed regulation, the Commission did not consider efficiency: the Commission proposed that the government establish an industry-specific regulator to set total service long-run incremental cost (TSLRIC)-based prices for a variety of Telecom's fixed-line services (excluding value-added services such as asymmetric digital subscriber line – ADSL), but ignored dynamic efficiency issues in its cost–benefit analysis of a new regulatory agency. Furthermore, distribution considerations – rather than efficiency considerations – prevailed in the Commission of Inquiry's recommendation that the costs of social obligations be borne by Telecom alone (echoing the earlier Court of Appeal ruling). Such inconsistency suggests pragmatic political interest trading, rather than principled economic analysis.

8.3.3 Introduction of Industry-Specific Regulation with Efficiency Objectives

The government chose to give legislative force to almost all of the Commission of Inquiry's recommendations, introducing them into the terms of the Telecommunications Act of 2001. The Act compromised the pursuit of efficiency objectives, to some extent, by pursuing other, largely distributional, objectives. It established a Telecommunications Commissioner within the Commerce Commission, and mandated TSLRIC pricing for 'designated services' ('retail-minus pricing' was imposed for lower-level 'specified services'). This pricing approach in part addressed the distributional objectives of Telecom's competitors, despite compromises to dynamic efficiency (Quigley, 2004).

However, the Act did include some pro-efficiency measures. For example, it ring-fenced the costs of the 'Kiwi Share' social obligations as the Telecommunications Service Obligation (TSO), a charge the Commissioner could levy on all industry participants. This helped to avoid some of the worst dynamic efficiency implications that would have arisen had the Court of Appeal decision on the *Clear* case not been overturned by the Privy Council. Furthermore, the Act created in Section 18(2) a specific obligation for the Commissioner to take account of efficiency when making decisions and recommendations:

> in determining whether or not, or the extent to which, any act or omission will result, or will be likely to result, in competition in telecommunications markets for the long-term benefit of end-users of telecommunications services within New Zealand, the efficiencies that will result, or will be likely to result, from that act or omission must be considered.

The Act also (potentially) safeguarded the Commissioner's independence from political influence by including the Office of the Telecommunications Commissioner within the Commerce Commission. Under the Crown Entities Act, the Commerce Commission is an Independent Crown Entity (ICE), and thus has the least obligation of any of the Crown Entities (statutory bodies outside of core government departments) to political imperatives. An ICE must take account of government directives only where it is specifically required to do so as part of its legislated duties (only the judiciary has greater independence from political direction). In the Commissioner's case, taking efficiency into account was a legislated duty under the Telecommunications Act.

This statutory obligation to consider efficiency suggests that, despite some of the more radical Commission of Inquiry recommendations, on balance the government believed that a balanced, total efficiency objective remained important for the telecommunications sector, albeit slightly diluted by the other provisions of the Act. Combined with the apparent independence of the Commission, it could be interpreted that the Telecommunications Act attempted to create a regulatory body, charged with pursuing efficiency, that was independent from risk of capture by vested (and especially political) interests. At the time, this was seen as an 'enlightened' form of industry-specific regulation.

8.3.4   Testing of the Regime with LLU Decision (Dynamic Efficiency Prevails)

The enforceability and political acceptability of decisions made using the legislated efficiency mandate were explicitly tested when the Commission undertook its statutory (Section 64) review into local loop unbundling (LLU) in 2003. The investigation was noteworthy internationally for employing a cost–benefit analysis based upon a total welfare decision criterion – the sum of consumer and producer surplus as opposed to consumer welfare alone (Hausman and Sidak, 2005) – albeit subject to criticisms regarding the treatment of investment in the model. The Commerce Commission (under the mandate of the Office of the Telecommunications Commissioner) explicitly applied dynamic efficiency principles in making the ultimate recommendation not to proceed with unbundling.

The Commerce Commission found that: 'The overall benefits from unbundling are not sufficiently persuasive to satisfy the Commission that a regulated solution is warranted'. It considered platform competition (for example, from wireless networks) likely to evolve and reduce the extent of Telecom's control of the bottleneck to access (Commerce Commission, 2003, para. 788). The Commission noted that, internationally, the experience of LLU was mixed in respect of increasing broadband penetration (para. 792). The Commission also cited high costs of mandatory unbundling, and made the critical point that the incentives for Telecom to invest would be substantially reduced under LLU, and that this would have very significant effects upon the potential welfare gains for consumers (para. 794).

In place of LLU, the Commission recommended accepting an offer by Telecom to make available a limited bitstream service. Given Telecom's imminent investment in a next generation network (NGN), a lesser form of unbundled access would enable a limited amount of service-differentiated entry, whilst preserving Telecom's investment incentives and limiting entrants' exposure to stranded assets in the event of the NGN resulting in the bypass of exchanges containing entrants' equipment (Gans and King, 2004; Covec, 2004; Howell, 2007).

The Minister of Communications accepted the Commission's recommendations, apparently endorsing both the methodology used and the conclusions reached. The decision was, however, received with considerable dismay by Telecom's competitors and other interested parties, who would undoubtedly have preferred a different outcome. The LLU decision illustrated to industry participants that industry-specific regulation constrained by an explicit efficiency obligation could result in an economically rational outcome, rather than bowing to pressures to meet specific competition or redistribution agendas. However, the decision also showed that, in the absence of public opportunities to deviate the regulator from legislatively mandated efficiency-based decision-making, lobbying efforts would need to concentrate on (less well-informed) decision-makers in the political arena.

8.3.5 Testing of the Regime with Mobile Termination Decision (Competition Trumps)

The ongoing supremacy of the total efficiency criterion and the balance between its static and dynamic components as interpreted in the LLU inquiry was further tested by the Commerce Commission's inquiry into mobile termination between 2004 and 2006. The initial inquiry was instigated by the Telecommunications Commissioner, who explained that 'features of the mobile termination market that give rise to concerns about the exercise of market power by mobile carriers' had led to 'complaints that lack of competition in the mobile termination market means charges for fixed-to-mobile calls in New Zealand are unreasonably high'.

Once again, dynamic efficiency considerations underpinned the Commerce Commission's cost–benefit analysis, which led to a recommendation in June 2005 that mobile termination charges for voice calls on 2G (second generation) networks, but not 3G networks, be regulated. The recommendation stated:

> the Act does not direct the Commission as to the weight that it should give to efficiencies, as opposed to other considerations. This is a matter for the Commission to consider. Where there are tensions between short-term allocative efficiency and long-term dynamic efficiency, the Commission takes the view that giving greater weight to the latter will generally better promote competition for the long-term benefit of end-users. (para. 28).

Notwithstanding, the Commission gave greater weight to distributional considerations in this recommendation than in the LLU case, with consumer welfare, rather than total welfare, providing the decisive criterion.

In August 2005, however, the Minister rejected the Commerce Commission's recommendation and ordered that a second review be undertaken, suggesting that a change in political priorities had occurred since the 2003 LLU inquiry. The second review was to reconsider 'definitional and implementation issues concerning 2G and 3G' and to take into consideration 'commercial offers made by Telecom and Vodafone following the Commission's final report'. The ministerial redirection appears to confirm that lobbying of politicians was now the preferred method of influencing regulatory decisions.

The Commerce Commission's second report, in April 2006, recommended that all fixed-to-mobile voice calls on all technology types be subject to regulation, on the basis that 'substantial net benefits to end users were likely to arise from making mobile termination a designated access service' (para. 32). The Commission defended using the redistributive consumer welfare decision criterion, explaining: 'where wealth transfers which are sustainable and not themselves conducive to inefficiency are likely to result from a measure promoting competition, the Commission ought to give weight to such transfers in the cost–benefit analysis' (para. 34). The Commission justified including 3G technologies in its regulatory prescription by noting that deployment had advanced considerably between the first and second decisions, to the extent that the Commission considered existing 3G investments to be irreversible.

More surprising than this shift in recommended regulatory scope, though, was the Commission's explicit rejection of the supremacy of long-term dynamic efficiency over short-term competition-based static efficiencies and distributive considerations in its decision-making criteria – a substantial change from the criteria emphasized in the first report.

The Commission argued that its statutory authority actually prioritized the pursuit of competition (not efficiency) as the prevailing sector objective. Specifically, the Commission claimed that the Telecommunications Act created a distinction between two provisions of the Commerce Act: Sections 36 and 47 on the one hand, which seek to promote competition by restricting the aggregation of market power and controlling its use, and Part IV on the other hand, which focuses on the regulation of existing market power. The Commission interpreted Part IV as requiring a focus on the net benefit to acquirers – that is, the Commission should take into account 'the wealth transfer that occurs in reducing the excessive profits of the regulated party' (para. 46) – an apparent acknowledgement of static efficiency and redistribution as the primary purpose of regulatory activity. The Commission deemed the Telecommunications Act to derive from the principles of Part IV of the Commerce Act, and thus was in no doubt that in addressing the tension between the promotion of competition (the means) and the pursuit of total, long-run efficiency (the end), the Telecommunications Act gave primacy to competition, stating: 'where there is a tension between the net public benefits and promotion of competition, the statutory context indicates that the primary consideration is the promotion of competition' (para. 47). A paragraph later: 'the Telecommunications Act is focused on regulating access to promote competition. It does not provide a mechanism that specifically allows for efficiency considerations to take precedence over the promotion of competition. Nor is there anything in the statutory scheme to suggest that this should be the case'.

That is, diametrically opposed to the LLU decision criteria, the static and dynamic efficiencies arising from increased competition on existing networks in the short run were deemed to take precedence over the longer-run efficiencies arising from investment in new network technologies in Commission decision-making. This suggests that, even if it could be demonstrated that restricting competition would result in greater total efficiency in the long run, and even ultimately the long-term benefit of consumers, the Commission must as its statutory duty promote short-run competition. The conclusion is that in this decision, the Commission is confirming a statutory obligation to place the short-term interests of competitors above even those of consumers.

This conclusion, therefore, begs the question of the purpose of Section 18(2) of the Telecommunications Act. For pursuit of competition to override pursuit of efficiency, 'consider' must constitute only an obligation to demonstrate that as some part of the decision process, efficiency issues were raised. The Commission evidently now deemed Section 18(2) to contain no statutory obligation that efficiency considerations should have any material effect upon the final recommendation if, in taking them into account, the pursuit of competition was impaired. If the Commission's 2006 view is accepted as legitimate, then it constitutes an admission that all of the previous Commission decisions prioritizing efficiency and weighing long-term dynamic considerations over short-term ones were based upon an erroneous interpretation of the government's intentions when passing the Act. In the historical context, this view is perplexing, given the apparent acceptance of all previous efficiency-based recommendations.

A plausible alternative explanation for the Commission's radical departure from previous decision-making precedents is that the prevailing political objectives changed between the 2003 LLU decision and the 2006 mobile termination decision. The 'independent' regulator had been perhaps too successful in prioritizing long-run total efficiency for political (and notably electoral) preferences which prioritize delivery of short-run changes. As an agent of political principals, the Commission became subject to pressures to resile from its previous prioritization of total, long-run efficiency in favor of a set of more politically acceptable short-run criteria. Evidence in support of this view includes the Governor General's (2005) 'Speech from the Throne' signaling that the incoming government would be 'advancing policies to ensure that the telecommunications sector becomes more competitive and that we achieve faster broadband uptake in line with our competitors' and the fact that the Minister, when rejecting the Commission's first mobile termination recommendation, instructed the Commission to take account of alternative offers made to the Minister by the firms facing regulation. An acknowledgement of the Commission's subservience to ministerial control is contained in its acknowledgement that ultimately 'the role of the Commission is to recommend and it is for the Minister to determine' (para. 53).

The above observations on the Commission's shift in decision-making processes suggest that competitor lobbying of politicians for competitive and redistributive interests to prevail successfully led not just to the change of objective but also to a fundamental change in the locus of control of New Zealand's regulatory decision-making. Ultimately, it was the Minister and not the Commissioner who began to exercise actual regulatory control. As further evidence of this shift, the Minister also rejected the Commission's second recommendation to regulate mobile termination rates, instead favoring a set of undertakings by the potentially regulated firms brokered by the Minister of Economic Development (the Minister of Communications having declared a conflict of interest as a consequence of a legal dispute with one of the companies facing regulation).

8.3.6 Ministry-Led 'Stocktake' in Interest of Increasing Competition

The pursuit of competition in the telecommunications sector became an increasingly important political issue as the incoming government began implementing its political agenda. In December 2005 the Ministry of Economic Development (MED) began a 'stocktake' of the telecommunications industry, with its primary focus being 'the broadband market and our broadband performance as a factor in economic performance' (MED, 2006). Given that the expertise to undertake the investigation lay principally in the Office of the Telecommunications Commissioner in the Commerce Commission, the use of a ministry to undertake an assessment of industry performance implies a lack of political confidence in the Commission and its processes. It also suggests a lack of certainty that Commission-led analysis 'taking account of' efficiency would deliver a set of recommendations consistent with the government's explicit competition agenda to which it was already committed. Unlike the Commission, the Ministry was not explicitly bound by a requirement to take account of efficiency issues in its analyses.

The stocktake report released by the MED in May 2006 is notable for its lack of principled economic analysis (see Howell, 2006 for a detailed discussion). In sharp contrast to the Commission's analyses, no cost–benefit analysis was undertaken. The Regulatory Impact Statement accompanying the stocktake proposals rejects the validity of the Commission's analysis in 2003, but specifically states that no new analysis was necessary to quantify the effects of the stocktake's recommendations. Rather, the stocktake justified full local loop unbundling and functional separation of Telecom primarily on the basis that it believed the broadband market had failed to meet a desired level of competitiveness (defined, arbitrarily, by the share of connections sold by competitors to Telecom), and that investment by Telecom had been 'insufficient' (on the basis of slippage from an investment schedule proposed in 2003 and an unscientific benchmarking against a handful of other OECD countries).

Compounding the severe analytical shortcomings of the stocktake, the process for reviewing its recommendations also lacked rigor and transparency. Submissions on the stocktake's proposals were heard by a select committee comprised of generalist politicians rather than a panel of expert commerce commissioners, and were not subject to the three-stage

Commission processes of a draft report, a quasi-court conference where the recommendations and written submissions on the draft report by all interested parties can be tested in a contestable manner, and a final report. Furthermore, the process was not subject to appeal or review on either process or substance.

8.3.7 Changes to the Telecommunications Act Increase Political Influence

Thus, with little alteration and considerable political support, the stock-take's recommendations were enacted in December 2006. Along with the provisions for full LLU and separation, a new section (19A) was added to the Telecommunications Act requiring the Commission to take account of any economic policies of the government that are communicated by the Minister in writing. This appears to place the requirement to take account of government economic policies on the same footing as the requirement to take account of efficiency in making recommendations as per Section 18(2). If treated as an imperative, as the efficiency obligation was from 2001 to 2005, it would appear to explicitly to compromise the Commission's independence from the government by making it an instrument via which government economic policy is enacted. Alternatively, if interpreted in the light of the second mobile termination inquiry, the new requirement could be subjugated, but only in favor of the pursuit of competition – which, incidentally, was an economic policy of the government that incorporated the requirement in the first place.

Rather than clarify the Commission's objectives, it appears that the new obligation compounds the problem of multiple priorities for the Commission, and reduces the possibility that any of the objectives will be satisfactorily delivered. Moreover, it creates a further tension if, in the future, government economic policies change again in a manner that brings them into conflict even with the competition imperative (for example, nationalization of assets).

Together, the 2006 stocktake and Telecommunications Act amendments signal that the New Zealand regulatory process no longer made any pretence of being independent. The Ministry and politicians were now firmly established, by a dint of formal legislative power, as the third party alongside the courts and the Commission in governing industry activities. In further confirmation of the strong government role in sector regulation, in May 2007 the Minister of Communications announced that he, rather than the Telecommunications Commissioner, would be overseeing a new process to functionally separate Telecom. The Minister (endorsed by the cabinet) justified his taking control of this process on the basis of 'the urgency attached by the government to the need to secure a clear outcome on this matter in the shortest possible timeframe. Because this is a major structural issue and not a matter of micro regulation, this was felt and is still felt to be the appropriate way forward' (New Zealand Government, 2007).

This declaration clearly indicated the government's view that the structure of the telecommunications industry, where almost all investment comes from the private sector, was the prerogative of political control, whilst the role of regulators was a micro one of carrying out politically determined policies. Ministerial supervision of Telecom's functional separation marked the third time since the 2005 election that the formal political power of the principal in the regulatory agreement was exerted in order to take back control of duties ordinarily in the ambit of a regulatory authority.

Such a view of political primacy is at considerable variance with the prevailing views espoused by the OECD and the International Telecommunication Union (ITU) that industry-specific regulation be insulated from political processes to the greatest extent possible precisely to guard against the pursuit of short-term self-interested agendas. However, under the present arrangements for sector governance in New Zealand, it is highly unlikely that future regulatory decisions will be driven primarily by economic efficiency imperatives, or even requires that an analysis taking account of the sectors' specific economic circumstances (for example, small-scale, high and sunk investment costs) be undertaken. Rather, recent history suggests that returns to political lobbying will more likely bear fruit (for both the incumbent and entrants) than economic debate with the regulator.

8.4   CONCLUSION

New Zealand telecommunications regulation from 1987 to 2009 offers a cogent example of the vulnerability of even those regulatory agencies designed with economic efficiency objectives in mind to capture by political principals in order to pursue alternative objectives. Over the course of less than two decades, the pursuit of competition and the achievement of short-term efficiency gains as a political priority prevailed over the pursuit of total efficiency in the long run – even though initially, total, long-run efficiency was explicitly articulated as the prevailing political objective. Reliance upon competition law alone was the starting point for the allocation of industry regulatory governance to achieve the stated objective. This was replaced by an industry-specific regulator, with a legislative mandate to consider total efficiency. However, over time this body has been politically influenced towards the pursuit of competition and politically motivated subjective short-term redistribution issues and notably competitor welfare, and its legislative mandate now requires that it consider political objectives in carrying out this function.

Whereas, typically, political capture of regulatory processes occurs informally as covert exertion of pressure and influence, the New Zealand case illustrates not only informal pressure, but also overt political action. Political dissatisfaction with both court adjudications and regulatory recommendations based upon total efficiency criteria has resulted in a 'third body' – the office of the Minister of Communications and the affiliated MED – assuming responsibility for a range of functions normally delegated to the purview of an industry-specific regulator. In the absence of legislative or constitutional formalization of this action, it is not clear whether the result is intended as competition for, or a complement to, competition law and regulation. What remains clear, however, is that current political decision-making has given very little weight to any efficiency considerations, has reduced industry certainty as it lacks precedents, and is subject to few of the checks and balances of either competition law or regulatory processes.

New Zealand's unique institutional structure appears to have evolved because the explicit attempts by governments in the past to create arrangements that prioritized the pursuit of total efficiency above other considerations, in a manner that reflects the competitive reality of the small New Zealand economy, have ultimately failed to withstand the threat of political capture. The government initially relied upon competition law alone to regulate the telecommunications sector, and subsequently, when introducing an industry-specific regulator, explicitly included efficiency objectives in the mandate given to the industry-specific regulator. However, subsequent governments diluted the force of the efficiency mandate, and ultimately have captured the regulatory process as a means of furthering a political agenda most likely driven by redistributive pressures. The erosion of the efficiency mandate began gradually, but has accelerated since 2005.

The chronology of erosion and capture illustrates the thesis that, despite the best of intentions, ultimately governance arrangements in the telecommunications sector are determined by those with the political power to make laws allocating responsibility for various tasks. Pursuit of efficiency is rationally justifiable from an economic perspective, and has been demonstrated to be the objective of (if not perfectly achievable via) competition law. However, making efficiency an explicit regulatory objective is not sustainable in the long run. An objective regulator would not heed the petitions of those seeking to capture the process for their own purposes, shifting pressure onto the lawmakers in government. The government will permit a continuing focus on efficiency only as long as this is congruent with its political objectives. But governments change and their objectives change with them. Eventually, in the absence of the ability to exercise informal control to capture the process, a successor government (which cannot be bound by its predecessors) may reverse earlier decisions, either changing the rules or taking over the regulatory task itself. The only reason that the courts administering competition law can avoid such capture is because their constitutional origins afford them more independence than agencies that derive their mandate from political processes.

Given the lack of ability to enforce an efficiency objective in the long run via industry-specific regulation, the only sustainable means of doing so would appear to be via competition law. Imperfect though competition law may be for the telecommunications sector, the New Zealand case suggests that, in the absence of constitutional protections for a regulatory agency also charged with the pursuit of efficiency, it may be the only sustainable institutional compromise.

NOTE

1. This chapter, prepared with editorial assistance from Seini O'Connor, is based on a working paper for the New Zealand Institute for the Study of Competition and Regulation ('The end or the means? The pursuit of competition in regulated telecommunications markets' see www.iscr.co.nz).

REFERENCES

Alleman, J. and P. Rappoport (2005). Regulatory failure: time for a new paradigm. *Communications and Strategies*, **60**, 105–21.

Armstrong, M. (2001). Access pricing, bypass and universal service. *American Economic Review*, **91**, 297–301.

Arnold, T., D. Boles de Boer and L. Evans (2003). The structure of industry in New Zealand: its implications for competition law. In M. Berry and L. Evans (eds), *Competition Law at the Turn of the Century* (pp. 11–18). Wellington, New Zealand: Victoria University Press.

Blanchard, C. (1994a). Telecommunications regulation in New Zealand: the Court of Appeal's decision in Clear Communications v Telecom Corporation. *Telecommunications Policy*, **18**(9), 725–33.

Blanchard, C. (1994b). Telecommunications regulation in New Zealand: how effective is 'light-handed' regulation? *Telecommunications Policy*, **18**(2), 154–64.

Blanchard, C. (1995). Telecommunications regulation in New Zealand: light-handed regulation and the Privy Council's judgement. *Telecommunications Policy*, **19**(6), 456–75.

Commerce Commission (2003). Telecommunications Act 2001 Section 64 review and schedule 3 investigation into unbundling the local loop network and the fixed public data network Final Report. Wellington, New Zealand. Available from www.comcom.govt.nz.

Connolly, S., and A. Munro (1999). *Economics of the Public Sector*. Harlow: Pearson Education.

Covec (2004). Benchmarking Telecom's UPC service. Retrieved from www.comcom.govt. nz.

Economides, N. and L. White (1995). Access and interconnection pricing? How efficient is the 'efficient component pricing rule'? *Antitrust Bulletin*, **40**(3), 557–79.

Evans, L., A. Grimes, B. Wilkinson and D. Teece (1996). Economic reforms in New Zealand 1984–1995: the pursuit of efficiency. *Journal of Economic Literature*, **34**(December), 1856–1902.

Feldstein, M. (1972a). Distributional equity and the optimal structure of public pricing. *American Economic Review*, **62**, 32–6.

Feldstein, M. (1972b). Equity and efficiency in public pricing. *Quarterly Journal of Economics*, **86**, 175–87.

Gans, J. and S. King (2004). Access holidays and the timing of infrastructure investment. *Economic Record*, **80**(248), 89–100.

Governor-General Dame Silvia Cartwright (2005). Speech from the Throne. Delivered to Members of the House of Representatives at the State Opening of Parliament on 8 November. Transcript available at www.beehive.govt.nz/node/24330.

Hausman, J. and G. Sidak (2005). Did mandatory unbundling achieve its purpose? Empirical evidence from five countries. *Journal of Competition Law and Economics*, **1**(1), 173–245.

Howell, B. (2003). Building best practice broadband: bringing supply and demand together. ISCR Working Paper. Available at www.iscr.org.nz.

Howell, B. (2006). Submission: Telecommunications Amendment Bill. Presentation to the New Zealand Finance and Expenditure Select Committee. Available at www.iscr.org.nz.

Howell, B. (2007). A pendulous progression: New Zealand's telecommunications regulation 1987–2007. ISCR Working Paper. Available at www.iscr.org.nz.

Kahn, A. (1970). *The Economics of Regulation*, Vol. I. New York: Wiley & Sons.

Kahn, A. (1975). *The Economics of Regulation*, Vol. II. New York: Wiley & Sons.

Ministry of Commerce and The Treasury (1995). Regulation of access to vertically-integrated natural monopolies. Wellington, New Zealand: Ministry of Commerce, The Treasury. Available at www.med.govt.nz.

Ministry of Economic Development (MED) (2000). Ministerial inquiry into the telecommunications industry: final report. Wellington, New Zealand: Ministry of Economic Development. Available at www.med.govt.nz/upload/30006/final.pdf, accessed 19 January 2010.

Ministry of Economic Development (MED) (2006). Promoting competition in the market for broadband services. Wellington, New Zealand: Ministry of Economic Development. Available at www.med.govt.nz.

New Zealand Government (2007). Delivering on telecommunications. Speech notes for address to TUANZ Telecommunications Day. Michael Fowler Centre, Wellington, New Zealand, 31 May 2007, http://www.beehive.govt.nz/node/29595 (official website of the New Zealand Government).

Peltzman, S. (1976). Toward a more general theory of regulation. *Journal of Law and Economics*, **19**, 211–40.

Quigley, N. (2004). Dynamic competition in telecommunication: implications for regulatory policy. Toronto: C.D. Howe Institute Commentary, February.

Schmalansee, R. (1981). *The Control of Natural Monopolies*. Lexington, MA: Lexington Books.

9. International regulatory comparisons: the evolution of IP-based fiber

Scott Marcus and Dieter Elixmann

9.1 INTRODUCTION

Technological and market forces are driving network operators and electronic communication service providers throughout the world to migrate their networks to Next Generation Networks (NGNs) based on the Internet Protocol (IP).[1] NGN access to the fixed network is in the process of being enhanced over time in many countries to provide higher speed using fiber-based technology. At an abstract level, one might imagine that a change in underlying technology would have little impact on regulation;[2] however, the evolution of the access network to IP-based fiber implies substantial challenges for regulators. Various regulators in various countries are finding somewhat different solutions to these challenges.

Two broad families of technical approaches exist. One approach is known as fiber to the home (FTTH) or fiber to the building (FTTB). The other is very high speed DSL (VDSL), which is associated with fiber to the curb/cabinet (FTTC) (because fiber is built out to the street cabinet but not all the way to the individual home or building). Whether the migration is to FTTC/VDSL or to FTTB/FTTH, traditional solutions to incumbent market power typically become difficult to apply. For FTTC/VDSL, the natural point of interconnection (PoI) for purposes of network access moves from the main distribution frame (MDF) to the far more numerous street cabinets; however, access to street cabinets is potentially difficult and costly, calling into question the practicality of local loop unbundling (LLU) as a competitive remedy. For FTTB/FTTH deployments to multiple dwelling units, many challenges exist regarding building wiring. As an additional problem for point-to-multipoint FTTB/FTTH, there are uncertainties as to the practicality of unbundling passive optical network (PON) solutions. All of these challenges call into question the practicality of LLU as a pro-competitive regulatory remedy.[3]

Various countries have considered these issues, but the pace of regulatory proceedings has been conditioned in each case by developments and market evolution in that country. In both the Netherlands and Germany, the incumbents (KPN and DTAG) are seeking to move rapidly to replace the traditional fixed access network with a VDSL-capable network. In France, the incumbent as well as its competitors are moving quickly to deploy FTTx in a number of major metropolitan areas. In Japan, fiber deployment has been very successful, with many of the deployments undertaken by third parties other than the incumbents (NTT East and NTT West). In the US, incumbent operators AT&T and Verizon have made substantial investments in fiber for their respective customers. Our empirical analysis reflects the state of the world as of the first half of 2008.

Section 9.2 discusses the regulatory challenges relevant to FTTC/VDSL and FTTB/FTTH solutions. Section 9.3 discusses regulatory solutions that have been attempted in various countries. Section 9.4 provides a comparative assessment, in terms of the degree to which competition has been preserved, and the extent of fiber-based deployment and adoption. Section 9.5 provides concluding remarks and a few recommendations.

9.2   REGULATORY CHALLENGES

This section explores the various regulatory challenges associated with migration to IP fiber-based Next Generation Access. Section 9.2.1 explores challenges in an FTTC/VDSL context. Section 9.2.3 discusses problems that are unique to an FTTB/FTTH environment. Finally, section 9.2.3 reviews problems that are common to both. Our focus in this chapter is on challenges at the access level of the network; consequently, we disregard a broad array of other regulatory concerns raised by the migration to NGN access from copper to fiber, notably including possible interactions with the interconnection regime (which has more to do with the NGN core than with the access network).

9.2.1   FTTC/VDSL

In this section, we consider the challenges the migration to FTTC/VDSL poses to pro-competitive regulation. First, we consider challenges in general; then we review specific challenges to the business model that have been studied in the literature. Finally, we consider the degree to which bitstream access, which is largely unperturbed by the migration from ADSL to FTTC/VDSL, could substitute for both LLU and shared access.

The deployment of FTTC/VDSL poses challenges to local loop unbundling (LLU)

The first-mover for FTTC/VDSL deployments has often been the incumbent. Examples include Belgium (Belgacom),[4] Germany (Deutsche Telekom), Italy (Telecom Italia)[5] and the Netherlands (KPN).[6] With an (all-IP) FTTC/VDSL strategy, fiber is deployed throughout the access network up to the street cabinet. Wherever fiber infrastructure is deployed, copper infrastructure is removed (or taken out of service). VDSL access devices (DSLAMs – DSL access multiplexers) are generally deployed at the street cabinet, in order to be close enough to the end-user to achieve suitably high bandwidth; consequently, the main distribution frames (MDFs) typically no longer serve as active network nodes. This means, as previously noted, that the point of interconnection (PoI) moves from the MDF to the street cabinet. This shift implies multiple challenges (in comparison to copper-based ADSL – Asymmetrical Digital Subscriber Line) for would-be competitors who seek to replicate the incumbent's FTTC/VDSL infrastructure:

- There are far more nodes to connect to, because there are far more street cabinets than MDFs.
- There are far fewer end-users per street cabinet.
- There are likely to be limitations with the physical characteristics of the street cabinets, including (alleged) lack of space, heat dissipation, and so on.
- There is no realistic possibility to provide separate, additional street cabinets for competitors due to the resistance of city governments and home owners (whose premises are potential locations for a separate street cabinet).
- There are likely to be high barriers or expenses in connecting to the existing street cabinets, because the competitor is unlikely to have a cost-effective alternative to the duct infrastructure that the incumbent has inherited from monopoly times.

Thus, incumbent FTTC/VDSL deployments are likely to pose serious challenges for existing wholesale services based on local loop unbundling (LLU), including shared access models.

Sub-loop unbundling and the business case for FTTC/VDSL

Three major studies have found that the business case for sub-loop unbundling (SLU) is likely to be problematic.[7] A study for the Dutch regulator OPTA (Analysys, 2007a) found that the use of sub-loop unbundling (SLU) by an alternative provider would be economically viable as an alternative to continuing to use LLU only under very limited conditions. Some combination of: (1) high market penetration; (2) higher average revenue per user (ARPU) than has typically been achieved to date in Europe; or (3) a substantial reduction in wholesale prices charged by the incumbent would be required. These requirements are collectively and individually unlikely to be fulfilled. For example, a mass-market competitor could be viable with both: (1) a market share greater than 55 percent of all broadband lines (including cable) in areas served; and (2) an increase in ARPU across all broadband users of 10 euros per month. Alternatively, an SLU competitor with a more limited footprint could be economically viable with: (1) a 10 percent market share of all broadband lines in areas served, corresponding to about 1000 of the largest street cabinets in the densest urban areas; in conjunction with (2) a 50 percent reduction in the interconnect and wholesale tariffs from KPN for SLU line rental, co-location and links to the street cabinets; and (3) an increase in ARPU of about 9 euros. Again, these conditions are unlikely to be fulfilled.

A second study that was conducted for the Irish national regulatory authority ComReg (Analysys, 2007b) reached similar conclusions. SLU is subject to much stronger economies of scale than is LLU. SLU is not as commercially attractive as LLU, even when the competitor concentrates on large street cabinets (more than 300 lines). The largest costs are the line rental charge, cost for the street cabinet and the backhaul link to the MDF. These are the points to be addressed first by regulatory means.

A third study (J.P. Morgan, 2006) compared the incremental investments and costs associated with FTTC/VDSL to those of ADSL. Both were assessed using a greenfield approach, which is to say that all investments are assumed to be new. Compared to an ADSL competitor, a VDSL alternative operator incurs:

- additional costs for unbundling at the street cabinet;
- additional costs to deploy fiber between the MDF and the street cabinet;
- additional costs of the backhaul network due to the greater capacity requirements implied by VDSL access.

The overall incremental cost depends in part on the assumed network topology, star or ring. In the worst case, an alternative operator with a market share of 25 percent incurs additional costs of 11 euros per access line per month. Like the two previous studies, J.P. Morgan (2006) shows that FTTC/VDSL deployment is economically feasible for an alternative operator only under very strong conditions. Even with a 40 percent market share, the VDSL alternative operator's profit per subscriber decreases by 2 euros per month compared to the comparable ADSL case.

Bitstream access to FTTC/VDSL
Given that SLU is viable only under (very) limited conditions, bitstream access must be considered as a practical alternative. The economics of bitstream access for the incumbent will change somewhat with the migration from ADSL to FTTC/VDSL (due, for example, to changes in the number of aggregation points), but the viability of bitstream overall does not appear to be threatened. A troubling question for the regulator is whether bitstream access alone is sufficient. Europe has sought to maintain a ladder of investment with four rungs: resale (which provides no service differentiation), bitstream access, shared access, and full LLU. The move to FTTC/VDSL apparently breaks two out of three rungs of the ladder. Is bitstream alone enough to ensure effective competition?

9.2.2   FTTB/FTTH

This section reviews the business and regulatory challenges posed by FTTB and FTTH and present an analysis of their implications for business models.

The deployment of FTTB/FTTH poses different challenges
One can find numerous examples of fiber to the building (FTTB) or to the home (FTTH) in Asia (Hong Kong, Japan and Korea), Europe (France, Sweden and Italy), and the United States. While FTTC/VDSL players are usually incumbents, FTTB/FTTH players can be either incumbents or competitors (or sometimes both, even in the same country).

   The two common architectural options are (1) passive optical network, or PON (which is a point-to-multipoint infrastructure containing no active electronic equipment where a single fiber path branches to several houses); or point-to-point (which is comprised of a single dedicated fiber path from the provider's network node to the consumer's building, that is, between the optical line terminal (OLT) and the optical network unit (ONU). Civil engineering costs and other difficulties (digging up the streets, getting the permissions and rights of way) dominate the cost of deployment. Where aerial deployment is permitted, however, or where existing ducts can be used, the cost of running fiber can be greatly reduced. Key regulatory challenges include:

   ● In-house wiring: for multiple dwelling units (MDUs), running fiber to individual apartments implies rewiring the building. Few buildings are suitable for more than one set of fiber. This adds a new dimension to the last mile problem.

- Sharing of PON infrastructure: for point-to-multipoint PON systems, it is unclear how to provide an LLU equivalent to competitors. Sharing would appear to be expensive if not impossible.

The business case for FTTB/FTTH

The J.P. Morgan (2006) study found that the investment required to pursue an FTTB strategy can be up to 12 times higher than those required for traditional ADSL access, and up to 2.5 times higher than for VDSL access. If, however, existing ducts can be used for the deployment of fiber, then the investment required to deploy FTTB is only slightly higher than in the VDSL case. The generic greenfield business case developed by JP Morgan reflects the following baseline model assumptions, which from our perspective seem to be plausible:

- The model is calibrated on a 25 percent market share.
- The operator has no access to duct infrastructure.
- There is no ARPU or market share gain due to FTTB/FTTH.
- For each FTTH customer acquired the operator saves 13 euros per month regarding LLU and interconnection.
- There are no changes regarding other operating expenditure (OPEX) components due to the FTTH roll-out.
- The weighted average cost of capital (WACC) is equal to 8 percent, the tax rate is equal to 35 percent and the depreciation period is equal to 20 years.

Under these baseline model assumptions, the payback period for the investments in FTTB/FTTH is 16 years. The net present value (NPV) of the investments is nonetheless negative, at -500 euros per subscriber. A market share of just 40 percent generates a positive NPV. The investments can, however, be economically feasible with a market share of 25 percent if the investments can be reduced to a level of 2000 euros per subscriber, if additional ARPU is possible, or if additional savings are possible. Regarding the latter, very important profitability effects arise with savings due to the deployment of fiber. Provided that 50 percent of the deployment costs to and within the building can be saved (by utilizing existing ducts and in-house sharing), then the investment costs can be reduced to a level of 1500 euros per subscriber. In this case, the payback period decreases from 16 years to ten, and the investments become profitable with an NPV of more than 200 euros per subscriber. A similar result would be obtained if the ARPU were increased by 6 euros per customer per month.

9.2.3   The FTTC/VDSL and FTTB/FTTH Deployments: some Common Issues

This section reviews a range of issues common to both, the FTTC/VDSL and FTTB/FTTH deployments. Particularly, it deals with the problems of market definition, the issues of network topology and the challenges they pose to the traditional regulatory models.

Market definition
The identification of markets is fundamental to the European regulatory framework. Consistent with numerous communications from the European Commission, we have treated the various forms of fiber-based IP access in this chapter as being part of a single continuous market that also includes conventional broadband (such as ADSL and cable). It is possible that applications that are heavily dependent on the bandwidth that only fiber can offer will become more prevalent over time. At that point, it might be appropriate to treat fiber-based access as a distinct market segment from conventional broadband, inasmuch as conventional broadband will no longer be an adequate substitute for fiber-based access.[8]

Change of network topology
The migration to Next Generation Access (NGA) will result in changes in the number and location of points where competitors can get access to the network of the incumbent (henceforth also called Points of Interconnection, or PoIs). The phasing out of MDFs[9] for competitive access is a key point. More generally, IP-based NGNs will be flatter than today's networks, and will therefore offer fewer points of interconnection as well. These changes in network topology are likely to lead to stranded investments (equipment that was deployed in the past, is not yet fully depreciated, but is no longer needed) both for the incumbent and for the alternative operator. KPN's approach in the Netherlands suggests that the incumbent can benefit by selling off the real estate associated with facilities that are no longer needed. These benefits can offset the cost of stranded incumbent investments. Competitive operators are less likely to benefit from such changes.

Stranded investments are most likely to be important in the transition period from switched network to IP-based NGN. They need to be reflected in the regulatory environment of the transition period. Transparency regarding the network deployment plans of the incumbent is crucial. The less transparency on the part of the incumbent, the greater will be the risk of stranded investments on the part of competitors.

Changes in regulated costs

Numerous regulatory decisions rest on assumptions about the cost of the network. The migration to fiber-based NGN access will, at a minimum, require regulators to upgrade their cost models. This migration potentially raises more complex issues than the implementation of ADSL, because the investments in fiber are so much larger, and also because they are often associated with fundamental changes in the core network.[10] The overall migration to fiber-based IP requires massive investment at the outset, but presumably results in lower ongoing operating costs when the transition is complete. At the same time, the network as a whole may incur higher operating costs during a period where the circuit-switched public switched telephone network (PSTN) and the packet-switched NGN must operate in parallel. Ongoing operating costs should eventually decline, but initially they are likely to increase. What is an appropriate regulated rate of return under these circumstances?

Ofcom explained analogous issues in developing a weighted average cost of capital (WACC) for British Telecom.[11] The WACC provides a measure of what might constitute a reasonable return on BT's investment of capital. Ofcom found it advantageous to disaggregate the WACC, computing different WACCs for different parts of the business in order to accommodate different levels of risk (beta). In the end, it did not specifically address the risk associated with the migration to NGN access, but it left open the door to possible future use of more innovative approaches, including the application of REAL OPTIONS.

9.3 REGULATORY RESPONSES: COUNTRY-SPECIFIC APPROACHES

Different regulatory approaches are visible in different countries. Note that the Netherlands and Germany are (chiefly) characterized by FTTC/VDSL roll-outs, while France and Japan are primarily characterized by FTTB/FTTH roll-outs. In the US, both are present (with AT&T prominently following a path of FTTC/VDSL, and Verizon a path of FTTB/FTTH).

9.3.1 Netherlands

KPN (as of 2008), is about to deploy a nationwide FTTB/VDSL infrastructure.[12] OPTA, the regulator for the Netherlands, initially was open to KPN's all-IP deployment plan and planned to find pertinent regulatory rules to shape the migration period. However, the Analysys study (see section 9.2.1) called into question the economic viability of sub-loop unbundling from the street cabinet. Several rounds of discussion and negotiations followed. It has now (as of mid-2008) been concluded that the current MDF related services will be maintained until the middle of 2010, that there will be an improved offering regarding high-quality wholesale broadband access, and that a substantial portion of MDF locations will remain usable for unbundled access. Moreover, OPTA has asked market participants also to discuss alternatives like Ethernet access on fiber and interconnecting leased lines.

9.3.2   Germany

In February 2007, the European Commission launched an infringement procedure against Germany in response to a specific clause in the new German telecommunications law that opened up the possibility of a regulatory holiday for DTAG's FTTC/VDSL deployment (Marcus and Elixmann, 2008). As a result of the ongoing market definition and market analysis process, the BNetzA (the German regulator) imposed an obligation on DTAG in June 2007 to open up the ducts between the MDF and the street cabinet to competitors. In the event that access to ducts is not possible (for example, due to technical reasons or limited capacity), DTAG must offer competitors access to dark fiber.

Competitors in Germany view this as a step in the right direction; however, they claim that access to ducts (dark fiber) is only one of the elements that make up a viable business case. Equally important, they argue, is unbundled access to fiber and copper within the street cabinet. Moreover, they are demanding (bundled) access to the hybrid local loop, consisting of copper and fiber, at the MDF. It is clear that there are still substantial information asymmetries between DTAG and the competitors as regards duct availability and space.

On 13 May 2008, the BNetzA issued a decision regarding the price of the IP bitstream access that DTAG is obliged to offer to its competitors. With IP bitstream access, DTAG sells a competitor a regulated wholesale service whereby data is transmitted to and from the end-user's access line over DTAG's concentrator network, and is handed off to the competitor at a designated DTAG PoI. This is the first time that IP bitstream has been meaningfully available in Germany (although Germany has well-established LLU rules for copper-based LLU). This effectively opens a new rung on the ladder of investment, which may partly compensate for the lack of access to fiber-based IP access.

9.3.3 France

ARCEP, the French telecommunications regulator (Autorité de Régulation des Communications Électroniques et des Postes), finalized two consultations in September 2007, focusing on access to ducts, and sharing of in-house infrastructure in an FTTB/FTTH environment. Regarding access to ducts, ARCEP argued that competitors must have access to FT's infrastructure in order to establish fair competition in the high-speed broadband market. Hence, the objective of regulation should be to set appropriate incentives for investments in local loop infrastructure. A possible implication could be more symmetric regulation. Moreover, ARCEP sees no need to impose functional separation as has been done in the UK.

ARCEP notes that it cannot be efficient for each fiber-based network operator to deploy its own fiber and optical connectors in each building and each apartment. Moreover, residents do not benefit if there is inefficient and excessive deployment within their buildings and apartments. Finally, ARCEP favors the principle that end-users have the opportunity to switch their broadband network operator without the need to move from one location to another. Thus, they see the need for infrastructure sharing.

In June 2007, Free/Iliad had sued France Telecom (FT) before the antitrust authority over FT's allegedly anticompetitive behavior for refusing to give its competitors access to its civil engineering infrastructures. The Conseil de la Concurrence issued an important decision on 12 February 2008. It found that FT's holding of civil engineering infrastructures is likely to give the company a particular responsibility, notably including not to distort the play of competition on the very high-speed budding markets in keeping for itself the use of the infrastructures and refusing to allow its competitors to use them, or giving them discriminatory access. FT started to deploy optical fiber in its civil engineering infrastructures with the copper local loop, while postponing its response to competitors' request for having access to the same infrastructures. The Conseil has not implemented any immediate remedial measures, considering that there was no serious or immediate infringement to the sector; nonetheless, the Conseil decided to carry on with the investigation on the merits of the case. Meanwhile, France Telecom made commitments to ARCEP to create an operational offer for access to its infrastructure. The Conseil was generally satisfied with the reactions of alternative operators to the first tests.

9.3.4 Japan

Japan is the country with by far the highest FTTB/FTTH penetration in the world. There were more than 11 million subscribers at the end of 2007.

The most popular broadband service in Japan is still ADSL. On the other hand, there has been a remarkable shift to FTTH/FTTB over the past four years (2006–09).[13] The Next Generation Broadband Strategy 2010 formulated in August 2006 by the Japanese Government aims at eliminating the non-broadband areas (Taniwaki, 2008). By fiscal year 2010, broadband service should be available to 100 percent of the population; and super high-speed broadband (FTTH) should be available to 90 percent of the population.

In Japan, there is infrastructure competition on the very last mile, at least in the largest metropolitan areas. Important drivers of this kind of infrastructure competition are the density of the population, the fact that aerial deployment of fiber on the very last mile is permitted in Japan (in contrast to much of Europe), and regulation. Even so, NTT's market position with respect to fiber deployment is very strong. As of the end of March 2007, the combined FTTB/FTTH market share of NTT East and West by number of lines was 78.9 percent, while the FTTB/FTTH share by revenue was equal to 69.0 percent according to Taniwaki (2008).[14] For many years, the Japanese broadband market has been built around an open access regime. Take-up of ADSL took off rapidly once collocation and unbundling rules for the access networks of NTT East and West were established in the middle of 2000. Although unbundled access continues to be an obligation for fiber going forward, these obligations have been less effective than the previous copper-oriented regime; thus, the role of unbundling going forward in Japan is unclear.

Japan is on the verge of changing its policy due the anticipated transition to full IP-based networks. In October 2006, the Japanese government launched a New Competition Promotion Program 2010. Japanese policy appears to be moving from a world of *ex ante* regulation to a world of *ex post* regulation. This shift appears to imply substantial easing of price and tariff regulations, and the introduction of a competition review mechanism. Currently, the fiber access of NTT East and West is unbundled, and its charge is calculated on the basis of the estimated actual cost of fiber deployment from 2001 to 2007 (that is, for seven years). On the basis of an exchange rate of 160 yen per euro, Katagiri (2008) reports a current price of 31.7 euros per fiber.[15] Katagiri points out that this price regime might be unfavorable for NTT's competitors. The reason is that FTTB/FTTH is usually provided via a passive optical network architecture, and in NTT's network one fiber is split to eight users. Given that the NTT's market share is more than 70 percent (see above), the cost per user would be quite different for NTT in comparison to its competitors.

9.3.5 The United States

Regulation – or rather deregulation – of fiber-based access in the United States is relevant both to VDSL and to FTTB/FTTH deployments. The United States during the years 2001–08 has consistently pursued a policy of deregulation of last mile access, possibly in the hopes of stimulating deployment of fiber-based solutions. This reflects a dramatic change from prior US communications policy (Marcus, 2005).

The most significant change in regard to fiber-based access came in 2003 with the Federal Communication Commission's (FCC) Triennial review (FCC, 2003). This decision exempted fiber from loop unbundling obligations. In a series of related decisions, the FCC abolished non-discrimination obligations for telephony incumbents in regard to IP-based services; eliminated obligations to implement shared access (line sharing); and exempted broadband services delivered over cable television from pro-competitive regulatory remedies. It is important to remember that the market in the United States differs from that of most other countries in that cable operators provide a larger fraction of broadband access than do telephone companies. This has established a significantly different market dynamic than that of other countries.

9.4 COMPARATIVE ASSESSMENT OF THE REGULATORY APPROACHES

In this section we discuss Next Generation Access policy in the context of a trade-off between investment and competition, and we consider the degree to which regulatory approaches have been effective.

9.4.1 Is there a Trade-off between Investment and Competition?

What are the likely implications of regulation for competition, for deployment and adoption of fiber-based access, and for consumer welfare? How will regulation impact on the incentives and viability of business models in the market? Does regulation negatively impact incentives to invest, for incumbents and for competitors? How will it impact on the intensity of infrastructure-based and service-based competition?

These questions necessarily involve complex trade-offs, the resolutions of which are not clear. What is clear is that regulators are interested in both competition and innovation, and that there is sometimes a tension among these objectives. For Europeans, this tension is manifested in Article 8 of the Framework Directive, which simultaneously calls on regulators to ensure 'that users . . . derive maximum benefit in terms of choice, price, and quality', and 'that there is no distortion or restriction of competition in the electronic communications sector', but at the same time to '[encourage] efficient investment in infrastructure, and [promote] innovation'.[16] What is the 'poor' regulator to do if these goals are somewhat at odds with one another?

The arguments in favor of pro-competitive regulation are well established in network industries. There are key assets that would not be cost-effective for competitors to replicate. In the absence of regulatory intervention, the industry would tend towards monopoly, resulting in a loss of consumer choice, in a transfer of surplus from consumers to suppliers, and typically in deadweight social loss inefficiencies. If sufficient infrastructure-based competition were present, it would be appropriate to withdraw last mile regulation and to leave matters to the unregulated market. Today, these conditions do not appear to be satisfied anywhere, including, in our opinion, North America. Thus, our sense is that procompetitive interventions on last mile access will need to be maintained for the foreseeable future, even if that were to imply some cost in the speed of fiber-based access deployment. It is by no means certain that maintenance of regulatory remedies on last mile access has a dramatic adverse effect on deployment. In the case of broadband, Europe has achieved near parity in penetration with the United States, despite intensive regulation, a dearth of cable television infrastructure, and a simply enormous head start for the United States. A number of studies have analyzed empirically the interrelationship of investing in broadband infrastructure/broadband penetration and regulation. However, there is no unique conclusion regarding the issue of whether willingness to deploy is conditioned on lack of regulation (Friederiszick et al., 2007; Wernick, 2007).

Having argued that it is appropriate in all or nearly all cases to maintain regulation on the last mile, we now come to the question of how it should be implemented. The two primary alternatives would appear to be: (1) imposition of conventional regulatory remedies as is practiced in the European Union; or (2) functional separation, as has been implemented in the UK by Ofcom and British Telecom (BT). The functional separation model is attractive, but it is still a bit soon to assess its true effectiveness. Since BT has not implemented fiber-based solutions yet, there is no working example to point to (and indeed, one must ask whether functional separation has somehow discouraged the deployment of fiber in the UK). These questions are too large to deal with in this chapter, but they will unquestionably occupy the attention of experts in the years to come (Kirsch and von Hirschhausen, 2008).

9.4.2 How Effective have Current Approaches been in Promoting Investment and Competition?

This sections aims at illuminating how effective current regulatory approaches have been in promoting investment and competition. To this end, we focus on Germany, France, Japan, the Netherlands and the US. We are addressing both the deployment and adoption of fiber and the competitive structure in these countries.

Germany
As of May 2008, DTAG has deployed FTTC/VDSL infrastructure in 27 cities, and ADSL 2+ infrastructure in about 750 cities, thus, reaching around 17 million households. By the end of 2008, a further 13 cities will be connected to the VDSL network, and ADSL 2+ will be available in about 1000 cities. Thus, high-speed DSL broadband access (fast enough for Internet Protocol television – IPTV, for example) will be potentially available to 20 million households (about half of all households in Germany). As of now, there is no publicly available information about phasing out of MDFs. Conventional wisdom suggests that FTTC/VDSL is only a step on the road to a long-term deployment of FTTB/FTTH.

Although its overall market share still is considerable,[17] there are at least two main driving forces for DTAG's FTTB/FTTH deployment activities. First, the speed of market losses regarding telephony access lines is considerable: In the period 2006–2008, DTAG has lost about 2 million access lines per annum to competitors. Second, there is fierce price competition in Germany which has changed more and more to an overall flat rate regime for triple play services (VoIP telephony, broadband access, Internet access). As a result, DTAG has had to respond to capital market expectations in order to sell a convincing new growth story. In this respect, however, DTAG has not been that successful. Its share price has gone down from an all-time high of more than 100 Euros in 2000 to a level on or below the first IPO quotation price of around 14 Euros since 2005.

Regional competitors in Cologne, Munich, and Hamburg have launched (or are planning to launch) FTTB/FTTH infrastructure. These ventures, however, are concentrated on densely populated areas. An important factor for their business case is savings from LLU wholesale services currently purchased from DTAG. Thus, a lower LLU price would have made these investments of the competitors less likely. Cable modem access still plays a minor role in the German broadband market, that is, intermodal competition is still relatively low; however, the cable operators have upgraded their networks in order to offer HFC (hybrid fiber coaxial based) triple play services. Even though cable represents a small fraction of the installed base for broadband in Germany, it represents a quite substantial fraction of the new customers added. As we have mentioned in section 9.3.2, the economic conditions for the IP bitstream access service offering of DTAG are now clear; however, it is far too early to assess how this service is going to reshape the German competitive landscape.

France

In France, there are ambitious FTTB/FTTH deployment plans by the incumbent and competitors alike (although geographically mainly focusing on big cities with very dense population and a high percentage of MDUs). A key driver for this development is the presence of man-high sewage channels leading into many buildings, a circumstance that represents a special advantage for Paris and certain other French cities.

The regulator in France is very active and successful in convincing city authorities to charge competition-friendly prices for providing access to their infrastructure (for example, sewage channels). Policy-makers in France are also monitoring the discussions with housing companies in order to establish a win–win situation as regards access to in-house infrastructure by fiber deployers.

Japan

A far-reaching unbundling policy (for copper and more recently for fiber) has been a decisive factor for the rapid adoption and diffusion of xDSL. Moreover, aerial deployment on the final part of the loop brings costs down. ADSL competition was fierce in Japan. This, in turn, has led to fiber-based infrastructure competition, although the two incumbent companies (NTT East, West) still account for an FTTB/FTTH market share of more than two-thirds.

Netherlands

Unlike in other European countries, there is substantial publicly available information regarding the deployment plans of the incumbent in the Netherlands. The regulator is in principle open to the next generation fiber-based deployment. The originally very ambitious deployment plan by the incumbent (phasing out six out of seven MDFs by 2010) has been postponed by at least 2–3 years due to regulatory decisions. Thus, there will be no major impact on competitors in the short and medium term.

The United States

The withdrawal of pro-competitive regulation in the United States has had more or less predictable effects. On the one hand, there is a reasonably good roll-out of fiber-based access to the home, using a mix of VDSL and of FTTH/FTTB technologies.[18] On the other hand, there has been the predictable negative effect on intramodal competition (that is, competition among telephone companies for a given customers). For traditional copper-based broadband, intramodal competition is negligible, having declined since the Triennial Review to some 3.1 percent of ADSL lines. Intramodal competition over cable was never mandated, and is negligible.[19] For fiber deployment, 91 percent has been undertaken by incumbents large and small, and only 9 percent by competitors (CLECs).[20]

Advocates for the US approach will argue that consumers have many access options. They may, for instance, point to FCC statistics that show robust deployments of wireless broadband. Our sense is that these arguments are not well founded, for a number of reasons. The FCC counts as broadband many services that nobody else would count as broadband, including services that are less than 200 Kbps in the slower direction, and mobile services where it is not clear that the subscriber is using the broadband and possibly does not view it as a substitute for a wired connection. For the great majority of Americans, the realistic choices are broadband from one cable operator or broadband (over copper or fiber) from one telephone company.

In assessing the overall effect of the FCC's deregulatory approach, it is difficult to say what the net effects have been. It seems clear that there has been far less broadband deployed and adopted than one might have expected given the long and early lead that the United States enjoyed, and also the widespread availability of cable television. At the same time, the robust roll-out of fiber is definitely an undeniable strength. How one assesses the relative balance of the two necessarily reflects complex judgment calls: What is the true value to the consumer of the greater bandwidth available with fiber? What is the impact on the consumer of the lack of intramodal competition? One must also bear in mind that the lack of intramodal competition does not quite imply a re-monopolization of the US market. Most Americans can choose cable broadband as an alternative to broadband from the telephone company – that is, two decent options are available in most cases. It has led, not to monopoly, but rather to an effective duopoly.

9.5   CONCLUDING REMARKS

The regulatory approaches in the countries that we considered can be broadly grouped into three categories:[21]

- Maintenance of pro-competitive remedies (France, Japan, Netherlands).
- Implementation of functional separation (UK).
- Radical deregulation (the United States, and to some extent Germany).

Can we say anything about the relative merits of the approaches? Which approach is best? If there is anything that emerges clearly from comparative assessment of different regulatory approaches in different countries, it is that they started in different places, and that they have achieved different results. The process of fiber-based NGN access deployment can be viewed as an enormous feedback loop, beginning with the intrinsic characteristics of each country prior to the initial deployment of NGN. The countries we evaluated initially differed from one another (and continue to differ from one another) in multiple dimensions, including: availability of alternative last mile infrastructure, notably cable television; demography; geography (topology); and societal and governmental attitudes toward industrial policy.

These factors can be viewed as inputs to the formulation of overall public policy, which drives regulatory policy. The regulatory policy that results does not, however, uniquely determine the pace of adoption and deployment; rather, companies base their deployment decisions on business models that reflect their expectations in regard to profitability. The regulatory framework is a factor in those expectations, but not the only factor and not necessarily the most decisive one (Wernick, 2007). The regulatory framework and the pace of deployment then shape the levels of intramodal and intermodal competition going forward. Over time, the competitive environment itself becomes an additional input to the policy formulation process, which can be viewed as a feedback loop.

What does all of this mean? It means that at least on a worldwide scale it is too early to expect a one-size-fits-all approach to the regulation of fiber-based NGN access. Different countries are likely, for now at least, to arrive at different solutions to the range of regulatory challenges that we have identified. These different approaches do not mean that the regulators are wrong; rather, it (primarily) indicates that the circumstances of their respective countries are different from one another.[22] In other words, there is no 'magic touch' here, no 'silver bullet'. There is no dominant strategy, no solution that is clearly superior in all cases. Not yet, at least. This also implies that a regulatory approach to the deployment of IP-based fiber cannot be simply and cavalierly carried over from one country to another. In particular, US-style deregulation cannot be effective in a country that does not have sufficient alternative last mile infrastructure to build on.

The choice among the three alternatives identified at the beginning of this chapter (regulation, functional separation or deregulation) must reflect the market circumstances of the country or region to which it pertains.

NOTES

1. IETF, Internet Protocol: Darpa Internet Program Protocol Specification, RFC 791, September 1981.
2. Indeed, technological neutrality is ostensibly a core tenet of European regulation.
3. Bitstream access seems to be much less problematic in connection with VDSL and FTTx solutions. Also point-to-point fiber introduces fewer problems with unbundling than PON.
4. See, for example, Van Heesvelde (2007).
5. See, for example, Amendola and Pupillo (2008) and Pileri (2008).
6. See Marcus and Elixmann (2008) for the deployment plans of Deutsche Telekom and KPN, respectively.
7. The issue of crucial factors for the viability of an FTTx business case is also addressed in Avisem (2007a, 2007b), ERG (2007a, 2007b) and OECD (2008a, 2008b).
8. Kirsch and von Hirschhausen (2008) consider this case. They also find that NGN access is unlikely to be replicated, that is, infrastructure competition is unlikely.
9. The issue of a potential dismantling of MDFs has been addressed in Reichl and Ruhle (2008).
10. A number of papers and reports have evaluated these costs. See Amendola and Pupillo (2008) and J.P. Morgan (2006).
11. Ofcom's approach to risk in the assessment of the cost of capital: Ofcom (2005).
12. See Marcus and Elixmann (2008) and Kirsch and von Hirschhausen (2008). KPN has also formed a joint venture with rival telco Reggefiber under which the pair will amalgamate their FTTH activities in a new partnership named Reggefiber FttH. The two companies had earlier agreed to cooperate on a FTTH project in Almere, designed to roll out fiber-optic links to 70000 households. The Almere project will be rolled up into the new partnership. See *TeleGeography* 23May 2008.
13. The number of NTT customers taking broadband services via its ADSL platform dropped from 5.3 million (as of 31 March 2007) to 4.6 million (as of 31 March 2008). The number of conventional main lines in service fell to 39.2 million from 43.3 million in this period; see *TeleGeography*, 'NTT fiber users up, ADSL subscriptions down', 15 May 2008.
14. NTT reports that the number of people signed up to its FLET'S Hikari fiber-optic service reached 8.78 million as of 31 March 2008, up from 3.2 million a year earlier; see *TeleGeography*, 'NTT fiber users up, ADSL subscriptions down', 15 May 2008.
15. In October 2007, the Japanese government started to examine the interconnection charge for the fiber access from fiscal year 2008. They decided to keep the current method, but to recalculate the fiber cost based on high-demand estimation. As a result, fiber charges will go down to 28.8 euros/fiber (of NTT East).
16. *Framework Directive*, European Parliament and Council of the European Union (2002, p. 38).
17. DTAG has a share of 48 percent with respect to the overall (nominal) telecommunications service market volume in Germany. With respect to telephony access lines DTAG has a market share of more than 80 percent (81.4 percent). DTAG has a share of slightly below 50 percent (48.6 percent) in the overall number of DSL access lines; see BNetzA (2007).
18. 1.4 million lines as of 30 June 2007 according to FCC statistics, and growing rapidly.

19. An obligation was imposed on AOL/Time Warner as a merger condition, but it proved to be ineffective and the FCC declined to take steps to enforce it.
20. The latest FCC data are as of 30 June 2007. This time lag (nearly a year) is normal for FCC data.
21. This breakdown generally follows Kirsch and von Hirschhausen (2008).
22. European regulation explicitly recognizes the principle that regulation exists to deal with market imperfections, notably with market power, and that regulatory remedies should appropriately differ from one country to another to the extent that market power differs.

REFERENCES

Amendola, G.B. and L.M. Pupillo (2008). The economics of next generation access networks and regulatory governance: towards geographic patterns of regulation. *Communications and Strategies*, **69**, 85.
Analysys (2007a). The business case for sub-loop unbundling in the Netherlands. Study prepared for OPTA, January.
Analysys (2007b). The business case for sub-loop unbundling in Dublin. Study prepared for Comreg, December.
Avisem (2007a). Étude portant sur les modalités de déploiement d'une boucle locale fibre optique. Study prepared for ARCEP, June.
Avisem (2007b). Étude portant sur les spécifications techniques des infrastructures de génie civil susceptibles de supporter des réseaux d'accès FTTH. Partie 2 de l'étude: Eléments de spécifications des infrastructures. Study prepared for ARCEP, September.
Bundesnetzagentur (BNetza) (Germany's Federal Network Agency) (2007). Jahresbericht 2007, http://www.bundesnetzagentur.de/enid/2.html and http://www.bundesnetzagentur.de/media/archive/13212.pdf.
ERG (2007a). ERG opinion on regulatory principles of NGA. October.
ERG (2007b). Supplementary document. October.
European Parliament and Council of the European Union (2002). Directive 2002/21/EC of March 7 2002 on a common regulatory framework for electronic communications networks and services (Framework Directive), (*Official Journal* (OJ) L108, 24.04.2002, pp. 33–50). Brussels: European Parliament and Council of the European Union.
FCC (2003). *In the matter of reviewing of Section 251 unbundling obligation of incumbent local exchange carriers*, Triennial review order, 21 August.
Friederiszick, H.W., M. Grajek and L.-H. Roeller (2007). Analyzing the relationship between regulation and investment in the telecom sector. European School of Management and Technology, expert report commissioned by Deutsche Telekom, Berlin. November.
J.P. Morgan (2006). The fiber battle. December.
Katagiri, Y. (2008). Recent regulatory reform in Japanese telecommunications. Slide presentation at the International WIK Conference, Review of the European Framework for Electronic Communications. Bonn, Germany, 24–25 April.
Kirsch, F. and C. von Hirschhausen (2008). Regulation of NGN: structural separation, access regulation, or no regulation at all. *Communications and Strategies*, **69**, 63–83.

Marcus, J.S. (2005). Is the U.S. dancing to a different drummer? *Communications and Strategies*, **60**, 4th quarter.

Marcus, J.S. and D. Elixmann (2008). Regulatory approaches to NGNs: an international comparison. *Communications and Strategies*, **69**, 19–40.

OECD (2008a). Public rights of way for fiber deployment to the home. DSTI/ICCP/CISP(2007)5/FINAL, April.

OECD (2008b). Developments in fiber technologies and investment, DSTI/ICCP/CISP(2007)4/FINAL, April.

Ofcom (2005). Ofcom's approach to risk in the assessment of the cost of capital. Final statement, 18 August.

Pileri, S. (2008). Technologies and operations evolution, telecom italia analyst and investor briefing, 2007 Results and Strategic Guidelines. Retrieved from http://www.telecomitalia.it/TIPortale/docs/investor/Pileri_070308.pdf.

Reichl, W. and E.-O. Ruhle (2008). NGA, IP-interconnection and their impact on business models and competition. *Communications and Strategies*, **69**, 41–62.

Taniwaki, Y. (2008). Broadband competition policy in Japan. Presentation by Ministry of Internal Affairs and Communications (MIC), March.

Van Heesvelde, E. (2007). Approaches to VDSL regulation in Belgium. Slide presentation at the WIK Conference, The Way to Next Generation Access Networks. 21 March, Königswinter.

Wernick, C. (2007). Strategic investment decisions in regulated markets: the relationship between infrastructure investments and regulation in European broadband. Wiesbaden.

PART IV

Structural separation and regulation of the telecommunications industry

10. Diffusion of broadband Internet and structural separation

Arata Kamino and Hidenori Fuke

10.1 INTRODUCTION

One of the most frequently discussed topics in industrial organization theory is vertical integration. However, the competitive effects of vertical integration are two-sided. When a vertically integrated firm with market power in the upstream market is also active in the downstream market, concerns regarding competition give rise to fears that the firm might abuse its market power. Abuse might take the form of obstructing competitors with a refusal to deal or by raising rivals' costs. To prevent this kind of abuse by the vertically integrated firm, both conduct regulation and structural separation are brought into consideration. When conduct regulation is deemed insufficient to solve the problems, structural separation is viewed as an alternative option. However, structural separation sacrifices economic efficiency such as economies of scale and scope. In recent studies other inefficiencies inherent in vertical separation such as 'coordination' between separated firms and 'hold-up' problems are pointed out.

In the case of the telecommunications industry, which is characterized by dynamic technological changes, structural separation might risk rapid transition to new technologies and preserve the market structure already shaped by regulation. Structural separation assumes the market power of a vertically integrated firm in the upstream market. However, the scope of market power is likely to change with the introduction of new technologies. Market structure is also bound to undergo change and it becomes difficult to define the scope of activities which exhibit bottleneck characteristics. This is especially true in the current telecommunications industry that is experiencing a rapid transition from plain old telephone service (POTS) to broadband. It is important, therefore, to conduct a cost–benefit analysis of structural separation.

Discussions on structural separation in the telecommunications sector originated at the time of traditional POTS, an era based on the sharp distinction between long-distance and regional communications. Although

211

this distinction is becoming obsolete with the diffusion of the Internet, structural separation is still on discussion tables in many of the advanced countries except for the US. In the US, with AT&T broken up in 1984 and divested Regional Bell Operating Companies (RBOCs) and the former AT&T reintegrated in the mid-2000s, the issue of vertical separation is no longer being raised. On the other hand, in Japan, where NTT was reorganized into several companies in 1999 – two regional, one long distance, one mobile and others – all under a holding company – ownership separation of these NTT subsidiary companies is still under consideration despite the rapid diffusion of broadband. In the EU, meanwhile, functional separation of incumbents became a hot topic during the EC 2007 telecommunications regulatory reform debate.

This chapter tries to survey the differences in recent discussions on the structural separation of incumbents in Japan, the US and Europe and explain them from the perspective of broadband market structure in these regions. We focus especially on Japan where ultra-fast broadband has been developed and where the traditional approach to vertical separation needs to be reconsidered. In addition, to make our conclusions more robust, the chapter also reviews existing market structure and discussions on structural separation in the US and Europe. Though the situation varies a lot among European nations, we will examine the UK, which has introduced functional separation of British Telecommunications (BT), as well as Germany and France as nations taking a negative position against functional separation of their incumbent carriers.

This chapter is organized as follows. First, we will briefly survey the discussions on broadband market structure and structural separation. Second, we will put forward an evaluation of the history and background of the treatment of structural separation in Japan. Third, we evaluate recent discussions on structural separation in the US and Europe. Finally, based on the points above, we will explain the differences between structural separation policies in Japan, the US and Europe in terms of the broadband Internet market structure and consider lessons for future progress on the structural separation issue.

10.2   BROADBAND MARKET AND STRUCTURAL SEPARATION

One of the important issues for policy attempting to govern competition is to constrain the abuse of market power by vertically integrated firms toward the downstream market, whenever a dominant company in the upstream market is also active in the downstream market. The measures to secure fair competition in these circumstances are generally classified into 'structural separation' and 'conduct regulation'. As incumbents in the telecommunications industry historically dominated local facilities, securing fair competition between incumbents and new entrants became a big issue. Consequently, topics such as separating an incumbent's local businesses from its long-distance businesses (that is, structural separation) and introducing conduct regulation mainly by way of interconnection policies have been widely discussed. In the telecommunications sector, in addition to this type of narrowly defined structural separation, functional separation that does not include ownership separation is also a matter of debate. Various interpretations of vertical separation also coexist, such as the Organisation for Economic Co-operation and Development (OECD) distinction between 'structural separation' and 'operational separation' (OECD, 2006), or Cave's (2006) definition of eight types of separation ranging from 'accounting separation' to 'ownership separation'. The European Commission (EC), in its 2007 telecommunications regulatory reform program, defines separation with ownership unbundling as structural separation and separation without it as 'functional separation' (EC, 2007). We will follow this EC definition hereafter in this chapter. In addition, we will use 'vertical separation'[1] as including the notions of both 'structural separation' and 'functional separation'.

Vertical separation in that sense is based on the assumption that the merits of promoting competition through regulatory measures outweigh losses in the economies of scale and scope resulting from the subdivision of an incumbent's organization and/or businesses. Vertical separation, however, raises the fundamental issue that certain forms of conduct regulation are required as long as a discreted operating company holds market power in the upstream market. Furthermore, vertical separation is accompanied by the risk of preventing flexible responses to market changes such as the shift from traditional POTS to Internet Protocol (IP) services. Numerous studies of theoretical industrial organization exist but since there are various problems in applying conclusions derived from them, policies instituted differ significantly between countries. The US divested AT&T in 1984, and Japan reorganized NTT in 1999 by establishing a holding company overseeing a long-distance and two regional companies, a mobile operating company, and other subsidiaries. Direct or virtual types of structural separation as such were adopted only by few nations in the twentieth century.[2] This was to assure new entrants of fair interconnection of long-distance networks with incumbent local networks used for legacy fixed voice services. In Europe, since the market liberalization of the 1980s–1990s, there have been no cases of this kind of vertical 'long-distance and regional business' separation being introduced by regulators.

It is widely recognized that the separation of long-distance and regional businesses is not possible and makes no sense in the recent broadband Internet architecture. As a result, no one discusses re-enactments of the AT&T type of divestiture. However, some European and Oceanian countries find it desirable to separate functionally the access business of incumbents in order to remove problems with competition resulting from the bottleneck nature of access networks sustaining broadband services. Functional separation has already been introduced in the UK (BT), New Zealand (TCNZ)[3] and some other nations. In the US, on the other hand, there are currently no active views on the vertical separation of incumbents, including functional separation of access units. In Japan, it has been decided to resume the review of the organizational structure of NTT in 2010. The need for resumption is based on the common recognition among the related parties that the current divisions between local and long-distance businesses of NTT are inappropriate in the broadband era.

Why do debates on the separation of dominant incumbents in the telecommunications industry differ in Japan, the US and Europe? We suggest that differences between countries derive from variations in stages and phases of recent broadband diffusion and competition, as shown in Table 10.1.

Several official broadband statistics are used in this chapter, including the ITU Internet Report (ITU, 2006) and OECD Broadband Statistics (OECD, 2008), providing figures on a global basis. To begin with, we will briefly examine the OECD Broadband Statistics (Table 10.2) and summarize the world broadband status.

Table 10.2 reveals the characteristics of broadband diffusion in the countries covered by this paper as follows:

- Penetration rates per 100 inhabitants are almost the same among Japan, the US, the UK, Germany and France (around 25 percent).
- Digital subscriber line (DSL) is the most popular technology except in the US and Japan. Inferring the percentages for DSL as a segment of all broadband services, we get 95 percent for France, 93.8 percent for Germany, 78.6 percent for the UK and 41.7 percent for Japan.
- Cable ties up the majority share of all broadband services in the US at 52.8 percent, while DSL occupies 40.4 percent.
- Japan shows the highest (fibre to the x) (FTTx) share with 44.3 percent among all broadband access services.

(Note: These figures vary slightly from those in Table 10.1 mainly due to different time of records taken.)

Table 10.1   Status of broadband penetration and competition in Japan, the US and Europe

Japan	
Penetration	DSL services are the fastest and cheapest in the world (ITU, 2006) and the number of subscribers to FTTx was the largest with 14.42M at the end of 2008. The shares of FTTx, DSL and cable modem at that time were 47.9%, 38.5% and 13.5%, respectively (MIC, 2009a).
Competition	Due to intense service-based competition, NTT's share of retail DSL services was 36.0% as compared to 38.2% of Softbank Corp. at the end of 2008. NTT's share of retail FTTx services at the time was 73.7% (MIC, 2009b). The FTTx market is characterized by a mixture of service-based and facilities-based competition. KDDI, potentially the largest facilities-based competitor to NTT, is utilizing assets acquired from electric power companies and its own subsidiary cable TV companies.
Regulation	Unbundling obligation has been imposed on both DSL and FTTx loops of NTT.
The US	
Penetration	Until several years ago, the transmission speed of DSL services was rather slow and rates were expensive compared to Japan and Korea (ITU, 2003).[1] The total number of residential high-speed lines[2] excluding mobile wireless came to 64.87M at the end of 2007. At that point, ADSL occupied a share of 26.48M (40.8%), cable modem 35.33M (54.5%), and fiber 1.68M (2.6%) (FCC, 2009). Verizon and AT&T are recently accelerating FTTx deployment, and cable companies such as Comcast are upgrading the transmission speed of their cable modem services.
Competition	Incumbent telecommunications carriers and cable companies are competing actively on a facilities base. 'Triple play' competition is intense.
Regulation	FCC abolished both line sharing obligation for copper loops and unbundling for fibers excluding voice 64 Kbps path in 2005.
Europe (EU)	
Penetration	Penetration rates vary considerably among EU member states, from the Slovak Republic with the lowest rate at 8.9% to

Table 10.1   (continued)

	Denmark with the highest at 36.7%. However, many of the major member states show a penetration rate almost the same as in Japan (23.0%) and in the US (25.0%). It is 27.6% in the UK, 26.4% in France and 26.2% in Germany (OECD, 2008). Except for a few nations where cable modems are dominant and Sweden where FTTx is widely deployed, the dominant broadband technology is DSL. However, commercial FTTx services are starting to be launched in Germany and France.
Competition	Competition in many of the member states is limited to intramodal service-based DSL. To promote service-based competition, local loop unbundle (LLU) and bit stream access (BSA: a kind of wholesale product) are being implemented. The shares of incumbents in retail markets and the usage rates of LLU and BSA vary significantly among member states.
Regulation	The EU has shifted from the unified application of The LLU Regulation (EC, 2000) to the introduction of LLU and BSA obligations, a process which was based on market analysis in member states under the 2003 regulatory framework. Most states impose LLU and BSA obligations on incumbents. As for FTTx, many countries are conducting a regulatory review to decide how to treat FTTx. Remedies such as open access to duct as the first step, with dark fiber provision as the second step and sharing of intra-building wire are under discussion.

Notes:
1   Prices per 100 Kbps were $0.09 (Japan), $0.25 (Korea) and $3.53 (the US) at the time.
2   Over 200 Kbps in at least one direction.

Table 10.2   *Broadband penetration rates per 100 inhabitants by technology in major countries as of June 2008*

Penetration	DSL (%)	Cable (%)	FTTx/LAN (%)	Others (%)	Total (%)
Japan	9.6	3.1	10.2	0.0	23.0
The US	10.1	13.2	0.9	0.8	25.0
The UK	21.7	5.9	0.0	0.1	27.6
Germany	24.6	1.6	0.0	0.1	26.2
France	25.1	1.3	0.0	0.0	26.4

Source: OECD (2008).

10.3 PROGRESS ON BROADBAND COMPETITION AND STRUCTURAL SEPARATION IN JAPAN

10.3.1 Telecommunications Liberalization and Competition Policy

In Japan, the liberalization of the telecommunications market and the privatization of NTT (Nippon Telegraph and Telephone Public Corporation) were implemented simultaneously in 1985. However, as Fuke (2007) points out, there was no sufficient debate on whether conduct regulation should be imposed on NTT or whether structural separation should be chosen in order to secure fair competition. Though many European countries arranged interconnection rules and rate rebalancing at the time of market liberalization around 1998, interconnection rules were not well provided for in Japan until 1997 when the Ministry of Internal Affairs and Communications (MIC) implemented new rules by revising the Telecommunications Business Law. One of the reasons for the difference between Japan and the EU is the fact that market liberalization occurred 13 years earlier in Japan than in the EU, with very few international benchmark rules applicable at that time. Since the initial arrangement of conduct regulation was insufficient in Japan as already mentioned, views have continued to be canvassed on the structural separation of NTT into long-distance business and regional business as a tool to promote fair competition. Thus the structural separation of NTT has been under consideration in parallel with the arrangement of conduct regulations in the 1990s. The persistence of these discussions has been one of the particular characteristics of the Japanese telecommunications market to this day.

As a result, Japan introduced an indirect but practical structural separation of NTT. NTT was reorganized under a holding company structure in 1999, while adopting the same level of conduct regulations as the EU. However, it became clear that such a 'regional versus long-distance' type of structural separation could not keep up with massive telecommunications changes from POTS to the Internet and to mobile communications. In order to resolve this problem, MIC (2006) issued a ministerial report in 2006 that proposed the removal of line of business control of NTT regional companies. This was to be achieved by instituting additional fair competition rules including the ownership separation of NTT group companies by way of disbanding the NTT Holding Company in 2010. However, due to a lack of a concrete vision for NTT after reviewing its organization and given the need for further conduct regulations, the Cabinet and ruling party (that is, the Liberal Democratic Party) at the time decided to suspend the MIC (2006) panel proposal and agreed to postpone the discussions on NTT reorganization until 2010. The agreement said that:

'as for the organizational structure of NTT, we shall resume the review in 2010, taking into account the status of broadband penetration and NTT Group's Medium-Term Management Strategy'.[4] Conclusions are supposed to be reached as soon as possible following the future review.

10.3.2   Service-Based Competition and Facilities-Based Competition

Since the end of the 1990s, Japan has been implementing the most rigid open network policy among major developed countries to promote service-based competition. For example, unbundling obligation has consistently been imposed on both copper and fiber loops until now. There is no bit stream access (BSA) type of broadband wholesale scheme because unbundled-based competition relying on very low local loop unbundling (LLU) charges (especially for line sharing) has flourished from the beginning and there has been no room for Internet service providers (ISPs) to adopt BSA-based competition. How does one explain that the US and European countries, which implemented more or less the same kind of copper LLU as Japan, did not experience the same degree of DSL competition as Japan? There might be no single answer to this question (Ikeda, 2005). As the LLU charges were set very low, new entries by service-based carriers were encouraged. Furthermore, there have been competitors who, driven by a kind of 'animal spirit' (Akerlof and Shiller, 2009), continued to provide very cheap alternative DSL services despite long-standing fiscal losses during early 2000s.

Table 10.3 shows that line sharing charges were set very low during that period in Japan. Nevertheless, Softbank Corp., the largest DSL competitor, incurred a huge loss (Table 10.4) because the company set a disruptively cheap retail rate for DSL and, in the short term, faced massive marketing costs in capturing the majority market share (Kamino, 2007). As the International Telecommunication Union (ITU) (2003) observed, a strategy like this contributed to the realization of 'the world fastest and cheapest DSL'.

Table 10.5 summarizes characteristics of DSL competition in Japan. Though these characteristics have both merits and demerits, it is probable that demerits exceed merits in some cases. Further analysis will be required to make a full assessment of the balance between them to judge the result of DSL competition in Japan.

Eventually, Japanese competitive broadband providers would not have been able to sustain losses in FTTx businesses in the same manner in which DSL services had initially recorded substantial losses (Table 10.4). Consequently, unwilling to fall into the same money-losing trap, they requested regulation enabling very low wholesale charges for fiber loops.

Table 10.3   Comparison of LLU charges (October, 2004)

	Full unbundling	Shared access (line sharing)
Japan	NTT East: 1,256 yen	NTT East: 158 yen
	NTT West: 1,318 yen	NTT West: 165 yen
The UK	8.76 pound	2.26 pound
	(1,752 yen)	(452 yen)
Germany	11.80 euro	2.43 euro
	(1,652 yen)	(340 yen)
France	10.50 euro	2.90 euro
	(1,470 yen)	(406 yen)

Note: Calculated on the basis of the exchange rate at the time (1 pound = 200 yen, 1 euro = 140 yen).

Source: NTT tariff, BT Price List, Ofcom (2004) and RegTP (2004).

Table 10.4   Financial result of Softbank Corp.'s broadband infrastructure (million yen)

Fiscal year	2003	2004	2005	2006	2007
Operating revenue	40 007	128 906	205 306	268 451	258 824
Operating profit	−96 204	−87 597	−53 747	20 672	26 809

Source: Softbank Corp.'s financial data

It is precisely this kind of regulatory interference of setting very low wholesale prices that is likely to lead to a regulatory preference for a specific technology (that is, FTTx), which, in turn, might impair the development of other broadband technologies.

10.3.3   Status of Broadband Penetration and Competition

The Japanese broadband market entered the FTTx era prior to other countries. Given the rapid expansion of the market, the number of FTTx subscribers exceeded that of DSL in the first fiscal quarter of 2008 (April–June) (Figure 10.1).[5] As noted above in terms of intramodal competition within DSL, the total NTT market share of FTTx (73.7 percent) is considerably higher compared to that of DSL, with some exceptional competitors such as K-Opti.com, a subsidiary communications company of Kansai Electric Company in Osaka, holding the majority FTTx share in several prefectures[6] (MIC, 2009b). In a situation where even NTT itself is often asked

Table 10.5   Characteristics of DSL competition in Japan

Characteristics	Merits	Demerits
(1) Very low LLU charges	Intense LLU-based competition developed.	Neither intramodal facilities-based competition nor intermodal competition developed.
(2) Destructively cheap DSL retail prices	Only large companies with financial abundance could bring sustainable competition. FTTx retail price was set very low because cheap DSL prices had become benchmark for broadband prices.	Only competitors with large financial base that could bear deficits over several years were able to enter the market. Margin squeeze regulation applied to FTTx wholesale charges restricted the development of facilities-based competition (Fuke, 2007).
(3) Fierce DSL competition	NTT shifted its broadband strategy from DSL, in which it has less market share and profit, to FTTx. This, in turn, accelerated FTTx deployment and penetration.	While DSL competitors faced a net decrease in the number of DSL subscribers due to customer shift to FTTx, they started to produce profits in DSL businesses. (DSL customer numbers hit the peak in March 2006 and kept on declining after that.)

the reason why it is making huge investments in currently less profitable FTTx deployment, it is certainly difficult for competitors to shift wholly into deficit-making service-based FTTx competition, just as it had been the case with DSL (Table 10.4). While facilities-based competition between NTT and electric companies that have large access fiber assets was once thought to be feasible, electric companies, with the exception of some rare cases such as K-Opti.com, are reluctant to keep on promoting FTTx business as they had been doing before. In spite of this, with KDDI's purchase of a subsidiary communications company and other fiber access businesses from Tokyo Electric Company (TEPCO), signs are emerging that facilities-

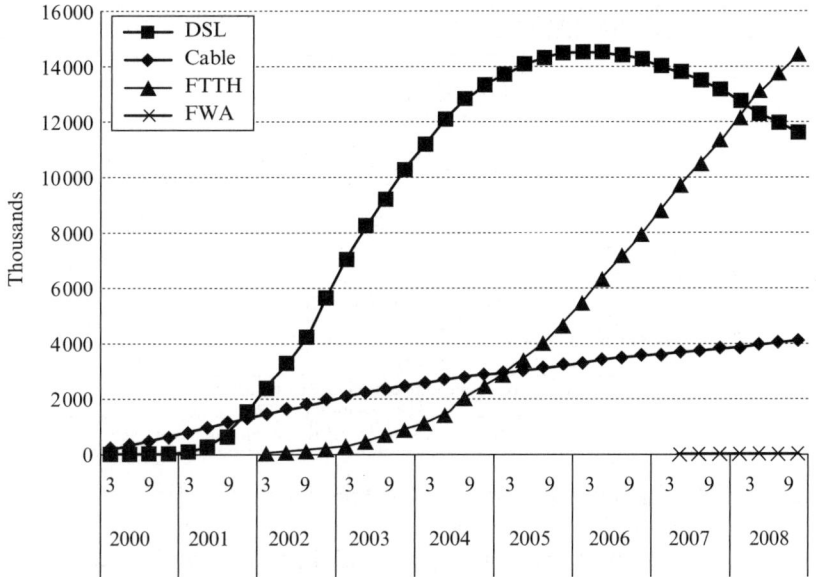

Source: MIC (2009a).

Figure 10.1   Growth of broadband services in Japan

based competition between NTT and other telecommunications providers such as KDDI is occurring in some densely populated areas.

As for cable broadband, which is widely operated in some countries including the US, the household penetration rate for cable TV in Japan soared to 33 percent around 2003 and then kept rising to 42 percent by the end of 2007. Nonetheless, cable broadband market share within the whole range of broadband technologies is slackening, at around 15 percent between 2005 and 2007. In summary, the characteristics of the Japanese cable industry can be listed as follows (MIC, 2008):

- Deregulation for multiple system operator (MSO) restriction was delayed until 1993, and MSOs less large than expected have appeared since then.[7] Though Jupiter Telecommunications Co. (J:COM), the largest MSO, was established in 1995, its subscribers came to only about 2.18 million by the end of February 2008. This number appears considerably small compared with 24.1 million at the end of 2007 for Comcast, the largest MSO in the US.
- Although the household penetration rate for cable TV itself has been increasing, overall cable broadband penetration per 100 inhabitants

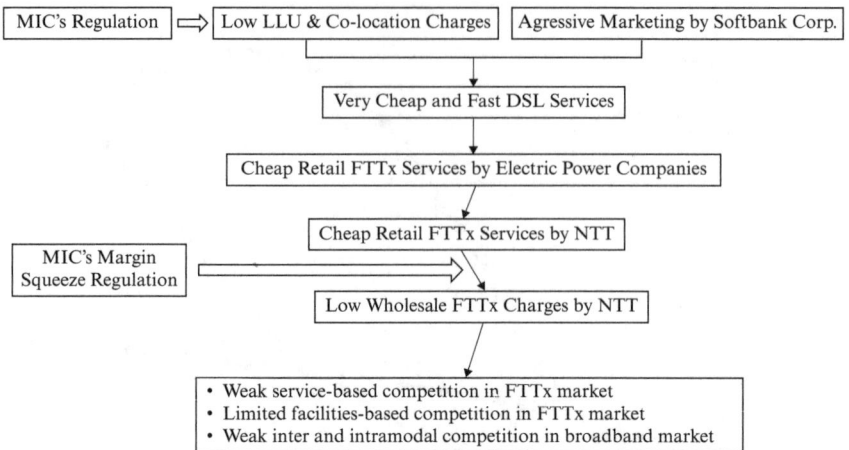

Figure 10.2   Influence of severe DSL competition on FTTx

has been decreasing slightly, from 15.4 percent to 13.5 percent over the three-year period 2005–07.

● As cable broadband does not play a significant role in triple play competition, the number of subscribers to IP multicast-type broadband broadcasting (that is, Internet Protocol television – IPTV) remained sluggish with just 240 000 subscribers at the end of 2007.

As we have seen, it was unfortunate for the cable industry that service-based DSL competition was so intense that there was little room for cable broadband to expand. Development of the Japanese broadband market is summarized in Figure 10.2. Both MIC's regulatory policy to enforce low LLU charges and the aggressive market entry of Softbank Corp. contributed to the rapid deployment of cheap and fast DSL services.[8] NTT, having been left behind by Softbank Corp. in the DSL market, has tried to catch up by focusing on FTTx services. However, it was forced to set retail prices of FTTx markedly low to enable it to compete against Softbank Corp.'s cheap DSL as well as against the FTTx services of newly entered electric power companies. By setting retail prices low, NTT was then obliged to set wholesale charges of FTTx below any profitable level due to margin squeeze regulation (Fuke, 2007).

Even with these low wholesale FTTx charges, service-based competitors have not been able to expand their retail FTTx. It has been difficult for them to invest in it due to the financial burden of their DSL businesses, where they have not fully recovered accumulated losses. Thus NTT's share in the FTTx market has continued to rise from 57.5 percent to 73.7 percent through the years 2005–08 (MIC, 2009b), with some competitors again arguing for the structural separation of NTT.

10.3.4   Current Progress on Vertical Separation

As shown below, progress on structural separation of NTT in Japan has aspects quite different from those in other countries:

- Current discussions in Japan followed the debates on ownership separation of long-distance and regional NTT businesses, originally started in the 1980s.
- Fundamental issues have not yet been resolved even following 1999 when a compromise was reached to reorganize NTT under a holding company structure.
- Reviews undertaken during 2005–06 to resolve these issues were suspended as a result of political compromise.

As was pointed out by the MIC Minister's panel report (MIC, 2006), NTT has been constrained by line-of-business control regulations that do not accommodate current technological and market environments. The problem arose because current laws and regulations designed for the reorganization of NTT in 1999 did not take into account rapid technological innovation. It appears unreasonable to maintain the current regulations, which virtually separate long-distance and regional NTT businesses (MIC, 2006).

Recent progress in the broadband market, especially that of DSL and FTTx services that were reviewed above, might support the argument for reorganization (that is, structural separation) of NTT. However, it must be noted that this market situation is the result of MIC's regulation in favor of service-based competition. If we want to encourage the development of facilities-based competition in the broadband market, structural separation is not a unique option. We should rather revise the regulatory policy that is much biased toward service-based competition.

10.4   CURRENT DISCUSSIONS ON VERTICAL SEPARATION IN THE US AND EUROPE

10.4.1   The Case of the US

There is little debate on vertical separation in the US as of early 2009. Large mergers and acquisitions (M&A) cases between RBOCs took place around 2005–06 and since then AT&T and Verizon have come to dominate the telecommunications market. The main reviewers of those merger plans were the Department of Justice (DoJ) and the Federal Communications Commission (FCC). The FCC conducted an analysis balancing 'potential harms and benefits to public interests' in its merger approval procedures. The FCC then judged that merger benefits outweighed harms if certain attached conditions were implemented. As will be seen below, FCC's analytical framework is built upon the balancing theory of the post-Chicago School approach (Yoo, 2002).

On the one hand, during the proposed SBC–AT&T and Verizon–MCI merger reviews, the specific markets (for example, 'enterprise/mass market retail services' and 'the Internet backbone market'; FCC, 2005) that might have been harmfully affected by anticompetitive effect were observed. On the other hand, four potential benefits were considered by the FCC, as shown below:

- enhancements of national security and government services;
- efficiencies related to vertical integration;
- economies of scope and scale;
- cost synergies.

Considering vertical separation, it should be noted that the second and third factors were regarded as balancing factors promoting public interests. As Riordan (2008) pointed out, the FCC drew this conclusion based on a more realistic 'give and take negotiation theory' of the post-Chicago School thinking that the FCC had been adopting since the News Corp. and DirecTV merger case.

No discussions on vertical separation are currently taking place. If we try to find any sort of debate on vertical separation, it will involve the type of questions associated with merger cases such as: Would the costs of further strengthening vertically integrated incumbent carriers outweigh its benefits? This approach lies within the scope of post-Chicago School theory. However, even under a Republican-led FCC, as shown in the AT&T–BellSouth merger review, commissioners from both Democrat and Republican circles voiced considerably different opinions on balancing merits and demerits of the merger. As a result, additional voluntary commitments such as augmented net neutrality promise are required for the merger plan applicants (AT&T–BellSouth).

Generally speaking, there is a widespread tendency in the US to conduct analysis by balancing merits and demerits of vertical integration while taking into account particular situations. This trend is clearly found in recent reviews of the US M&A cases (Riordan, 2008). We should point out that the non-existence of arguments for vertical separation in US telecommunications businesses in recent years is a reflection of the development of facilities-based competition in the broadband Internet market between telecommunications companies and cable operators. This unique market structure is also the result of the regulatory environment, including 'finsyn' (financial interest and syndication) rules, which applies to the broadcast and media industry.

10.4.2   The Case of the European Union (EU)

Pros and cons of functional separation have been widely discussed in Europe (Waverman, 2006). It is true that LLU-based competition had not progressed much in the UK's DSL market at the time Ofcom recommended functional separation of BT (Ofcom, 2005) and accepted BT Undertakings (BT, 2005) in its Telecommunications Strategic Review (TSR). It is equally true that taking up of LLU has been accelerated since BT established its Openreach access division. However, Germany and France, which had not introduced any kind of functional separation, had already experienced explosive use of LLU when Ofcom initiated TSR in April 2004 (Table 10.6). Following the reorganization of NTT in 1999, the Japanese DSL market started to show spectacular growth from 2000 when LLU and collocation were formally mandated.[9] These facts lend little support to the assumption that functional separation itself promotes LLU usage. Therefore, it would be reasonable to think that necessary and sufficient conditions for the promotion of broadband competition are brought about not by functional separation but by equivalence of input (EOI) and fair rate settings for wholesale inputs.

Since functional separation was imposed on BT, the number of countries considering the introduction of the same kind of functional separation has been increasing. The EC is the most active supporter of this

Table 10.6   Take-up of LLU in the UK, Germany and France (end of June, 2004)

Country	Full LLU			Shared access (line sharing)	Total LLU
	DSL use	Voice use	Total		
UK	7 580	0	7 580	6 270	13 850
Germany	650 000	978 699	1 628 699	0	1 628 699
France	13 066	0	13 066	717 654	730 720

Source: ECTA (2004).

trend and took up functional separation as a 'remedy of last resort' in the proposals for its revised directives in the 2007 telecommunications regulatory reform (EC, 2007). However, key related parties opposed structural separation and the EC itself revealed an opinion recommending functional separation rather than structural separation in an explanatory note of the proposed EU directives, concluding that: 'given this experience and the high level and non-revocable intervention involved, very significant benefits of mandated structural separation in terms of gains from achieving equality would have to be demonstrated for it to be a suitable remedy in the telecommunications sector' (EC, 2007). Even among EC commissioners, some voices opposed to functional separation pushed by Telecom Commissioner Ms Reding could be heard before the proposals were submitted to the European Parliament and the Council. This led Competition Commissioner Ms Kroes and Enterprise and Industry Commissioner Mr Verheugen to criticize strongly Ms Reding's idea according to articles from the *Financial Times* excerpted below:

> Neelie Kroes, EU Competition Chief, and Günter Verheugen, the Industry Commissioner, have attacked a proposed overhaul by fellow Commissioner Viviane Reding of laws governing the bloc's €289bn-a-year ($407bn, £201bn) electronic communications sector. They warn that the proposed review could create more bureaucracy and harm investment . . . Ms. Kroes' officials argue that this move – known as 'functional separation' – risks hitting investment in the sector, especially in new, ultra-fast broadband networks. An internal document written by her department, and seen by the FT, said the measure 'is not only superfluous but also damaging. Functional separation does not prevent discrimination of alternative operators'.[10]

The European Regulators Group (ERG) announced acceptance of the functional separation proposal in October 2007 with the provision that the final decision on the introduction of functional separation is solely that of each member state (ERG, 2007). Even so, some NRAs (national regulatory authorities) such as ARCEP (France) or CMT (Spain), who participated in the ERG announcement, expressed their intention to refuse the adoption of functional separation in their own countries. Again, some government officials such as two vice-ministers in the German Ministry of Economics and Technologies disallowed the functional separation remedy itself. On the other hand, several countries such as Sweden and Italy are progressing towards the introduction of functional separation. Tensions are arising there between the EC, NRAs and incumbents because the targeted incumbents (Telia Sonera and Telecom Italia) voluntarily established new access business units to avoid regulatory mandated functional separation.

Though the EU scheduled the completion of deliberations on the 2007 telecommunications regulatory reform by the end of 2009 at the latest,[11] it was expected that many EU countries would suspend further consideration on functional separation until then. They will also draw conclusions on the effectiveness of functional separation from the case analysis of BT, given that it has already been more than three years since BT established Openreach in January 2006.

Incumbent carriers still hold strong market power in many of the European countries, a fact which reflects the rather late introduction of full competition in telecommunications in 1998. Broadband services started to evolve before competition had fully unfolded in the POTS market. Facilities-based competition is not likely to develop in these countries in the near future. In evaluating discussions on vertical separation, it is necessary to take into account this stage of market development.

Here it is useful to study Cave's suggestion, from the perspective of market reality, that:

> it is important that this remedy [of vertical separation] be applied proportionately. This requires that the detriments resulting from non-price discrimination exceed the costs of imposing an operational separation remedy, where those costs are not only those of changing the incumbents' business processes, but also [those] of any chilling effect on investment in new assets, by both the incumbent and competitors. (Cave, 2006)

At the same time, we should remember experiences made in Japan and France, which indicate that effective conduct regulation for mandating LLU and collocation, rather than functional separation, can lead to rapid growth in the DSL and total broadband market.

10.5 CONCLUDING ANALYSIS AND FURTHER ISSUES

10.5.1 Concluding Analysis

Given the above analysis of the broadband market and regulatory policy affecting it, what implications does vertical separation policy have for the three regions under consideration, Japan, the US and Europe? Our assessment is that current differences in the discussions on vertical separation are a reflection of differences in the market structure of broadband and that the market structure itself is formed by regulatory policy. In assessing the current status of deliberations on vertical separation, it is therefore necessary to understand the interaction between market structure and regulatory policy.

In the US, political and regulatory confusion deterred the deployment of DSL from the start and non-regulated cable modem services were able to spread ahead of DSL. To make up for this 'hold-up', the integrated incumbent operators focused on FTTx deployment alongside DSL, and such activities led to facilities-based competition in the broadband market. As a result, it became unnecessary to implement vertical separation of incumbents. In other words, we should like to make the point that the failure of regulatory policy resulted in an unexpected outcome which rendered the need to impose regulation unnecessary.

Though NTT was reorganized in 1999 to promote competition in traditional POTS, contradictions have been coming up from that time with the rapid growth of broadband Internet. NTT regional companies' launch of commercial NGN services in March 2008 may serve as an example. The way NTT provides NGN reveals the contradiction inherent in the regulation limiting NTT regional companies' business areas within intra-prefecture communications. As the Internet has developed as a global seamless network, there ought to be no boundary between long-distance and regional communications. When NTT regional companies tried to offer NGN, they were required to offer the intra-prefecture parts of the service themselves while leasing inter-prefecture parts from other carriers. That is to say, regulation forced an artificial division on the Internet. How should we deal with 'vertical separation problems' based on these market trends? It is important to recognize that the current market status is also a product of regulation. While it is true that LLU obligation mandated by MIC contributed to the rapid growth of the DSL market, it is service-based competition that relied on local loops of NTT regional companies. Though it seemed that facilities-based competition would grow in the fiber to the home (FTTH) market initially, MIC's policy of setting dark fiber charges at a low level restrains the development of facilities based-competition not only in FTTH itself but also in cable broadband and wireless services such as WiMAX (Worldwide Interoperability for Microwave Access) (Fuke, 2007).

In this situation, competition in the Japanese broadband market might remain service-based and no significant facilities-based competition may materialize. In general, ownership separation of an incumbent carrier or functional separation of its bottleneck unit would not lead to facilities-based competition even if it promoted service-based competition. Clearly, facilities-based competition has an advantage over service-based competition in the sense that it stimulates incentives for technological innovation while ownership separation on its own will not lead to facilities-based competition.

Finally, how should Europe treat vertical separation policy given the lessons of Japanese and the US experiences? Our chapter takes the view that the final goal in this debate is to materialize facilities-based competition. It is effective to impose LLU and lower rate setting for LLU in order to promote competition in the short term. However, as shown above, vertical separation as such is not effective. It can be anticipated that incumbent operators would show reluctance to accept a strict unbundling regulation. In this case, it may be more effective to show intentions of introducing vertical separation as a kind of 'stick', in turn inducing incumbents to accept regulation.

Although this kind of remedy would lead to service-based competition, it might hamper facilities-based competition as shown in the case of Japan. Hence, it should be regarded as a temporary remedy until an adequate degree of competition develops in the broadband market. In any case, regulators should hesitate before enforcing vertical separation without taking into account its inherent demerits.

10.5.2   Further Issues

The introduction of FTTx and NGN built on FTTx raises additional issues in the discussion of vertical separation in telecommunications industry. The first issue is related to the 'hold-up' and 'coordination' problem. In the case of POTS, technology is mature and it is easy to define bottleneck parts of the network. However, FTTx is being newly built and technology is going through innovation. For example, NTT is utilizing various technologies for FTTx including fiber to the building (connecting homes with copper intra-building wires) and shared use of a fiber core from a local exchange to a nearby cabinet shared by several users. In these cases, it is difficult to tell which part of a fiber might constitute a bottleneck. Further, the build-out of fibers and the adoption of a particular technology depend on the forecast of user demand in terms of service quality and quantity. However, the firm in charge of maintaining access facilities cannot obtain the kind of user information collected by the firm in charge of providing retail services relied on wholesale products. Even if the upstream firm invests in facilities based on a particular technology, there is no guarantee that the firms in the downstream market will lease the wholesale facilities. In these circumstances, the upstream firm will become too conservative in new investment and is likely to hamper the smooth deployment of new services. When we discuss vertical separation in telecommunications industry, we should therefore keep in mind this kind of 'hold-up' and 'coordination' problem.

The second issue is related to content and application providers. Current discussions on vertical separation are mainly focusing on fair competition between vertically integrated telecommunications carriers and non-integrated telecommunications carriers. Since non-integrated competitors usually lack access assets, current discussions on vertical separation are heating up on the topic of functional separation of access bottlenecks. In the age of core NGNs and NGA (next generation networks and next generation access), however, competition between telecommunications carriers and content providers many of whom do not own network facilities (that is, ultimately non-integrated providers) is expanding. In cases where free broadband content is subsidized with advertising revenue from content providers and is competing with content by new entrants from the telecommunications market, it is not clear at present that owning the network (that is, holding vertically integrated organization) is always accompanied by competitive advantages. In particular, it can be assumed from net neutrality discussions that, if incumbents are unable to change their current rate structure for broadband access provision, upgrading, maintaining access and combined backhaul, then operating a network may become burdensome for them. Therefore, future discussions on vertical separation should consider the perspective of competition that is emerging between traditional carriers and content providers on the basis of new business models.

NOTES

1. Theories on vertical separation in telecommunications industry are analyzed in detail by Sasaki (2007).
2. Though NTT keeps ownership relationships between long-distance, regional, mobile and other businesses under a holding company, such an organizational structure is the tentative result of deliberations on the structural separation of NTT that have continued since the privatization of the company. Each operating company is regulated separately to conduct businesses stipulated by NTT Law and by a guideline issued at the time of reorganization in 1999. Therefore, we should regard current NTT group companies as virtually divested companies.
3. TeliaSonera and Telecom Italia voluntarily established functionally separated access divisions in early 2008.
4. NTT published its Medium-Term Management Strategy in November 2004 and promised to shift 30 million customers to optical fiber access and next generation network services by 2010 (NTT, 2004). The promise has been modified several times after then.
5. Status of broadband diffusion and competition in Japan is described and analyzed in detail by Fuke (2008).
6. Japan's local administration is divided into 47 prefectures.
7. Since it was permitted for one cable operator to extend their businesses in several separated regions and to accept investment from foreign companies, MSOs started to emerge.
8. See Kamino (2007) for prehistory of Softbank Corp.'s entry into the broadband market.
9. See Fuke (2007) and Fuke (2008) for details. Metrical analysis by Akematsu and Tsuji

extracts regulatory policy, competition and technological development as factors for DSL progress in Japan (Akematsu and Tsuji, 2006).

10. It is cited from an article from the *Financial Times* titled 'Brussels divided over telecoms plan' (25 September 2007).
11. The reform was formally appraised in late November 2009 and entered into force on 19 December 2009.

REFERENCES

Akematsu, Y. and M. Tsuji (2006). *Deregulation or market competition, which has larger effect on Japanese ADSL development; panel data and AHP analyses.* Paper presented at ITS 17th Biannual Conference.

Akerlof, G. and Shiller, R. (2009). *Animal Spirits: How Human Psychology Drives the Economy, and Why It Matters for Global Capitalism.* Princeton, NJ: Princeton University Press.

BT (2005). Undertakings given to Ofcom by BT pursuant to the Enterprise Act 2002. 22 September.

Cave, M. (2006). Six degrees of separation: operational separation as a remedy in European telecommunications regulation. *Communications and Strategies*, **64**, 89–103.

EC (2000). The European Parliament and the Council of 18 December 2000. Regulation (EC) No 2887/2000 on unbundled access to the local loop.

EC (2007). Commission staff working document. Impact assessment. Accompanying document to the Commission proposal for a Directive of the European Parliament and the Council, {COM (2007) 697, COM (2007)698, COM (2007)699, SEC (2007)1473}, 13 November.

ECTA (2004). ECTA broadband scorecard end of June 2004. 28 September.

ERG (2007). ERG opinion on functional separation. ERG(07)44.

FCC (2005). In the Matter of SBC Communications Inc. and AT&T Corp. Applications for Approval of Transfer of Control, Memorandum Opinion and order, FCC 05-183, 17 November.

FCC (2009). High-speed service for Internet access: status as of December 31, 2007.

Fuke, H. (2007). *Info-Communications Policy in the Broadband Era.* Tokyo: NTT Publishing. (In Japanese.)

Fuke, H. (2008). Structural changes and regulatory challenges in Japanese telecommunications industry. In Y. Dwivedi, A. Papazafeiropoulou and J. Choodrie (eds), *Handbook of Research on Global Diffusion of Broadband Data Transmission Volume 1* (pp. 90–107). Hershey, PA: Information Science Reference.

Ikeda, N. (2005). *The Architecture of Information Technologies and Organizations.* Tokyo: NTT Publishing. (In Japanese.)

ITU (2003). ITU internet report 2003: birth of broadband.

ITU (2006). ITU internet report 2006: digital life.

Kamino, A. (2007). 'Realization of complementarities through full scale integration – M&A in telecommunications sector'. In H. Myiajima (ed.), *M&A in Japan* (pp. 259–82). Tokyo: Toyo Keizai. (In Japanese.)

MIC (2006). Report of Panel on Frameworks of Communications and Broadcasting. (In Japanese.)

MIC (2008). Status of cable TV as of end December 2007, February 2008. (In Japanese.)

MIC (2009a). Number of Broadband Service Contracts, Etc. (as of the end of December 2008).

MIC (2009b). Disclosure of Quarterly Data concerning Competition Review in the Telecommunications Business Field, Third quarter of FY2008 (as of the end of December 2008).

NTT (2004). NTT Group's medium-term management strategy. Press release, 10 November. Retrieved from http://www.ntt.co.jp/news/news04e/0411/041110d. html.

OECD (2006). Report to the council on experiences on the implementation of the recommendation concerning structural separation in regulated industries, (C(2006)(65).

OECD (2008). OECD Broadband Statistics, OECD Broadband Portal, June.

Ofcom (2004). Review of the wholesale local access market Explanatory statement and notification, 26 August.

Ofcom (2005). Final statements on the Strategic Review of Telecommunication, and undertakings in lieu of a reference under the Enterprise Act 2002, 22 September.

RegTP (2004). RegTP Newsletter. One-off local loop charges. 25 June.

Riordan, M. (2008). Competitive Effects of Vertical Integration. In P. Buccirossi (ed.), *Handbook of Antitrust Economics* (pp. 145–82). Cambridge, MA: MIT Press.

Sasaki, T. (2007). Perspective on structural separations or functional separations. Working paper No. 07-02, August. (In Japanese.)

Waverman, L. (2006). The challenges of a digital world and the need for a new regulatory paradigm at Ofcom. In Ofcom (edited by E. Richards, R. Foster and T. Kiedrowski), *Communications – The Next Decade, A Collection Of Papers Prepared for the UK Office of Communications*, London: Ofcom, pp. 158–75.

Yoo, C. (2002). Vertical integration and media regulation. *Yale Journal on Regulation*, **19**, 171–300.

11. Implementing functional separation in fixed telecommunications markets: the UK experience

Peter Curwen and Jason Whalley

11.1 INTRODUCTION

Since 2004 or so there has been considerable interest in the implementation of functional separation within fixed telecommunications markets. In these markets, which are largely but not exclusively to be found within the European Union, functional separation is seen as a way to resolve the tensions that exist between incumbent operators and those other service providers that require access to incumbents' networks to deliver their own services.

At the forefront of the implementation of functional separation is the UK. In late 2005, Ofcom and British Telecom (BT) agreed on a series of undertakings that culminated in the creation of a new company, Openreach, to run BT's local access network. Accompanying the establishment of Openreach was the imposition of key performance indicators and penalties for non-achievement. As Openreach is clearly an important landmark in the development of the UK telecommunications market, this chapter will focus on the implementation of the undertakings to date. With this in mind, the remainder of this chapter is divided into five sections. A brief overview of the different types of separation possible within the telecommunications industry is provided in section 11.2. Background information regarding the adoption of functional separation is detailed in section 11.3, while section 11.4 focuses on the implementation of the undertakings. The adoption of functional separation in the UK is appraised in section 11.5, and conclusions drawn in section 11.6.

11.2 LITERATURE

Any assessment of functional separation within the UK raises two questions: What is functional separation, and what has been the UK experience?

This section will address the former of these questions, while the remainder of the chapter will address the second question. A useful starting point when answering the first question is Xavier and Ypsilanti (2004). Through focusing on the separation of competitive from non-competitive services, the authors (ibid., p. 76) identify a range of separation measures that have been implemented within the telecommunications industry, namely:

- accounting, functional and corporate separation;
- separation into regional operators;
- separation of local from long-distance services;
- separation of local and mobile services;
- separation of local and broadband/advanced services;
- separation of an incumbent into smaller, vertically integrated carriers.

With respect to structural separation, four different approaches are identified (ibid., pp. 77–81). In the first of these, LoopCo, the incumbent divests its access business to form a new company while in the second, NetCo, an arm's-length relationship is established between the incumbent's access and non-access networks. The establishment of an alternative distribution company is the third approach suggested, and entails the collective ownership of the non-competitive assets by those operators present in the competitive parts of the market. The final approach suggested is that of voluntary suspension.[1]

Although all four approaches have been variously discussed, it is not altogether surprising that attention has focused on the LoopCo and NetCo proposals. Cave (2002), Dounoukos and Henderson (2003) and Xavier and Ypsilanti (2004) identify a range of factors that need to be taken into consideration regarding the viability of the LoopCo option. One such factor is the network scope of LoopCo, while another is whether coordination problems between LoopCo and other operators would result as a consequence of technological advances. With respect to the restructuring of BT announced in late 2000, Sandbach (2001) stated that the proposed NetCo inadequately addressed the local loop issues as it would cover both the copper and switched network. That is, the possibility remained for NetCo to favor its own services or those provided elsewhere in BT.

Cave (2002, p. 30) argues that with the NetCo option there is the possibility that the company will leverage the market power accruing from the monopoly part of the business into those that are competitive. In addition, he suggests that the pace of technical innovation will slow under the NetCo option as service providers would have to convince NetCo that sufficient demand existed for it to warrant making the necessary investment.

Due to the problems and uncertainties associated with both the LoopCo and NetCo proposals, it is not surprising that attention has focused on other ways of resolving the tensions that exist within the telecommunications market. Cave (2006a, p. 94), for example, identifies six alternatives as follows:

- creation of a wholesale division;
- virtual separation;
- business separation;
- business separation with localized incentives;
- business separation with separate governance arrangements;
- legal separation.

According to Cave (2006a, p. 94), accounting separation at the time of his writing was more or less the modus operandi of European incumbents with the exception of BT. In terms of the six identified alternatives, BT–Openreach falls under business separation with local incentives and consequently is closer to the ownership separation end than the accounting end. BT–Openreach is an example of functional separation. According to the European Regulators Group (ERG, 2007, p. 2), this involves the selective separation of those parts of the network that are difficult for other operators to replicate but which they need to access in order to provide their own services. Although this is frequently interpreted as the separation of the incumbent's wholesale and retail businesses from one another, this is slightly misleading. If the focus is solely on those parts of the incumbent's network that cannot be replicated then the scope of the functional separation may be narrower than is implied by the separation of wholesale from retail (ERG, 2007, p. 8).

Regardless of the extent to which functional separation is implemented, the result is to run and manage one part of the network separately from the rest. The separated part of the incumbent should be provided with local incentives so that it acts in the interests of all its customers, internal and external, and not in the interest of its parent company.[2] In addition to the use of local incentives, the ERG (2007, pp. 2f) suggests a range of 'key elements' that need to be provided if functional separation is to be effective.[3]

11.3 BACKGROUND

Before providing a brief overview of the events that culminated in the establishment of Openreach in late 2005, it is necessary to note that the decision to initiate the strategic review of telecommunication can be located at the confluence of three sets of drivers. The first of these drivers was the need to incorporate EU directives into the UK regulatory framework, while the second was the relatively recent establishment of Ofcom in 2003. Thus, the strategic review could be viewed as drawing a line under the old regulatory framework and providing a basis on which converged regulation could progress.

The review could also be regarded as being a reponse to a third set of drivers, namely, the failure of competition to develop as anticipated in the UK. Although some companies had invested in their own infrastructure, these networks lacked scale (Ofcom, 2004a, p. 53). The cable operators, which operated the most extensive networks geographically, collectively covered less than half of the population and their ability to compete was limited by their continued financial woes.[4] Service-based competition had been possible since the late 1990s (Ofcom, 2004c, p. 53) but had enjoyed only limited success because, it was alleged, BT had abused its dominant position in the wholesale market to enhance its retail competitiveness (Wilsdon and Jones, 2002).

The results of such anticompetitive behaviour can be seen with respect to broadband and local loop unbundling. Although towards the end of the 1990s many companies expressed an interest in offering broadband services, most subsequently left the market (Turner, 2003, p. 6). One consequence of this was that only a handful of companies emerged to compete against BT, while another was the limited uptake of local loop unbundling in the UK.[5] With this in mind, the strategic review could also be viewed as being driven by the desire to enhance competition within the broadband telecommunications market and to encourage greater adoption of local loop unbundling.

11.3.1   The Strategic Review of Telecommunications

At the end of 2003, Ofcom announced its intention to hold a review of the telecommunications market during the following year (Ofcom, 2004a). The initial consultation document was wide-ranging in nature, raising issues that were subsequently clarified in the second consultation document (Ofcom, 2004b). Central to the second consultation document was the identification of three regulatory options, the first of which was deregulation. Ofcom concluded, however, that this was not possible, not least because sector-specific regulation was faster and more precise than the alternatives.

The second option was a reference under the Enterprise Act (2002) to the Competition Commission. Such a reference would inevitably necessitate a wide-ranging review of the telecommunications market that could result in the eventual imposition of structural remedies. The third option, the one preferred by Ofcom, was termed 'real equality of access' and would enable those companies purchasing wholesale products from BT to do so on the same terms as BT's own retail operations. Thus, wholesale customers would have access to (Ofcom, 2004c, p. 14):

- the same or a similar set of regulated wholesale products as BT's own retail activities;
- at the same prices as BT's own retail activities; and
- using the same or similar transactional processes as BT's own retail activities.

Two different types of equivalence were proposed, outcome and input, and a range of products identified where it could be applied (Ofcom, 2004b, p. 68). In the case of equivalence of outcome, wholesale customers receive products that are comparable to those offered to BT's own retail operations but the underlying processes would not be the same. In contrast, where equivalence of input is applied, wholesale customers receive the same products as BT's own retail operations using the same set of underlying processes (Ofcom, 2004c, pp. 67f).

Recognizing that a range of issues had been identified by many of BT's wholesale customers during the consultative process, which these customers believed placed them at a competitive disadvantage relative to BT (ibid., p. 70), there was also a behavioural dimension to equivalence. While the range of issues highlighted was broad, two areas in particular – namely the incentives for inappropriate behaviour and transparency – were singled out as areas where action could be taken. This said, Ofcom did note that BT had, in the past, devoted considerable effort and resources to addressing the complaints raised by its competitors.

11.3.2 Undertakings in Lieu of a Reference under the Enterprise Act 2002

In June 2005, Ofcom (2005a) announced that it was launching a consultation to determine whether it should accept the undertakings offered by BT to bring an end to the strategic review. Rather than trigger a reference under the Enterprise Act 2002, BT agreed to a series of legally enforceable undertakings (ibid., p. 2). BT agreed to create an access service division that would (Odell, 2005, p. 23):

- control the 'last mile' of the telecommunications network;
- be operationally independent of BT while remaining under its ownership;

- be branded differently from BT;
- have its own five-member board, headed by a non-executive director of BT;
- incorporate 15 000 of BT Wholesale's 28 000 employees.

In addition, BT also agreed to a schedule for equivalence for legacy products as well as stating the principles on which the company's next generation network (NGN) would be developed (Ofcom, 2005a, pp. 2ff). For its part, Ofcom stated that it would revisit issues such as leased lines and retail price controls in the near future (ibid., p. 5).

In September 2005, Ofcom accepted the undertakings offered by BT (Ofcom, 2005b). In total, 236 undertakings were made by BT. These governed the operation of the access service division to ensure that those wholesale customers reliant upon access to deliver their own products and services were treated no differently from BT's own retail operations (Ofcom, 2005c). At the same time as Ofcom agreed to accept the undertakings, BT rebranded its access service division as Openreach (Ofcom, 2005b).

11.4   IMPLEMENTING THE UNDERTAKINGS

Given the magnitude of the undertakings, it is no surprise that their implementation has been carefully monitored. During 2006 and 2007 Ofcom published two evaluations of the impact of the telecommunications strategic review that detail the progress that BT has made in implementing the undertakings. In addition, five quarterly reviews as well as correspondence between Ofcom, BT and others have been published.[6]

A useful starting point for an understanding of how the undertakings have been implemented is the two annual evaluations that have been published by Ofcom. The first of these, which was published in October 2006, acknowledged the effort that BT had invested into meeting the undertakings before identifying a range of areas where implementation had been less than satisfactory (Ofcom, 2006a). Eight areas where further action was required were identified. It is, perhaps, no surprise that these were broad in their scope, ranging from the need to resolve boundary issues between BT Wholesale and Openreach to agreeing how Openreach's management information systems (MIS) and operational support systems (OSS) could be separated out from the rest of BT.[7]

The second annual report suggested that further work was required to separate Openreach from the rest of BT and to develop, and subsequently deploy, equivalent products (Ofcom, 2007b). In addition, the report also stated (on p. 4), somewhat vaguely, that more effort was required if the full benefits of functional separation were to be achieved.

Both annual reports highlighted the difficulties to be faced in separating Openreach from the rest of BT, noting in particular the information system-based difficulties being encountered. The three information systems in question are the MIS, OSS and the equivalence management platform (EMP). Openreach is required to separate its OSS from the rest of BT in a logical manner and to separate them physically from one another by June 2010 (Ofcom, 2007b). However, Ofcom and BT placed a different interpretation on logical separation, with the consequence that clarification was required (Ofcom, 2007d). Although this inevitably resulted in some delays, it also produced a clear timetable for the migration of users to physically separate systems.

Additional time was also sought by BT to separate the MIS between Openreach and the rest of the company. While Ofcom did agree to this request, BT was required to assist users to restrict access and to define the subsequent separation process. This has largely been achieved although Ofcom does note (2007a, p. 48) that risks still remain with those systems that draw on BT-wide initiatives. The delivery of equivalence is supported through the use of the EMP,[8] the implementation of which has been less than satisfactory since it was first introduced in early 2006. The delivery of the initial system was delayed and subsequent versions were released with reduced functionality (OTA, 2006a, 2006b). Perhaps more importantly, concerns have been raised as to the stability of the EMP (OTA, 2007a) and the extent to which the service is unavailable (OTA, 2007b, 2007c). Both of these have caused problems for the telecommunications companies using the EMP. Although these issues have been tackled with varying degrees of success, they have engendered a degree of uncertainty regarding the platform's robustness and reliability.

The correspondence published by Ofcom highlights some of the difficulties that have been encountered in the implementation of the undertakings.[9] BT has, on more than one occasion, sought more time to implement the undertakings. Although Ofcom has invariably granted these requests, it is worth noting that the extensions are temporary and not open-ended. This correspondence is relatively brief, which is in contrast to the consultations surrounding the range of exemptions and variations that BT has sought since June 2006.

To bring the strategic review to a swift conclusion, it was agreed that Ofcom and BT could consider at a later date the equivalence of inputs needs of some products (Ofcom, 2006e). In June 2006, BT sought exemptions and variations in 15 different areas, with 13 requiring consultation due to their complexity (Ofcom, 2006f). As a result of this consultation,

Ofcom agreed to nine of the requests. The remaining four requests required additional consultation as they involved products relying on fiber (Ofcom, 2006e).

In turn, this additional consultation resulted in three out of the four requests being granted. Ofcom granted a temporary extension until December 2007 in the case of the fourth request while further consultation was undertaken (Ofcom, 2007e, p. 2). Another set of exemptions and variations granted to BT was published in October 2007 and was once again the outcome of a consultation process that began in July of the same year (Ofcom, 2007f). The variation request made in May 2008 granted BT and Openreach more time to implement the undertakings (Ofcom, 2008b). All 18 granted variations are shown in Table 11.1.

The published correspondence, as well as the exemptions and variations consultations published by Ofcom, draw attention to boundary issues. Boundary issues arise where the distinction between Openreach and the rest of BT is blurred. One area where boundary issues have arisen was noted above, namely to ensure that the information systems that linked Openreach with the rest of BT were altered so that the two were separate from one another.

A second area where boundary issues have emerged is that of access to engineering resources. BT has sought to move engineers between Openreach and BT Wholesale as circumstances dictate (Ofcom, 2006b, p. 3). For example, BT requested permission from Ofcom to move engineers between the two divisions in the aftermath of the floods that swept the south of England during 2007. This was, however, a temporary measure that addressed a particular series of events.

The quarterly reports published by Ofcom draw attention to the need to ensure that the 'Chinese Walls' between Openreach and the rest of BT are maintained (Ofcom, 2006b, 2006d). A separate Openreach head office has been established, and some of the earlier concerns that the 'Chinese walls' were unsatisfactory due to organizational changes within BT Wholesale have been addressed.[10] The second report on the implementation of the undertakings notes that Openreach is reliant on other parts of BT for access to space and power within exchanges (Ofcom, 2007b, pp. 46ff). As a consequence of this, it was felt that Openreach does not have adequate control over the products that it delivers.

Also highlighted by the quarterly reports are the concerns expressed by other telecommunication operators as well as by Ofcom regarding product development. The former have noted that a gap existed early on between what BT announced and what was delivered (Ofcom, 2006b, p. 11), while more recently their interaction with BT has become an issue (Ofcom, 2006c, pp. 9ff; Ofcom, 2006d, p. 10). It was alleged that this interaction

Table 11.1   Variations to the undertakings

Variation	Scope	Date
1	Products & services supplied by Access Services (Openreach); share schemes and BT Group Deferred Bonus Plan; EAB report to OFCOM; EAB Summary Annual Report	March 2006
2	Equipment location	April 2006
3	Products & services supplied by Access Services (Openreach)	August 2006
4	OSS separation	September 2006
5	OSS separation	October 2006
6	Products & services supplied by Access Services (Openreach)	December 2006
7	Information flows & system separation	December 2006
8	Products & services supplied by Access Services (Openreach)	April 2007
9	OSS separation	June 2007
10	Incident management processes	October 2007
11	Extensions to OSS and EOI timetables	November 2007
12	Changes to sections 2.1 (definitions), 5 (access services) and 6 (management & structure of BT Wholesale)	December 2007
13	Products & services supplied by Access Services (Openreach)	December 2007
14	Provision of equivalent products and services – changes to section 3.1.1, section 3.1.2 and annex 1 of the undertakings	December 2007
15	Products & services supplied by Access Services, changes to section 5.46.2	May 2008
16	NGN, Space and Power and OSS separation	October 2008
17	IPStream in certain geographic markets and Wavestream National	December 2008
18	Changes to sections 2.1 (definitions) and 5.46.2 (date of effect of undertakings)	December 2008

Source: www.ofcom.org.uk/telecoms/btundertakings/exemptionsandvariations, accessed 1 June 2009.

was insufficient, and that in some cases the ability of other telecommunication operators to influence product specifications was limited. To this, Ofcom (2007b) adds that the pace of product development has been slow, before acknowledging that this may be due to developments elsewhere in BT.[11]

11.5   THE IMPACT OF FUNCTIONAL SEPARATION

The previous section has demonstrated that the separation of Openreach from the rest of BT has not been straightforward. While Openreach was established relatively swiftly in the aftermath of the undertakings being agreed, the actual separation of Openreach and BT has been more problematic in terms of both products and processes. As the implementation of the undertakings has progressed, BT has sought exemptions, variations and extensions. Such requests were perhaps inevitable given the unprecedented nature of the undertakings and thus may simply reflect the inherent difficulties of separating Openreach from the rest of BT.

Notwithstanding the difficulties that have been experienced, the Openreach model of functional separation has been praised. In a speech to the ERG, Commissioner Reding argued that functional separation had contributed to the rapid rise of unbundled lines and increased network investment (Reding, 2007). When the undertakings were announced in September 2005, the number of unbundled lines stood at 123 000 (OTA, 2005). In contrast, the most recent figures published in May 2009 show that there are now 5.9 million unbundled lines (OTA, 2009). Although this growth, which is shown in Figure 11.1, is undoubtedly impressive, it has not been without its problems as we have shown above. Monthly updates from the Office of the Telecommunications Adjudicator (OTA) show that although progress has been made in implementing the undertakings, with many key performance indicators improving after initial

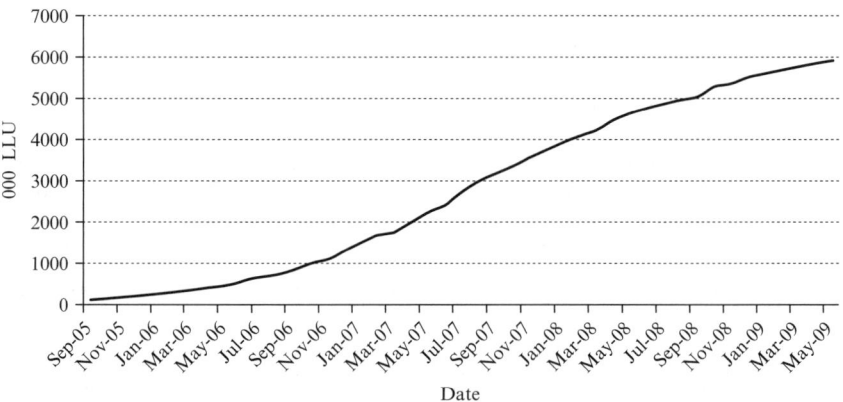

Source: Monthly OTA updates, available at www.offta.org.uk.

Figure 11.1   Unbundled local loops, September 2005–May 2009

disappointments, problems continue to emerge such as the delays in publishing the Wholesale Line Rental 3 (WLR3) roadmap and providing the necessary functionality to other service providers (OTA, 2008a, 2008b, 2008c).

There can be no doubt that BT has invested to deliver the undertakings. According to its 2008 annual report, BT invested £35 million in a 'proactive maintenance programme' that reduced the number of faults experienced (BT, 2008, p. 21). The total amount invested by BT in the financial year ending 31 March 2008 was £53 million, which was an increase on the corresponding figure for 2007 (£30 million) but less than in 2006 (£70 million) (ibid., p. 105). This is, however, slightly misleading as these figures relate to the costs of establishing Openreach and meeting the undertakings.[12] In other words, they do not include capital expenditure. Over the same period, capital expenditure remained steady at more or less £1.1 billion per annum.[13] It is not clear, of course, how much of this would have occurred regardless of whether Openreach was established, and how much is specifically due to the implementation of the undertakings.

There are other service providers in the marketplace. One such operator is Cable & Wireless (C&W). Prior to the undertakings being agreed, C&W stated its intention to invest in local loop unbundling (LLU) (Cable & Wireless, 2004) and acquired Bulldog, a broadband provider, to strengthen its position in the marketplace. After failing to control costs, C&W announced in June 2006 that it would stop offering retail products (Stafford, 2006) although it has continued to use LLU to deliver services to business clients (Cable & Wireless, 2008).[14]

Another operator providing broadband services is BSkyB. As of 31 March 2008, BSkyB had 1.428 million broadband subscribers, an increase of 229 000 subscribers over the quarter (BSkyB, 2008). In addition to acquiring Easynet at the end of 2005 for £211 million (Wray and Milmo, 2005), BSkyB has also made subsequent investments in infrastructure. In the nine months ending 30 April 2008, BSkyB invested £127 million in its residential broadband and telephony business (BSkyB, 2008). While it is not clear how much of this investment was targeted towards broadband, it is worth noting that it has been claimed that this represents the peak of its broadband-related investment cycle (Edgecliffe-Johnson and Fenton, 2008).

A third operator in the marketplace is Carphone Warehouse. Since announcing its intention to use LLU in November 2005, Carphone Warehouse has grown to become a significant player in the market. The economics of LLU will be further enhanced when Carphone Warehouse completes its £236 million acquisition of the UK operations of Tiscali that was announced in May 2009 (BBC, 2009). The company's growth has been driven by innovative new products such as TalkTalk Free

Broadband, launched in April 2006, which gave free broadband to those customers purchasing voice and line rental (Carphone Warehouse, 2007, p. 6). Although this was highly successful, attracting more than 500 000 customers, losses associated with the product widened from £20 million in April to £70 million in October 2006 (Parker and Braithwaite, 2006). Accompanying the launch of TalkTalk Free Broadband was the intention to invest in 1000 unbundled exchanges by July 2007 (Carphone Warehouse, 2007, p. 6). One motive for investing in LLU was that delivering services to customers this way was profitable, unlike the situation when wholesale products were used (Pratley, 2006).[15]

The acquisition of AOL UK from Time Warner in December 2006 further expanded the company's customer base.[16] This not only served to increase the number of subscribers but would make the economics of LLU more attractive to Carphone Warehouse (Parker and Braithwaite, 2006, p. 23). As a consequence, it is perhaps no surprise that Carphone Warehouse has switched the bulk of its subscriber base onto LLU and expanded the number of exchanges in which it has invested (Carphone Warehouse, 2008, p. 5).[17] Although the expansion of the business would not be possible without infrastructure investment of one sort or another, it is not clear how much the company has actually invested as the relevant figures are consolidated with other investments in its annual report.[18] This said, Charles Dunstone, the Chief Executive of Carphone Warehouse, has recently been quoted as saying that the company has made large expenditure commitments towards unbundling (Parker, 2008a, p. 19).

Dunstone has also raised the issue of the fees that Carphone Warehouse pays Openreach. Charge ceilings for WLR and LLU services were set by Ofcom between December 2004 and January 2006.[19] Since then, Openreach has failed to achieve the 10 percent return that it is permitted, with the consequence that BT would like to raise these charges (Parker, 2008e, p. 18). BT has argued that the aforementioned charge ceilings do not reflect the underlying costs of providing services and that they need to be amended to reflect the changing nature of the product portfolio being delivered.[20] Unsurprisingly Carphone Warehouse has a different view, stating that it would unfair to 'change the game' (Parker, 2008a, p. 19).

Notwithstanding the complexities of determining rates of return and charges that Ofcom (2008a, 2008d) highlights, the implications are clear: if Ofcom agreed with BT, the costs for other service providers of using Openreach's network would increase. In mid-May 2009 Ofcom announced that Openreach would be allowed to increase the prices that it charges other operators for both metallic path facilities and shared metallic path facilities (Ofcom, 2009).[21] The price of metallic path facilities would rise to £86.50 and that for shared metallic path facilities to £15.80 (Ofcom, 2009,

p. 4). Significantly these increases were towards the lower end of the range suggested in the consultation document (Ofcom, 2008d), leading BT to state subsequently that it may challenge this decision (Parker, 2009).[22]

BT is also quoted as noting that the lower charges were a 'real disincentive for future investments' (Parker, 2009). In its 2009 annual report, BT states that the regulatory environment must encourage investment and provide a rate of return commensurate with the risk involved (BT, 2009). Whilst BT does not say what constitutes a reasonable rate of return, it does note that regulatory clarity and certainty are important factors when determining whether the company will invest or not. It is arguably the case that Ofcom has provided regulatory stability through outlining how prices will increase in the coming years, and clarity through extensively explaining how it has arrived at its conclusions. Although Ofcom (2009, p. 30) does acknowledge that the price changes will have an impact on telecommunication companies, it then goes on to state that it is more interested in providing a stable and predictable regulatory framework than guaranteeing the returns of individual companies within the market.

Within such an environment, the inevitable question is: Why has BT continued to invest in Openreach? One reason is that Openreach is profitable and that the investments that have been made are intended to maintain this profitability.[23] In other words, it is in BT's economic interest to invest in Openreach. Thus, the complaints that BT has made regarding the rate of return that it is allowed to make on such investments could be interpreted as being part of a 'game' that is played between Ofcom, other telecommunication companies and itself. Another explanation could be that notwithstanding the undertakings that have been implemented, BT still derives sufficient advantages from owning the last mile of the network to warrant retaining ownership of, and investing in, Openreach. It may also be the case that BT has continued to invest in Openreach because it believes that through doing so it will be treated more favorably in other markets.

In the process of assessing the costs and benefits of functional separation, Amendola et al. (2007) state that functional separation enhances competition. One measure of this is market share while another is the price and speed of broadband products. As can be seen from Figure 11.2, BT remains the largest Internet Service Provider (ISP) whereas the market shares of Orange, Tiscali, Virgin Media and smaller ISPs have all fallen. In contrast, the market shares of Carphone Warehouse and BSkyB have both risen since the undertakings were agreed.

Ofcom (2008a, p. 14) states that broadband prices have fallen whereas speeds have increased. While there is clear evidence that prices have fallen,[24] there has been much discussion of late regarding broadband

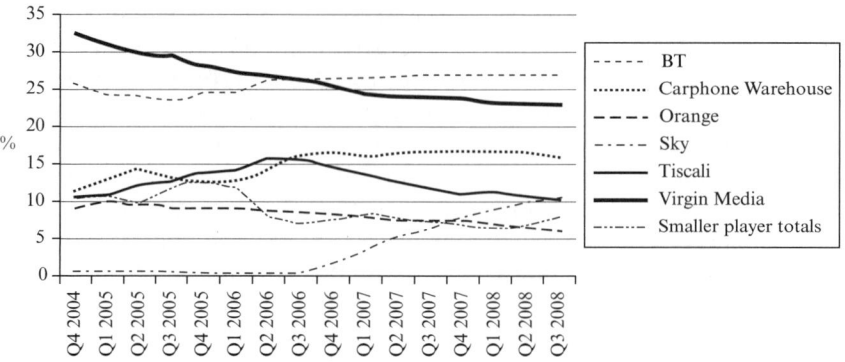

Source: www.broadbanduk.org.

Figure 11.2   Market share of ISPs

speeds. Average download speeds vary considerably across the UK, from 4.5 Mbps in London to 2.3 Mbps in Northern Ireland (BBC, 2008). In addition, the relationship between the advertised and the actual broadband speeds has been questioned, with a discrepancy being noted between the two. According to Point Topic (2008), as advertised speeds increase the proportion of subscribers receiving such speeds declines. The discrepancy between advertised and actual broadband speeds has prompted Ofcom (2008c) to issue a voluntary code of conduct containing eight principles.

11.6   CONCLUSION

This chapter has focused on functional separation in the UK. The September 2005 decision to opt for functional separation represents a milestone in the regulation of the UK telecommunications market, and in the process imposed a range of undertakings on BT. The process of unraveling the numerous undertakings that bound Openreach to the rest of BT has proved to be more difficult and protracted than anticipated. Not only were extensions, variations and exemptions sought, but Openreach also took longer than anticipated to meet several key performance indicators. If nothing else, these difficulties suggest that implementing functional separation is by no means straightforward.

Nevertheless, the period subsequent to the adoption of functional separation has seen significant broadband competitors emerge in the form of Carphone Warehouse and, to a lesser extent, BSkyB. For both Carphone Warehouse and BSkyB, LLU plays a central role in their strategies, not least because LLU is a more profitable way to service their customers than WLR. While LLU does not appear to have spurred other service providers into investing in other parts of the 'ladder of investment' as suggested by Cave (2006b), they do appear to have invested relatively large sums to deliver LLU-underpinned broadband services. With the emergence of LLU as a vehicle for the deployment of broadband services, not only has the importance of BT in the marketplace been elevated, but so too has the necessity of ensuring that the relationship between BT and those companies using its network is functioning as intended.

Functional separation was adopted to resolve the tensions that existed between BT and those companies wishing to access its network to provide their own services, with the undertakings providing a framework for assessing the state of this relationship. Looking back over the period since the undertakings, it could be argued that the relationship between BT and other service providers has improved as the number of unbundled lines has substantially increased. However, the relationship could sour if BT continues to argue that the rate of return that it is allowed to earn is unsatisfactory. If the relationship between BT and other operators does sour and undertakings are breached, Ofcom could refer the matter to the Competition Commission. The unpredictable outcome of such a referral may mean that Ofcom is unwilling to make such a reference and will adopt a more pragmatic approach to implementing and monitoring the undertakings.

NOTES

1. This is not to suggest that this is the only feasible categorization of structural separation. Dounoukos and Henderson (2003, pp. 44f), for example, distinguish between 'actual' structural separation and 'internal' separation. With respect to the former, they identify four alternatives – club or joint ownership, operational separation, separation into several vertically integrated companies and separation of the non-competitive components into several parts; while three 'internal' separation alternatives – accounting, functional and corporate – are suggested. A discussion of structural alternatives based on a broader range of regulated industries than just telecommunications can be found in OECD (2001).
2. While referring to wholesale and retail, the OECD (2006) states that incentives may be given to wholesale managers that conflict with those provided to retail managers. This reinforces the suggestion that incentives should be localized and not tied to the overall profitability of the operator.
3. The European Regulators Group (2007) provides a summary of the measures undertaken in countries where functional separation has been implemented under the three headings of functional, employee and information separation. Interestingly, not all measures have been implemented in all cases, and some are only feasible in conjunction with others.
4. In 2004 just 46 percent of UK homes were passed by broadband-enabled cable (Ofcom,

2004c, p. 38). Subsequent communication market reviews have shown that in 2005 and 2006, digital cable was available to 45 percent of the UK population (Ofcom, 2007a, p. 15). For a discussion of the financial woes of cable operators see, for example, Curwen (2004).

5. See, for example, de Bijl and Peitz (2005) for a discussion of unbundling that highlights the relatively slow uptake of local loop unbundling in the UK compared to other European Union member states.

6. For details of the issues raised by the quarterly reports, see Ofcom (2005a, 2006b, 2005d, 2007c). These reports, as well as the other material published by Ofcom relating to the telecommunications strategic review and the implementation of the undertakings, can be found at www.Ofcom.org.uk/telecoms/btundertakings.

7. See Ofcom (2006a, p. 2) for a full list of the eight areas identified that require further action.

8. The EMP is an information system that supports the delivery of the products offered by Openreach. Given the anticipated large volumes of some of these products, the system is designed to be automated to ensure that services are provided as demanded.

9. See www.Ofcom.org.uk/telecoms/btundertakings/exemptionsandvariations for a full list of the correspondence between BT and Ofcom that has been published.

10. BT Wholesale has been reorganized with two management units – BT Wholesale Core Network Services and BT Wholesale Value-added Network Services – being established (Ofcom, 2006b). The concern expressed by some telecommunications operators was that this would complicate the implementation of the 'Chinese walls' that were established. However, Ofcom stated that more time should be given before making a judgment as to whether the 'Chinese walls' were being breached.

11. Two developments are noted, namely, the implementation of the undertakings and the development of the company's twenty-first century network.

12. It is worth noting that the annual report states that these figures are estimates of the incremental and directly attributable costs incurred as a consequence of establishing Openreach on the one hand and meeting the undertakings on the other (BT, 2008).

13. For the year ended 31 March 2007, capital expenditure was £1108 million while the figure for the following year, albeit an estimate on the part of BT, was £1100 million (Ofcom, 2008a, p. 20). Although capital expenditure has remained more or less steady since Openreach was established, the 2007 figure did represent a slight increase on the previous year's level of £1038 million (Ofcom, 2008a, p. 20).

14. Cable & Wireless does not break up capital expenditure by its European, Asian and US businesses by geography or product. Thus, it is unclear how much the company has invested in LLU in the UK. Having said this, capital expenditure by this business did decrease from £235 million in 2006/07 to £221 million in 2007/08 (Cable & Wireless, 2008).

15. According to Pratley (2006), Carphone Warehouse would make a profit of £7 per month per customer when using LLU whilst it would lose £5 per month when using wholesale products.

16. With the acquisition of AOL UK, Carphone Warehouse controlled 16 percent of the broadband market (Carphone Warehouse, 2007). While not without its risks, the acquisition has also been regarded as strategic in nature as it increased the presence of Carphone Warehouse in this market (Parker, 2008b, 2008c). Carphone Warehouse sought to expand its presence in the UK broadband market by bidding to acquire the UK operations of Tiscali, which had 1.8 million broadband subscribers at the time. Although Parker (2008d) reports that C&W is no longer in contention, other reports published around the same time contest this.

17. At the end of 2006, 19 percent of the company's broadband subscribers were located in unbundled exchanges (Carphone Warehouse, 2008). This figure had increased to 61 percent by the end of the following year. Over the same period, the number of unbundled exchanges increased from 569 to 2457. Only part of this increase – 924 – was due to the acquisition of AOL UK.

18. Total investment increased from £351.1 million in 2006 to £562.2 million in 2007 (Carphone Warehouse, 2007), with 'acquisition of property, plant and equipment' increasing from £89.4 million to £161.4 million over the same period (Carphone Warehouse, 2007). Some, but not all, of this presumably relates to LLU.
19. See Ofcom (2008a, pp. 1f) for details. The charge ceilings were as follows: £100.68 for residential WLR, £110.00 for business WLR, £81.69 for metallic path facilities (MPF) and £15.60 for shared mpf (SMPF).
20. The move away from WLR has, for example, contributed to a decline in Openreach's rate of return (Ofcom, 2008a, p. 18).
21. See, for instance, Ofcom (2009), for a detailed description of the new pricing arrangements for Openreach.
22. As part of this decision to set prices at the lower end of the range suggested in the consultation document, Ofcom declined to allow BT to use the Openreach's fees to reduce BT's pension deficit (Parker, 2009). For an assessment of BT's pension deficit see, for instance, Ralfe (2009).
23. Openreach is one of BT's four operating divisions. According to its 2009 annual report, three of these divisions – BT Retail, BT Wholesale and Openreach – maintained or increased their levels of operating profits for the year ending 31 March 2009 compared to 2008, whilst the fourth division – BT Global Services – incurred a substantial operating loss (BT, 2009). Between 2008 and 2009, Openreach increased its operating profit from £1050 million to £1209 million whilst BT Global Services saw its 2008 operating profit of £117 million turn into a loss of £2106 million in 2009.
24. See, for example, Ofcom (2006g, p. 120), Ofcom (2007a, p. 298) or Ofcom (2008a, p. 15).

REFERENCES

Amendola, G., F. Castelli and P. Serdengecti (2007). Is really functional separation the next milestone in telecommunications (de)regulation? Paper presented at the 18th European Regional ITS Conference, Istanbul, Turkey, 2–5 September.

BBC (2008). Towns triumph in broadband tests. 3 June. Available at news.bbc.co.uk.

BBC (2009). Carphone to purchase Tiscali UK. 8 May. Available at news.bbc.co.uk.

de Bijl, P. and M. Pietz (2005). Local loop unbundling in Europe: experience, prospects and policy challenges. *Communications and Strategies*, **57**, 33–57.

BSkyB (2008). 3rd quarter results. 30 April. Available at www.bskyb.com.

BT (2008). Keeping BT ahead of the game. Annual Report & Form 20-F. Available at www.bt.com.

BT (2009). Annual Report & Form 20-F. Available at www.bt.com.

Cable & Wireless (2004) Press release – Cable and Wireless PLC presentation on local loop unbundling. 16 September. Available at www.cwplc.com.

Cable & Wireless (2008). Results for the year ended 31 March 2008. 22 May. Available at www.cwplc.com.

Carphone Warehouse Group (2007). Annual Report 2007 – Delivering value to our customers and shareholders. Available at www.cpwplc.com.

Carphone Warehouse Group (2008). Preliminary results for the 52 weeks to 29 March 2008. Available at www.cpwplc.com.

Cave, M. (2002). Is LoopCo the answer? *info*, **4** (4), 25–31.

Cave, M. (2006a). Six degrees of separation: operational separation as a remedy in European telecommunications regulation. *Communications and Strategies*, **64**, 89–103.

Cave, M. (2006b). Encouraging infrastructure competition via the ladder of investment. *Telecommunications Policy*, **30**, 223–37.

Curwen, P. (2004). Markets vs regulators and the battle to determine market structure: evidence from the UK cable industry. *info*, **6** (1), 24–36.

Dounoukos, S. and A. Henderson (2003). Structural separation in telecommunications: a review of some issues. *Agenda*, **10** (1), 43–60.

Edgecliffe-Johnson, A. and B. Fenton (2008). BSkyB hits peak of its investment in broadband. *Financial Times*, 7 February, 20.

European Regulators Group (ERG) (2007). ERG opinion on functional separation. ERG (07) 44. Available at www.erg.eu.int.

Odell, M. (2005). New unit emerges from the shake-up. *Financial Times*, 24 June, p. 23.

OECD (2001). Structural separation in regulated industries. DAFFE/CLP(2001)11, Paris: OECD.

OECD (2006). Report to the Council on experiences on the implementation of the recommendation concerning structural separation in regulated industries. C(2006)650, Paris: OECD.

Ofcom (2004a). Strategic review of telecommunications – phase 1 consultation document. April, London: Ofcom.

Ofcom (2004b). The Communications Market 2004 – Telecommunications. August, London: Ofcom.

Ofcom (2004c). Strategic review of telecommunications – phase 2 consultation document. 18 November. London: Ofcom.

Ofcom (2005a). Telecommunications statement. June. London: Ofcom.

Ofcom (2005b). Ofcom accepts undertakings from Board of BT Group on operational separation. BT access service business – Openreach – formally established today. 22 September. London: Ofcom.

Ofcom (2005c). Final statement on the strategic review of telecommunications and undertakings in lieu of a reference under the Enterprise Act 2002 – Statement. 22 September. London: Ofcom.

Ofcom (2005d). Report on the implementation of BT undertakings – first quarterly report. 25 October. London: Ofcom.

Ofcom (2006a). Evaluating the impact of the Telecoms Review – an interim report one year on. 18 October. London: Ofcom.

Ofcom (2006b). Report on the implementation of BT undertakings – second quarterly report. 26 January. London: Ofcom.

Ofcom (2006c). Report on the implementation of BT undertakings – third quarterly report. 23 May. London: Ofcom.

Ofcom (2006d). Report on the implementation of BT undertakings – fourth quarterly report. 5 September. London: Ofcom.

Ofcom (2006e). Requests from BT for specified exemptions and agreements to its Undertakings under the Enterprise Act 2002 – Part 2. 21 December. London: Ofcom.

Ofcom (2006f). Requests from BT for specified exemptions and agreements to its Undertakings under the Enterprise Act 2002 – Part 1. 15 June. London: Ofcom.

Ofcom (2006g). The communications market. 10 August. London: Ofcom.

Ofcom (2007a). Communications market review. 23 August. London: Ofcom.

Ofcom (2007b). Impact of the Telecoms Strategic Review. Evaluation. 10 December. London: Ofcom.

Ofcom (2007c). Report on the implementation of BT undertakings – fifth quarterly report. 12 February. London: Ofcom.

Ofcom (2007d). BT OSS separation. 21 June. London: Ofcom.

Ofcom (2007e). Requests from BT for specified exemptions and agreements to its Undertakings under the Enterprise Act 2002 – Part 3. 19 July. London: Ofcom.

Ofcom (2007f). Requests from BT for specified exemptions and agreements to its Undertakings under the Enterprise Act 2002 – Part 4. 3 October. London: Ofcom.

Ofcom (2008a). A new pricing framework for Openreach. Developing new charge controls for wholesale line rental. unbundled local loops and related services. 30 May. London: Ofcom.

Ofcom (2008b). Variation of the undertakings given to Ofcom by BT pursuant to the Enterprise Act 2002 – variation number: 15. 20 May. London: Ofcom.

Ofcom (2008c). Voluntary code of practice: broadband speeds. 5 June. London: Ofcom.

Ofcom (2008d). A new pricing framework for Openreach. 5 December. London: Ofcom.

Ofcom (2009). A new pricing framework for Openreach. Statement. 22 May. London: Ofcom.

OTA (2005). Update for September 2005. Available at www.offta.org.uk.

OTA (2006a). Update for March 2006. Available at www.offta.org.uk.

OTA (2006b). Update for April 2006. Available at www.offta.org.uk.

OTA (2007a). Update for January 2007. Available at www.offta.org.uk.

OTA (2007b). Update for April 2007. Available at www.offta.org.uk.

OTA (2007c). Update for June 2007. Available at www.offta.org.uk.

OTA (2008a). Update for February 2008. Available at www.offta.org.uk.

OTA (2008b). Update for March 2008. Available at www.offta.org.uk.

OTA (2008c). Update for April 2008. Available at www.offta.org.uk.

OTA (2009). Update for May 2009. Available at www.offta.org.uk.

Parker, A. (2008a). Carphone hits at BT plans to raise line fees. *Financial Times*, 14 April, p. 19.

Parker, A. (2008b). Risky move into broadband looks as if it might pay off. *Financial Times*, 14 April, p. 19.

Parker, A. (2008c). Carphone deal opens up lines of expansion. *Financial Times*, 9 May, p. 19.

Parker, A. (2008d). Carphone off Tiscali buyer list. *Financial Times*, 13 May, p. 17.

Parker, A. (2008e). BT seeking increased landline charges. *Financial Times*, 16 May, p. 18.

Parker, A. (2009). Regulator keeps line open for BT. *Financial Times*, 22 May. Available at www.ft.com.

Parker, A. and T. Braithwaite (2006). Carphone is catapulted into the big league. *Financial Times*, 12 October, p. 23.

Point Topic (2008). How fast is your broadband? 12 March. Available at point-topic.com.

Pratley, N. (2006). Broadband disaster should turn to success. *Guardian*, 12 October. Available at www.guardianunlimited.co.uk.

Ralfe, J. (2009). Behind BT's deep, dark pension deficit. *Financial Times*, 24 May. Available at www.ft.com.

Reding, V. (2007). Better regulation for a single market in telecoms, Speech/07/624, Plenary Meeting of the European Regulators Group, Athens, 11 October.

Sandbach, J. (2001). Levering open the local loop: shaping BT for broadband competition. *info*, **3**(3), 195–202.

Stafford, P. (2006). C&W's Bulldog turns tail from broadband market. *Financial Times*, 9 June, p. 22.

Turner, C. (2003). Issues in the mass market deployment of broadband. *info*, **5**(2), 3–7.

Wilsdon, J. and D. Jones (2002). *The Politics of Bandwidth: Network Innovation and Regulation in Broadband Britain*, London: Demos.

Wray, R. and D. Milmo (2005). Watchdog clears BSkyB acquisition of Easynet. *Guardian*, 31 December. Available at www.guardianunlimited.co.uk.

Xavier, P. and D. Ypsilanti (2004). Is the case for structural separation of the local loop persuasive? *info*, **6**(2), 74–92.

12. Efficiency and sustainability of network neutrality proposals[1]

Toshiya Jitsuzumi

12.1 INTRODUCTION

Recent times have witnessed the phenomenon of increasing congestion on the Internet; this is characterized by the 'crowding out' of average users by peer-to-peer (P2P) users due to a rapid spread of bit-intensive applications (Cisco Systems, 2008a, 2008b, 2008c; Swanson, 2007; Swanson and Gilder, 2008), and further, it is sometimes accompanied by the possibly anticompetitive conduct of network operators. In an attempt to deal with this, 'network neutrality', a term coined by Wu (2003), has come to be a focus of discussion among operators, academics and telecom regulators. For example in parallel with a series of controversial events, for example, blocking of Voice over Internet Protocol (VoIP) by Madison River Communications LLC in March 2005 and the allegedly discriminatory treatment of P2P by Comcast since 2007,[2] the US Federal Communications Commission (FCC) has issued important statements. First, in 2004, the then FCC Chairman Michael Powell challenged the broadband network industry to preserve the four 'Internet Freedoms': freedom to access content, freedom to use applications, freedom to attach personal devices and freedom to obtain service plan information.[3] Secondly, on 23 September 2005, the following principles were incorporated into a policy statement.[4]

- Consumers are entitled to access lawful Internet content of their choice.
- Consumers are entitled to run applications and use services of their choice; subject to the requirements of law enforcement.
- Consumers are entitled to connect their choice of legal devices that do not harm the network.
- Consumers are entitled to competition among network providers, application and service providers, and content providers.

Moreover the FCC mentioned the necessity of preserving these principles in the conditions agreed upon for the October 2005 approval of both the Verizon–MCI and the SBC–AT&T mergers and for AT&T's concessions in its merger with BellSouth, which was finally approved on 29 December 2006. In addition the US Federal Trade Commission (FTC) issued a policy report in 2007 (FTC, 2007), and several bills have been proposed and were proposed and discussed in the 109th and 110th Congress.[5]

In Japan, the Ministry of Internal Affairs and Communications (MIC) issued a report in September 2007 regarding network neutrality (MIC, 2007b). Similar considerations are being made in the EU as well.

Since the concept of network neutrality encompasses many different aspects, a discussion on the related issues has thus far been quite complex. Therefore, this chapter divides network neutrality issues into economic and non-economic ones and focuses mainly on the former, which is a direct result of traffic congestion and negative externality thereof. I further categorize the network neutrality challenges into short- and long-term issues and discuss policy recommendations. Short-term solutions include a case-by-case approach and a re-evaluation of the 'smart market' concept proposed by MacKie-Mason and Varian (1998), while long-term solutions involve the need for a new business model and the introduction of a mechanism similar to a universal service fund (USF).

The remainder of this chapter is organized as follows. In section 12.2, the definition of network neutrality is analyzed. Section 12.3 discusses some basic concepts that make the network neutrality issue unique, and section 12.4 proposes policy recommendations for general cases. Section 12.5 concludes the discussion.

12.2 MULTIPLE ASPECTS OF NETWORK NEUTRALITY

The most important factor that has contributed to the complexity of the discussions on the topic under consideration is the lack of a standardized definition of network neutrality. Examining the related arguments that have been presented thus far, there are at least six dimensions to this concept, as discussed below. However, some of them are not strictly orthogonal from each other.

12.2.1 Economic Aspect versus Non-economic Aspect

Some discuss this issue from the viewpoint of optimal resource allocation or maximizing economic efficiency; others underscore the non-economic value of this 'neutrality'. For example, network users stress the value of the Internet as a medium and relate it with freedom of speech or democracy, while network operators emphasize the efficient use of network. However it is important to remember that these two aspects are not mutually exclusive. Rather the non-economic aspect works as an important factor modifying the naive outcome generated by a pure economic consideration. In fact the fundamental difference between network users and operators lies in how they perceive these two aspects.

12.2.2 Long Term versus Short or Very Short Term

Faced with congestion on the Internet, some people mention the importance of network investment or technological development in alleviating capacity constraints in the long term, while others focus on how to control peak demand, implicitly assuming that the network capacity is fixed in the short term. However, as Yoo (2006) and Peha (2007) point out, these windows are not necessarily mutually exclusive but merely different sides of the same coin.

Moreover, in the long term, prospects of future technological developments may become an important element of this discussion. It is important to bear in mind that a short-term solution that assumes constant technology can be mutually incompatible with a long-term solution that takes some technological development into consideration; in other words, the static-efficient solution may be detrimental to the dynamic-efficient one. For example a local loop unbundling (LLU) to incubate access competition may have different impacts on different time horizons.

12.2.3 Fixed Telecom Network versus Cable Network or Mobile Network

Congestion requires government intervention only when the resources in question are scarce and shared as 'common goods'. Thus, theoretically speaking, congestion can be a problem for the trunk/backhaul/backbone portion of the network and for the access network of cable and mobile services, while it does not pose a problem for telcos' access segments where certain capacities are usually reserved for individual users.[6]

12.2.4 Single-Sided Market versus Two-Sided Market[7]

On some occasions, a solution to network neutrality can be developed assuming that subscription revenue is the only source of an operator's revenue. Alternatively one can assume that network operators can expect additional revenues from content and/or application providers (two-sided market) and can therefore utilize the advertisement model as an integral part of a solution.

12.2.5 Unbundled Market versus Bundled Market

In reality, network operators are either vertically separated from content and application providers or vertically integrated with them. Since network operators usually possess a certain market power, vertical integration requires additional considerations of leveraging, extraction, discrimination or price squeezing.

12.2.6 Legal Use versus Illegal Use

A significant portion of Internet traffic reportedly comprises unlawful content,[8] and as a result, people who emphasize the legitimate use of the Internet and others who represent content creators would like to include anti-piracy or anti-indecency paragraphs in the network neutrality solution.

12.3 SOME BASICS CONCERNING NETWORK NEUTRALITY

The Internet is characterized by several distinctive features that require special attention when proposing network neutrality policies. This section will briefly outline them.

12.3.1 Congestion on the Internet

When a capacity of a shared portion of the network cannot meet the demand, it results in congestion and hence yields negative externality.[9] In this case, the market mechanism is unable to produce a socially optimal resource allocation. In order to improve the situation, some policy intervention can be justified from the viewpoint of social welfare maximization, provided the social benefit exceeds the cost.

The demand for network usage is increasing rapidly in the US (Swanson and Gilder, 2008) and, in fact, in the world at large (Cisco Systems, 2008a, 2008b, 2008c). In the case of Japan, the MIC (2008a) estimates that the overall traffic increased by 150 percent from November 2004 through November 2007 and reached more than 800 Gbps (monthly average, estimated figure) in November 2007; in fact the MIC suggests evidence of congestion over the trunk network (2007b).

Possible strategies for congestion management can be categorized into two groups: demand control and supply expansion. Demand control aims to restrict the demand within the capacity. Possible tools for doing so include a Pigovian tax, a 'congestion right' trading system, and quasi-property rights granted for certain capacity through auction. Supply expansion can involve direct investments by the government and financial incentives offered to private sectors. Such an expansion can materialize not only by installing more fibers, but also by introducing new technologies, including network cache systems (for example, Akamai) and a new protocol such as P2P or P4P.[10]

A problem may arise if a congestion management policy is poorly designed or used for anticompetitive purposes. If an Internet service provider (ISP) holds a dominant market position and employs such an ill-designed congestion management policy, the market cannot independently remedy the resulting inefficiency.

12.3.2 Efficiency in a Vertically Related Market

Since Internet access providers need content and applications to satisfy subscriber needs, it is not a priori apparent whether or not such firms utilize their market power in an anticompetitive manner. According to Farrell and Weiser's (2003) analysis, the monopolist does not have an incentive to jeopardize its complementary market at all times. Under the concept of 'internaliz[ing] complementary efficiencies' (ICE), they point out that:

> in choosing how to license interface information, certify complementors, and otherwise deal with developers, such a firm has a clear incentive to choose the pattern that will best provide it or its customers with applications. That is, a firm will *internalize complementary efficiencies* arising from applications created by others. (p. 101)

They also state that: 'if a platform monopolist integrates into an adjacent market, it will still welcome value-added innovations by independent firms' (p. 102). Similar concerns are shared by Baumol et al. (2007).

It is true that these arguments have clear limitations (Farrell and Weiser, 2003; van Schewick, 2007). For example if the complementary market is monopolized, the basic ICE framework cannot be applied in the first place since it assumes that a dominant power resides only on the bottleneck monopolist side. Moreover, as Farrell and Weiser (2003) claim, when the situation in the complementary market allows a network monopolist to execute price discrimination, ICE cannot produce socially efficient outcomes. Furthermore the original ICE concept does not consider the two-sided market structure. In the case where the downstream market faces Cournot competition, Economides (1998) shows that an upstream monopoly will ultimately decrease social welfare. Bakos and Katsamakas (2008) show that the integration of a monopolistic two-sided platform by one set of users on either side can serve to attain social optimum without policy interventions.

If the tendency to behave anti-competitively is confirmed, the possible network neutrality proposals must include an external mechanism that disciplines market players.

12.3.3   Quality as 'the Collectively Produced Commons'

A packet-switching system, in which a certain capacity of the Internet is 'shared' by every user, generates non-rivalry (until it becomes congested) and the non-excludability nature of the communication quality. Maintaining communication efficiency under the surge of traffic should be considered as constituting 'the commons', in which case an individual network user's decision-making could cause negative externality. Since this is a rather classic problem that was already pointed out by MacKie-Mason and Varian (1998), economic theory already has some solutions prepared in order to finance the optimal size of these goods, including the classical Lindahl mechanism as well as a recently proposed Falkinger (1996) mechanism.

Nevertheless, what complicates this simple and rather straightforward situation is the segmented ownership of the Internet. The Internet is a set of individual networks, and there exists no controlling authority or fund allocation system. In order to guarantee a certain level of communication quality on an end-to-end (E2E) basis, cooperation and cost clearance among ISPs is essential, which is also mentioned in FTC (2007). There are three kinds of contracting frameworks between network operators at present: 'interconnection', 'peering' and 'transit'. Among these, interconnection, which was originally applied between facility-based telephone service providers, has a bidirectional cost clearance system, while the other two do not. ISPs are connected only through peering or transit. A peering partner or an upstream ISP in the transit agreement need not bear the cost of usage accruing at the connecting network. This violates the fairness of cost sharing or the 'he who benefits ought to pay' principle unless the amount of packets is balanced in both directions, which is essential for an optimal fund distribution. To make matters worse, a rapid spread of bit-intensive application, especially Internet Protocol (IP) video content, will upset the balance between upstream and downstream ISPs. Therefore, it is currently very difficult to attain optimal fund distribution.

12.3.4 Other Considerations

There are other factors that should also be taken into consideration when making policy recommendations: providing broadband Internet as a universal service or improving the availability of broadband Internet, the cost of regulation and the ensuing market reaction towards such intervention. In addition some proponents in the US suggest that stipulating network neutrality rules and promoting innovation at the edge of the network is a good option for raising the country's Organisation for Economic Cooperation and Development (OECD) ranking.[11]

12.4 POLICY RECOMMENDATIONS

Although it is important to take non-economic aspects into account as previously mentioned, this section will basically discuss network neutrality solutions from an economic viewpoint for simplicity reasons, by first stating the issues from the short-term perspective and then from the long-term one.

The short-term solution to be discussed below implicitly assumes the dominance of existing network operators, whereas the long-term one does not. This may create mutually incompatible solutions in each time framework, and can explain a certain discrepancy in the network neutrality debates. For example a strong confidence in wireless innovation and uncertainty regarding the future industrial organization support Yoo's (2006) network diversity claim, whereas the pro-network neutrality side focuses more on the current status quo.

12.4.1 Short-Term Issues

In the short term, under which network capacity is considered to be fixed and new entry to the market is impossible, the issue can be phrased as follows: How to maximize the static efficiency? or, How to properly discipline incumbent network operators who control bottleneck facilities?'[12] In this case, we assume that some market power exists in the network market, while the content and/or application market is more competitive. For simplicity, my discussion will focus on a situation where ISPs are completely integrated into the network facility operators' business.

There are two major tasks to be dealt with in order to solve this short-term issue. First, Task 1 is to prevent anticompetitive behavior on the part of a network operator. Considering the cost of regulation and information asymmetry, competition is always the most desirable solution – that

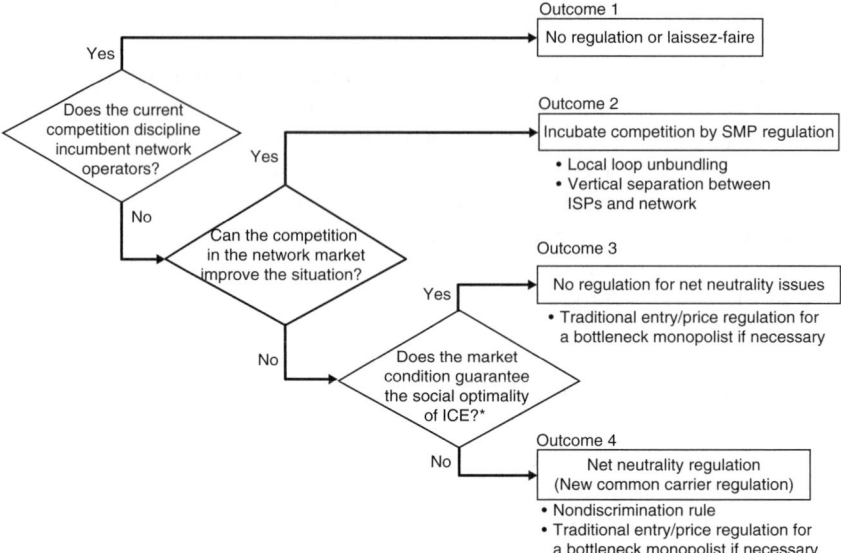

Note: * The counterexamples listed in Farrell and Weiser (2003) and van Schewick (2007) provide a good basis for implementing this decision box.

Figure 12.1   Flowchart for selecting an optimal policy mix for a short-term issue

is, government intervention must be kept to a minimum. In this case, the aforementioned ICE concept serves to determine the optimal policy mix. Since ICE assumes that a profit-seeking bottleneck monopolist acts to maximize its efficiency, it can be used as a foundation for policy-makers to build a least-regulated market. However the effective range of ICE has some limitations. Thus, it is important to distinguish the case where we can rely on ICE from that where we cannot. On the basis of this considera-tion, I propose a possible framework that regulators can rely on (Figure 12.1). For example this flowchart shows that when the situation leads us to 'Outcome 2', building up a competitive pressure in the ISP market through a significant market power (SMP) type of regulation is required.[13]

There may be some drawbacks in the use of the flowchart. The most significant one is that implementation requires a great deal of time and substantial costs. Due to the heterogeneity of the market structure, deci-sions must be made on a community-by-community basis. Moreover the speed of technological change requires a frequent review of such decisions. Furthermore information asymmetry may cause deterioration of the effi-ciency of the overall mechanism. Thus it is not practicable to prepare a complete set of policies on an *ex ante* basis. In other words, policy-makers should initially trust the initiative of market players and the overall market mechanism, and then start intervening only in the event of a market failure that cannot be remedied by private initiatives, and if the social benefit of a certain government intervention exceeds its cost. In this case, in order to minimize the risk of or the uncertainty about the future possible policy intervention, it is important to declare a general policy guideline beforehand. Since the primary policy target we are now considering is to prevent anticompetitive behavior of bottleneck monopolists, I believe it is appropriate for a government to declare a non-discrimination principle as a ground rule. Related authorities (an antitrust court or the quasi-judicial section or 'alternative dispute resolution' of telecom regulators) then monitor the market to see if any breach is happening. Once detected, they can follow the flowchart and build an appropriate policy tool.[14]

The second-biggest drawback is that taking the *ex post* antitrust mechanism into consideration, the use of the flowchart might be considered as 'gilding the lily'. In the testimony before the US Senate Committee on Commerce, Science, and Transportation (22 April 2008), Lessing pointed out that relying on an *ex post* remedy will stifle the incentive at present and choke future innovation. If this is the case, the next step that should be considered is determining the optimal mix of *ex ante* and *ex post* treatments.

The efficiency of the aforementioned proposal presupposes the efficient use of the current network capacity. Task 2, therefore, entails motivating network operators to utilize their existing resources efficiently through 'reasonable' network management. The majority of network neutrality debaters on both sides agree on the necessity of such management; they disagree, however, on what constitutes 'reasonable management'.

There is good reason to believe that it is very difficult to attain economically justifiable 'reasonableness' under the current market where the E2E transmission quality has the characteristics of a commons without a proper cost-sharing mechanism, and the monthly fixed charge contract for subscription is very popular. Because of these, ISPs and end-users do not have the socially optimal level of incentives for efficient use of network resources. Thus it is important to motivate ISPs properly to conduct economically justifiable reasonable network managements in the short run (and to invest sufficiently in the long run) and, at the same time, to introduce a pricing mechanism that charges a higher fee if the end-user requires better throughput.

Since the short-term issue has been defined above from the viewpoint of maximizing static efficiency, one of the theoretical answers to this is an auction-like mechanism, such as the concept of a 'smart market' put forth by MacKie-Mason and Varian (1998). In an ideal situation, such a mechanism will guarantee economic efficiency. Alternatively if the unit price level is somehow properly determined (possibly through trial and error), metered tariffs may be able to generate satisfactory results; however such tariffs incur high transaction costs, causing a serious negative impact on the broadband penetration. Kruse (2008) proposes a similar idea using four types of quality class, and MIT (2005) examines the possibility of modified flat rates (that is, 'pricing tiered by peak rate' and 'pricing tiered by traffic volume'). On the other hand when operators face enough competitive pressure and sufficient disclosing requirements, it might be better to take a laissez-faire approach and leave it to the market instead. It is important to note that all such suggestions implicitly require some clearing house mechanism on the basis of the 'sending party's network pays' principle (Kruse, 2008, p. 30).

12.4.2   Long-Term Issues

The challenge faced in the long term, under which it would be possible to alter network capacity, can be stated as follows: How to maximize dynamic efficiency? In other words, the challenge lies in determining how to motivate network operators to deploy the optimal network capacity while generating positive externality.

Capacity building of the network can be realized by employing various methods other than installing more fibers or routers. Operators can expand capacity 'virtually' by adopting new technology (for example, wavelength division multiplexing, IP multicast technology), better protocols (for example, P2P, P4P) and/or better network management. Regardless of the method actually deployed, it is still true that expanding capacity will require additional investment, and private operators will not be motivated without the appropriate incentives in place.

Some network neutrality proponents suggest that capacity expansions should be solely financed on the basis of the subscription model – that is, through additional monthly subscription revenues such as quality of service (QoS) surcharge from end-users, and not from content and/or application providers. Hermalin and Katz (2007) show that such a practice will degrade the transmission quality offered by ISPs. In addition, if the sum of subscription revenues is not sufficient, then this proposal will be utility-decreasing and unsustainable in the long term.

In order to answer the above question empirically, I conducted a survey of Japanese broadband users in 2007, using an e-mail and web-based system. Japan is clearly one of the frontrunners in the broadband policy development and has the most advanced and cost-effective broadband access environment around the world. Thus, I believe that analyzing the practicability of the above-mentioned pro-network neutrality proposal in Japan will provide certain guidelines for other nations. The survey period was from 27 to 30 July 2007; 912 valid responses were collected out of 3000 contracted monitors. The median age of the respondents was 37.3 years and the median annual household income was 696.4 million yen. On average, the respondents had 92.4 months of Internet experience and spent 4928 yen per month on Internet access. A total of 803 respondents used broadband (344 used fiber to the home – FTTH; 349 asymmetric digital subscriber line – ADSL; and 110 cable Internet) and had 50.1 months of experience of subscription. Two questions were specifically focused on estimating the respondents' willingness to pay (WTP) for improving the QoS over the Internet (Box 12.1).

Following Hidano (1999), the WTP figures were estimated by adopting a Weibull survival model. In equation (12.1), the survival function, $S(T)$, can be interpreted as a reduced-form description of the probability that an individual's WTP is at least as high as T:

$$S(T) = 1 - G(T) = \exp\left(-\exp\left(\frac{\ln T - \mu}{\sigma}\right)\right), \qquad (12.1)$$

Where $G(\cdot)$ = cumulative distribution function for T, and μ and σ are the parameters to be estimated.

When incorporating a sample's demographic features, the following survival function is applied:

$$S(T) = \exp\left(-\exp\left(\frac{\ln T - \boldsymbol{\beta}'\mathbf{X}}{\sigma}\right)\right), \qquad (12.2)$$

where $\mathbf{X}$ is a vector of demographic variables and $\boldsymbol{\beta}$ represents the parameters to be estimated.

Table 12.1 shows the responses for survey questions 31 and 32. Table 12.2 summarizes the results of the parameter estimations based on the maximum likelihood estimation method, which were calculated by using a software package called CVM 2002 (version 1.0).[15]

The estimation for equation (12.1), presented in the second column of Table 12.2, indicates that under perfect price discrimination, each respondent would be ready to spend between 452.1 yen ($5.00)[16] and 1064.6 yen ($11.77) on average as a monthly surcharge to improve the quality of communication. Assuming that this average WTP can be applied to broadband users in Japan – which totaled 26.44 million as of March 2007 – the total WTP was between 143.4 billion yen ($1.59 billion) and 337.8 billion yen ($3.73 billion) each year. This is equivalent to 6 percent to 14 percent of the major firms that invested in telecom equipment during FY2006.[17] However as the sixth column shows, such estimates are significantly

---

BOX 12.1   QUESTIONS FOR WTP ESTIMATION

Q31   Assume your ISP company is currently considering an investment plan for improving its backbone infrastructure to alleviate the congestion problem and make your broadband experience less stressful (i.e., faster downloading, uninterrupted streaming), which costs several billion yen and lasts several years. In order to finance this plan, your company intends to impose a surcharge of X yen per month. What is your most likely response?

A 1. I approve the plan and accept the required surcharge.
   2. I approve the plan but reject the required surcharge.
   3. I approve the plan but reject any surcharge.
   4. I disapprove of the plan.

Q32   (For those who chose Answer 4 in Q31) What is your reason for disapproving of the plan?

A 1. I do not require broadband content.
   2. I am satisfied with the current quality, and thus there is no need for further investment.
   3. I am not satisfied with the current quality, but there is no need for further investment.
   4. I do not understand the question.
   5. Other reasons.

Note: X is set as 500, 1000, 1500, 2000, 2500, 3000, 3500, 4000, 4500, 5000, 5500, or 6000 yen.

---

influenced by some demographic variables. The third, fourth and fifth columns show differences in WTP between subgroups.

On the other hand, 22.2 percent of the investments made by Japanese major telcos in 2005 were towards trunk facilities.[18] The traffic volume transmitted over the Internet in Japan increased 2.5 times over the three years following November 2004,[19] and is predicted to double every other year.[20] This prediction is based on a mere extrapolation of the past trends; however it is in line with the Internet traffic estimates of Cisco Systems (2008c). If this trend continues, even assuming the 20 percent annual increase of subscription,[21] the amount of possible WTP would run short

Table 12.1   Responses to Q31 and Q32

Responses to Q31

Surcharge	N	Would approve the plan and . . .			Would not approve the plan (%)
		Pay the required surcharge (%)	Not pay the required surcharge (%)	Not pay any surcharge (%)	
500	87	9.2	17.2	37.9	35.6
1000	85	5.9	23.5	42.4	28.2
1500	65	3.1	24.6	40.0	32.3
2000	77	1.3	18.2	50.6	29.9
2500	75	2.7	24.0	48.0	25.3
3000	75	2.7	21.3	49.3	26.7
3500	77	0.0	11.7	58.4	29.9
4000	74	2.7	16.2	37.8	43.2
4500	69	2.9	27.5	43.5	26.1
5000	74	2.7	21.6	41.9	33.8
5500	75	2.7	14.7	50.7	32.0
6000	79	1.3	15.2	35.4	48.1
N		29	178	407	298

Responses to Q32

Reasons for disapproval	N	%
I do not require broadband content.	106	35.6
I am satisfied with the current quality, and thus there is no need for further investment.	62	20.8
I am not satisfied with the current quality, but there is no need for further investment.	48	16.1
I do not understand the question.	69	23.2
Other reasons.	13	4.4

within only two to three years (Figure 12.2). Thus it is clear that the subscription-model proposal of network neutrality regulation supporters is not sustainable in Japan.

If technological developments cannot cover the difference,[22] ISPs must seek some alternate business models such as an advertisement model or a vertical integration model. Either scheme would ultimately entail heavy or bandwidth-intensive subscribers to bear the cost of traffic congestion; therefore both schemes appear socially fair.[23] It is also possible to rely on tax revenues, provided this is justified by the size of the positive externality of broadband Internet.

Table 12.2   Estimated parameters for equations (12.1) and (12.2)

Model	Equation (12.1)	Equation (12.1)	Equation (12.1)	Equation (12.1)	Equation (12.2)
Sample	ALL	Richer	Male	Male and rich	ALL
N	436	146	164	73	294
Log-likelihood	−102.16	−41.89	−55.39	−28.28	−68.83
A/C	208.32	87.78	114.77	60.56	151.66
Coefficients					
σ	4.124***	3.873*	3.210**	2.525**	3.061**
	(2.926)	(1.939)	(2.540)	(2.227)	(2.884)
μ	3.603**	4.304**	5.169**	5.999***	
	(2.510)	(2.352)	(4.968)	(7.079)	
In (household income)					1.365**
					(2.025)
Gender (M = 0, F = 1)					−1.569*
					(−1.904)
Marital status (married = 1, other = 0)					−0.803
					(−1.418)
Student (full time = 1, other = 0)					−1.782
					(−1.299)
FTTH (user = 1, non-user = 0)					0.600
					(1.058)
Constant					−3.180
					(−0.659)
Mean WTP	¥1064.6	¥1470.4	¥1381.8	¥1377.2	
Truncated mean WTP	¥452.1	¥593.9	¥729.5	¥923.7	
Median WTP	¥8.1	¥17.9	¥54.2	¥159.7	

Notes:
1 Numbers in the parenthesis are asymptotic t-values.
2 *** is less than 1%, ** is less than 5%, and * is less than 10%.
3 Truncated points are 0 and 6000 yen.
4 'Richer' sample includes respondents with above-median household income.

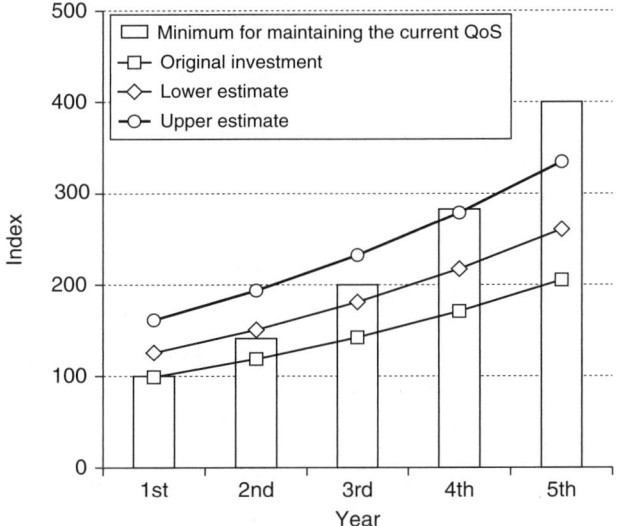

Figure 12.2   Simulations for subscription-model sustainability

When a new business model is adopted, the problem remaining is again how to allocate the raised resources to the appropriate network operators. For example if a single communication needs two 'peering' interconnections between three ISPs, the middle ISP cannot obtain any revenue from the other ISPs who can directly charge end-users. Since balancing accounts directly between numerous ISPs is very costly (Farnon and Huddle, 1997), it may be better to introduce an external and centrally controlled resource allocation system, such as a universal service fund-like external mechanism. In addition, as mentioned in section 12.3.3, it is necessary to provide each network operator with the appropriate incentives to upgrade the network capacity because of the common feature of Internet quality, which results in a more-than-optimal usage and a less-than-optimal investment. The challenges that will need to be addressed in the future are monitoring the necessity of each ISP's investment and its actual efforts and then compensating with a proper external fund injection.

12.5   CONCLUSION

This chapter shows that network neutrality is not a simple concept, but has at least six dimensions. Therefore before considering a possible solution to the network neutrality issues, it is important to specify the focus of

the discussion on this concept. Further, in considering policy recommendations, one must fully understand some of the peculiarities of the Internet
business, which include the characteristic of sharing some portions of the
network, the motivations of a monopolistic network operator in a two-
sided market, public good-like features of communication quality, and the
lack of a clearing house mechanism between ISPs.

One of the main findings of this analysis is that there exists no 'one size
fits all' solution when dealing with the short-term economic issue. For
example, in some cases, an SMP-type regulation should be introduced,
accompanied by a certain allowance of network management. To guarantee that such network management is reasonable, a 'smart market'
proposed by Mackie-Mason and Varian (1998) could be a good guideline.
Alternatively the market could be left to ascertain whether certain conditions have been satisfied. As for the long-term issue of capacity building,
my econometric analysis indicates that the subscription model may not
be sustainable without extraordinary high hopes for future technological
developments; it is necessary, rather, to introduce other funding sources to
finance network expansions. For both short-term and long-term cases, a
fund allocation mechanism must be established.

This chapter might generate more questions than answers. Among
them, three important issues are: (1) implementing these proposals in the
real world; (2) determining the accurate boundaries within which ICE can
guarantee overall efficiency; and (3) improving the accuracy and robustness of WTP estimations in other samples or settings. These constitute the
areas that need to be explored in future works.

NOTES

1. This research was assisted by a grant from the Abe Fellowship Program administered
   by the Social Science Research Council and the American Council of Learned Societies
   in cooperation with and with the funds provided by the Japan Foundation Center
   for Global Partnership. The survey was supported by a grant from the International
   Communications Foundation.
2. As Clark (2007) and others have already pointed out, the reason why the network neutrality issue has become the focal point of discussion in the US is that the government
   'ha[s] abandoned the idea of increasing competition through facilities unbundling' (p.
   704).
3. Keynote Address at the Silicon Flatirons Symposium, Powell (2004).
4. FCC (2005).
5. See Jordan (2007) for details on the seven bills submitted to the 2005–06 Congress. The
   historical development of this debate is well summarized by Atkinson (2007).
6. In the US, since the trunk segment of the Internet is relatively abundant in capacity
   compared to its access, partially due to the dot-com bubble, the locus of this problem
   seems currently limited within the cable's access portion. In Japan on the other hand,
   where the access segment has a much higher capacity than in the US, congestion may be

observed in the trunk segment. Of course, since NTT's FTTH is designed such that its access line is shared by a maximum of 32 users, there is a possibility of some congestion at the access line, if bit-intensive real-time applications are excessively used.

7. The discussion concerning the two-sided market is well summarized by Rochet and Tirole (2003, 2006), Eisenmann et al. (2006) and Armstrong (2006).
8. Svensson, P. (2008).
9. Excess demand is not the single cause of congestion. See Gupta et al. (2005).
10. 'P4P' stands for proactive network provider participation for P2P or provider portal for P2P. See Xie et al. (2008) for details.
11. OECD (2007).
12. These are the primary focuses of the MIC's discussion (2007b); the ideas presented in the report are fundamentally similar to the policy recommendations discussed below.
13. This particular outcome coincides with the situation that Jordan outlines in his article (2007).
14. Bauer (2007) proposes a similar 'wait and see' approach.
15. A software of the Research Institute for Community Policy Co., Ltd.
16. The currency conversion reflects the exchange rate in January 2009.
17. MIC (2007c).
18. This figure comes from investments by 'registered telcos' in trunk line facilities, switchers, routers and computers, coupled with investments by large 'notified telcos' in switchers, routers, multiplexers, line concentrators and computers (MIC, 2006).
19. MIC (2008b).
20. The presentation by Mori (2008) the then Vice-Minister of the Ministry for Internal Affairs and Communications (MIC) at 30th Anniversary conference of PTC '08.
21. This assumption is very optimistic. According to the MIC (2007a), Japanese broadband subscribers increased by 13 percent in 2006 and their growth rate has been decreasing.
22. A Japanese ISP, SoftBank, claims that the backbone cost per capacity has been decreasing at the rate of 20–30 percent per annum. (SoftBank, 2008). Assuming that this technological development continues at 20 percent annually, the broadband subscription increases at a more plausible rate of 10 percent annually, and the retail market condition remains constant, we would still need new revenue sources in the 7th to 10th year from the starting point of the calculation.
23. It is possible that the P2P or P4P protocol may create 'unconsciously' heavy users and raise some fairness concerns.

REFERENCES

Armstrong, M. (2006). Competition in two-sided markets. *RAND Journal of Economics*, **37**(3), 668–91.

Atkinson, R.C. (2007). Network neutrality overview. Presentation material from the 9th Annual Telecom, Cable and Wireless Conference, Austin, TX. Retrieved 14 August 2008 from https://www4.gsb.columbia.edu/null/download?&exclusive=filemgr.download&file_id=646279.

Bakos, Y. and E. Katsamakas (2008). Design and ownership of two-sided networks: implications for Internet platforms. *Journal of Management Information Systems*, **25**(2), 171–202.

Bauer, J.M. (2007). Dynamic effects of network neutrality. *International Journal of Communication*, **1**, 531–47.

Baumol, W.J., M. Cave, P. Cramton, R. Hahn, T.W. Hazlett, P.L. Joskow, A.E. Kahn, J.W. Mayo, P.A. Messerlin, B.M. Owen, R.S. Pindyck, V.L. Smith, S. Wallsten, L. Waverman, L.J. White and S. Savage (2007). *Economists' statement on network neutrality policy*. Retrieved 14 August 2008 from http://www3. brookings.edu/views/papers/litan/200703jointcenter.pdf.

Cisco Systems, Inc. (2008a). The Exabyte era. Retrieved 1 April 2008 from http:// www.cisco.com/en/US/solutions/collateral/ns341/ns525/ns537/net_implementation_white_paper0900aecd806a81a7.pdf.

Cisco Systems, Inc. (2008b). Approaching the zettabyte era. Retrieved 11 December 2008 from http://www.cisco.com/en/US/solutions/collateral/ns341/ns525/ns537/ ns705/ns827/white_paper_c11-481374.pdf.

Cisco Systems, Inc. (2008c). Cisco visual networking index – forecast and methodology, 2007–2012. Retrieved 11 December 2008 from http://www.cisco.com/ en/US/solutions/collateral/ns341/ns525/ns537/ns705/ns827/white_paper_c11-481360.pdf.

Clark, D.D. (2007). Network neutrality: words of power and 800-pound gorillas. *International Journal of Communication*, **1**, 701–8.

Economides, N. (1998). The incentive for non-price discrimination by an input monopolist. *International Journal of Industrial Organization*, **16**, 271–84.

Eisenmann, T., G. Parker and M.W. Van Alstyne (2006). Strategies for two-sided markets. *Harvard Business Review*, October, 92–101.

Falkinger, J. (1996). Efficient private provision of public goods by rewarding deviations from average. *Journal of Public Economics*, **62**(3), 413–22.

Farnon, M. and S. Huddle (1997). Settlement systems for the Internet. In B. Kahin and J.H. Keller (eds), *Coordinating the Internet* (pp. 377–403). Cambridge, MA: MIT Press.

Farrell, J. and P.J. Weiser (2003). Modularity, vertical integration, and open access policies: towards a convergence of antitrust and regulation in the Internet age. *Harvard Journal of Law and Technology*, **17**(1), 85–134.

Federal Trade Commission (FTC) (2005). Appropriate framework for broadband access to the Internet over wireline facilities. 20 FCC Rcd 14986 (policy statement). Retrieved 4 January 2009 from http://hraunfoss.fcc.gov/edocs_public/ attachmatch/FCC-05-151A1.pdf.

Federal Trade Commission (FTC) (2007). *Broadband connectivity competition policy: RTC staff report*. Retrieved 14 August 2008 from http://www.ftc.gov/ reports/broadband/v070000report.pdf.

Gupta, A., D.O. Stahl and A.B. Whinston (2005). Pricing traffic on interconnected networks: issues, approaches, and solutions. In S.K. Majumdar, I. Vogelsang and M.E. Cave (eds), *Handbook of Telecommunications Economics* (Vol. 2, pp. 413–39). Amsterdam: Elsevier.

Hermalin, B.E. and M.L. Katz (2007). The economics of product-line restrictions with an application to the network neutrality debate. *Information Economics and Policy*, **19**(2), 215–48.

Hidano, N. (ed.) (1999). *Kankyo to gyousei no keizai hyouka: CVM manyuaru* [Economic evaluation of environments and policy: manual for CVM analysis]. Tokyo: Keiso Shobo.

Jordan, S. (2007). A layered network approach to net neutrality. *International Journal of Communication*, **1**, 427–60.

Kruse, J. (2008). Network neutrality and quality of service. *Intereconomics*, **43**(1), 25–30.

MacKie-Mason, J.K. and H.R. Varian (1998). Economic FAQs about the Internet. In L.W. McKnight and J.P. Bailey (eds), *Internet Economics* (pp. 27–62). Cambridge, MA: MIT Press.

Ministry of Internal Affairs and Communications (MIC) (2006). *Survey on communications industry, August 2005: investment March 2006*. 17 February. Retrieved 6 January 2009 from http://www.johotsusintokei.soumu.go.jp/statistics/pdf/HS200513_001.pdf.

Ministry of Internal Affairs and Communications (MIC) (2007a). Information and communications in Japan. White Paper 2007. Retrieved August 2008 from http://www.johotsusintokei.soumu.go.jp/whitepaper/eng/WP2007/2007-index.html.

Ministry of Internal Affairs and Communications (MIC) (2007b). Network no churitsusei ni kansuru kondankai houkokusho [Report on network neutrality]. Retrieved 14 August 2008 from http://www.soumu.go.jp/s-news/2007/070920_6.html#bt.

Ministry of Internal Affairs and Communications (MIC) (2007c). Basic survey on the communications industry. 30 November. Retrieved 6 January 2009 from http://www.johotsusintokei.soumu.go.jp/statistics/pdf/HB200600_001.pdf.

Ministry of Internal Affairs and Communications (MIC) (2008a). Wagakuni no Intaanetto ni okeru torahikku no shukei keisan [Estimation of traffic volume on the Internet in Japan]. Retrieved 14 August 2008 from http://www.soumu.go.jp/s-news/2008/080221_3.html.

Ministry of Internal Affairs and Communications (MIC) (2008b). Tabulation and estimation of Internet traffic in Japan: announcement of tabulation results as of November 2007. 21 February. Retrieved 14 August 2008 from http://www.soumu.go.jp/s-news/2008/082221_3.html.

MIT (2005). The broadband incentive problem. A white paper prepared by the Broadband Working Group MIT Communications Futures Program (CFP) and the Cambridge University Communications Research Network. Retrieved 14 August 2008 from http://cfp.mit.edu/docs/incentive-wp-sept2005.pdf.

Mori, K. (2008). ICT policy in Japan. Presentation material at the 30th Anniversary conference of PTC '08, Honolulu, Hawaii, 13 January.

Organisation for Economic Co-operation and Development (OECD) (2007). OECD broadband statistics to June 2007. Retrieved 6 January 2009 from http://www.oecd.org/sti/ict/broadband.

Peha, J.M. (2007). The benefits and risks of mandating network neutrality, and the quest for a balanced policy. *International Journal of Communication*, **1**, 644–68.

Powell, M.K. (2004). Preserving Internet freedom: guiding principles for the industry. Keynote address at the Silicon Flatirons Symposium. Retrieved 4 January 2009 from http://hraunfoss.fcc.gov/edocs_public/attachmatch/ DOC-243556A1.pdf.

Rochet, J.C. and J. Tirole (2003). Platform competition in two-sided markets. *Journal of European Economic Association*, **1**(4), 990–1029.

Rochet, J.C. and J. Tirole (2006). Two-sided markets: a progress report. *RAND Journal of Economics*, **37**(3), 645–67.

SoftBank (2008). Dai yon kai intaanetto seisaku kondankai purezenteishon shiryo [Presentation material for the 4th meeting of panel on Internet policy]. 27 May. Retrieved 26 December 2008 from http://www.soumu.go.jp/joho_tsusin/policyreports/chousa/internet_policy/pdf/080527_2_si4-4.pdf.

Svensson, P. (2008). Peer-to-peer networks go legit. 14 March. Retrieved 23 December 2008 from http://www.msnbc.msn.com/id/23636873/.

Swanson, B. (2007). The coming exaflood. *Wall Street Journal*, 20 January. Retrieved 18 April 2009 from http://online.wsj.com/article/ SB116925820512582318.html.

Swanson, B. and G. Gilder (2008). Estimating the exaflood: the impact of video and rich media on the Internet – a 'zettabyte' by 2015? Discovery Institute. Retrieved 14 August 2008 from http://www.discovery.org/a/4428.

van Schewick, I.B. (2007). Towards an economic framework for network neutrality regulation. *Journal of Telecommunications and High Technology Law*, **5**, 329–91.

Wu, T. (2003). Network neutrality, broadband discrimination. *Journal on Telecommunications and High Technology Law*, **2**, 141–75.

Xie, H., A. Krishnamurthy, A. Silberschatz and Y.R. Yang (2008). P4P: explicit communications for cooperative control between P2P and network providers. Retrieved 14 August 2008 from http://www.dcia.info/documents/P4P_Overview. pdf.

Yoo, C.S. (2006). Network neutrality and the economics of congestion. *Georgetown Law Journal*, **94**, 1847–1908.

PART V

Interdependent innovations and regulatory
policies: mobile network deployment and
mobile Internet developments

13. Interdependent innovation in telecommunications: risk, standardization and regulation[1]

Bruno Basalisco, Andy Reid and Paul Richards

13.1 INTRODUCTION

This chapter is concerned with developments within and between three distinct strands of the telecoms industry, namely those of standardization, the development of technology policies and the regulation of telecom operators who are attributed some level of market power and subject to sectoral regulatory rules. It is taken as a self-evident fact that increasingly in the information and communication technology (ICT) sector, the commercialization of innovatory technologies which is the underlying aim of all parties to harness the benefits of technical progress, requires the coordination of activities across many parties and even different sectors. This chapter explores the issues arising for what might be described as good regulatory practice given the context of developments in standardization bodies and technology policies.

Regulation and innovation are not easy bedfellows. In fact, technological or business change adds much complexity to the dialogue between the state and the market which is the essence of regulation. The latter can have an indirect effect on firms' incentives to innovate via its impact on the level of competition in an industry.[2] Moreover, regulatory policy can directly affect the path of technological development – and be shaped by it.[3]

Because of the converging nature of technological change in the ICT industries, the interplay of regulation and innovation trespasses the boundaries of industries subject to regulatory oversight. Regulators, especially those showing an interest in more efficient forms of regulation (also known as re-regulation), are increasingly aware of models of innovation in network industries (Farrell, 2003) and are also keen to understand the functioning of highly innovative unregulated industries. On the other hand, innovations in technologies and businesses across the ICT industries have long been shifting the boundary between competition and regulation, which influences the whole set of public policies aimed at redressing market failures (Farrell and Weiser, 2003).

Industry players and public policy alike benefit from the highest possible degree of regulatory certainty. This implies that regulatory practice has to operationalize the innovation dynamics and its impact on competition in regulated markets. A key procedure in telecommunications regulation is the definition of markets, which underlies any analysis of market power and the potential remedies to address the latter.[4]

The concept of emerging markets has been introduced by regulators in order to account for the interplay of innovation and regulation in the procedure for market definition. It is a subject of intense debate among academics and practitioners (Dobbs and Richards, 2004; Crocioni, 2008). Innovations in technologies and business foster new products and services, which can provide a challenge for quantitative market definitions (informed by demand substitutability, which suffers in this case from high levels of uncertainty). In fact, regulators often acknowledge that the development of these innovations could be artificially shaped by regulation. As a result, regulators at times opt to forebear from these innovative products and services, labeled as part of emerging markets. We believe that the issue of emerging markets is of high relevance because it lies at the interface of two fields: the specific context of applying *ex ante* market power remedies in telecommunications regulatory regimes, but also the bigger picture of the technology policy discourse on the promotion of investment and innovation.

Specifically, the chapter discusses how regulatory policy-making could stand to benefit from a framework for emerging markets which accounts for the extent to which each innovation depends on the strategic interplay of many parties such as network operators, equipment manufacturers, regulators and policy-makers. Our analysis stresses that current regulatory theories and practice underappreciate the importance of the interdependent nature of innovation (in its risks and benefits). We perform a brief analysis of a set of new products and services spurred by communications network innovations and discuss the relative nature of innovation. In light of this and in a spirit of constructive criticism, we argue that there is considerable merit in operating the regulatory framework in such a way that it signals that different scenarios based on the nature of innovation will indeed be treated differently in regulatory terms guided by the degree to which the innovation is risk-bearing within and between different parties.

The chapter is structured as follows. Section 13.2 below sets the scene of the nature of interdependency of innovation for individual firms. Section 13.3 discusses the merits and shortcomings of two existing responses to the challenge of interdependent innovation in the communications industries: standardization and technology neutrality. Section 13.4 reviews developments of some examples of innovation of network technologies and their implications on communications services. Section 13.5 concludes by drawing some implications of our perspective on regulatory practice and specifically in the context of market reviews and the assessment of market power.

13.2 THE INTERDEPENDENT NATURE OF INNOVATION

Many firms need to vault over company boundaries in order to improve their products and services – a phenomenon popularized as 'co-opetition' (Brandenburger and Nalebuff, 1996). Even large, dominant industry players have realized those size and market shares are not sufficient to be able to leverage innovation processes in their industry. Some firms effectively manage innovation platforms. This is for instance the case of Intel's efforts in orchestrating innovation across its industrial ecosystem through carefully crafted policies of entry and intellectual property rights (IPR) sharing in complementary markets (Gawer and Henderson, 2007).

When considering the role of interdependence in affecting innovation efforts, an important aspect is whether systems present modularity in their components (Baldwin and Clark, 2000; Langlois, 2002; Farrell and Weiser, 2003). Many firms incorporate technological change in their competitive strategies. Henderson and Clark (1990) highlight that some firms exploit their knowledge of the very architecture of systems by leveraging linkages between core concepts and components. They thus suggest taking a more nuanced perspective of the traditional incremental versus radical innovation dichotomy by accounting for architectural and modular innovation.

The competitive advantages of a firm may depend on innovation processes involving its co-opetitors – defined as their network of suppliers, users and complementors (Afuah, 2000). These can be a source of innovation, lead users themselves of the technology, and providers of information useful to refine innovation. Moreover, many firms engage in collaborative ventures with other industry actors with the intent of coordinating the development of innovation – such as industry fora, joint ventures and collaborative agreements. As the rate and uncertainty of technological change vary across sectors, different industrial ecosystems are denoted by distinct forms of co-opetition. Each of these presents elements which show how firms' relationships may have adapted to the extent of innovation interdependence to deliver maximum benefit to their participants (Robertson and

Langlois, 1995). Telecommunications firms setting improved standards via the International Telecommunication Union (ITU), and Intel promoting better designs in the computer industry, are both instances of investment strategies which minimize the overall absolute level of risks of the entire production chain and not just of one part.

In fact, in industries such as telecommunications, network, equipment and protocol design influence the degree of standardization and interconnectivity. Further to that, regulatory decisions can potentially contribute to freeze the industry ecosystem and business models. On the one hand, the presence of path-dependence implies that successful innovation by one or more player(s) is likely to influence future technological choice across the industry. On the other, the existence of direct and indirect network effects provides a great economic incentive for different networks to interconnect. All of the above add considerable complexity to the exploitation of network-based innovations.

While international agreements between telecommunications firms serving different markets have been commonplace, on the other hand, the relationships between players in regulated industries have traditionally been adversarial, as new entrants accessed key incumbents' assets through regulatory supervised processes. Nonetheless, technology-driven changes in industry architecture and regulatory reform have fostered an increase in co-opetition.[5] Cooperation at a highly granular level between hitherto rivals is therefore increasingly acknowledged as necessary for industry-wide innovation and even mandated as part of sectoral regulation. Thus, the changing role of standardization will be analyzed in greater detail in section 13.3.1.

Fransman (2007) analyzes ICT ecosystems across some of the world's leading regions (US, Japan, Europe). He argues that the dominant paradigm in telecommunications regulation conceives the process of innovation as an endogenous but separate process from the competitive dynamics analyzed for the purpose of regulatory decision-making. In general, while the consequences of innovation are frequently taken into account by regulators, a failure to acknowledge how that came about in the first place leaves a key gap in the analysis.

Specifically, the telecommunications industry presents several challenges to welfare-maximizing policy-makers. Static and dynamic economies of scale and scope favor oligopolistic structures over atomistic competition.[6] Moreover, large players' investment decisions can have spillover effects towards future choices (as frequent in network industries), leading to path-dependence. Furthermore, this is compounded by technology interdependence – as discussed throughout this chapter.

In practice, interoperability and complementarity can be pivotal in allowing a technology to succeed. To take a recent example, Gunasekaran and Harmantzis (2008) show that the commercial viability of Wi-Fi could be materially enhanced by the development of what might have been regarded as competing technologies such as WiMAX (Worldwide Interoperability for Microwave Access). Frequently it is not apparent even at the stage of technical development and even of preliminary deployment, which effect will dominate – complementarity or substitution – and whether the descending services will be a net substitute or expander of market demand (Richards, 2007). Thus, regulatory decisions (on spectrum, access, and so on) can affect the path of technological development. It is evident that regulators are well aware of these complexities and some of them have adopted an approach which is intended to be 'technology-neutral'. Section 13.3.2 presents how this policy principle was introduced in European Union (EU) law and discusses its implications.

13.3 BUSINESS AND REGULATORY RESPONSES TO THE INTERDEPENDENT NATURE OF INNOVATION IN COMMUNICATIONS TECHNOLOGIES

13.3.1 The Importance of Risk Mitigation for Operators and Regulators

A key feature of the interdependent risk faced by innovators is the lack of information about how other players will act in response to a specific innovation. Where there is a strong mutual interdependency, this lack of information can escalate the amount of risk involved. This is especially true in telecommunications where interconnection is at the heart of the industry.

The key challenge for telecommunications firms attempting to innovate is how to assess project-specific risk, considering that information about future returns is unknown at the time of the decision to invest. This is essentially the definition used in the real options approach to the analysis of risk.[7] As a consequence, it is natural for players in the telecommunications industry, if possible, to defer investment until the reaction of other players is better known.

A discussion of risk mitigation necessarily falls into the more general debate on the nature of uncertainty and risk, informed by the seminal work of Knight (1921). While a comprehensive discussion of risk and its theories is beyond the scope of this chapter, our study aims (with an eye to the context of regulatory practice) to highlight a dimension of the innovation process – its interdependence – which informs the nature of the uncertainty and the extent of risk associated to it. Because we take a fundamentally qualitative approach, in what follows we will use the term 'risk' in its broadest meaning and this is also in line with the definition of risk which is currently used in regulatory parlance (for example, see Ofcom, 2005a).

This section discusses the effectiveness of standards in reducing interdependency risk between players and the attempt by regulators to set policy which is independent of specific technical innovations. While the former is a long-established approach, the latter is a relatively recent phenomenon.[8] First, section 13.3.2 analyzes how, throughout the long history of telecommunications, technical standardization has been a mechanism aimed at mitigating the investment coordination problem and highlights how shifts in the industry architecture are curtailing the effectiveness of this approach. Second, with the introduction of regulated competition into the industry in the 1980s and 1990s, the reaction of the regulators to innovation has been another unknown for players deciding on innovation investment. In response, EU regulators have therefore attempted to make regulatory policy stand above specific technology innovation by being technology neutral, which is discussed in section 13.3.3.

13.3.2   Response by Firms: the Case for Standardization

Many sectors of the economy present innovation which is associated with some elements of standardization. Even though this is the case, often products or services do not directly impinge on each other to be commercially or functionally viable. For example, car manufacturers may need to meet emission standards and conditions on technical performance but each car can achieve these goals to a large extent as a stand-alone system.[9] The telecommunications industry is characterized by features which mark it apart from many other sectors, even other utility sectors such as energy: the strength of network effects and their global scale create a huge economic incentive for interconnection between the telecommunications networks of different organizations. However, as we have shown in the previous sections, this introduces a considerable extra complexity in the exploitation of an innovation based on the application of new technology.

Firms attempt to mitigate mutual risks through the process of standardization. Historically, fora such as the ITU have played a strong part. This allows many parties to precommit to technical standards which will lessen the risks of failure for networks to interconnect. Transparency and commitment to standards also means that purchasers of services are less likely to be reluctant to contract with suppliers.

Industry evolution created a further motivation for agreeing technical standards. There is now a widespread split between equipment manufacturing and network service provision. Technical standards therefore serve an additional purpose in negotiating this vertical split in the production chain, allowing a reasonably open market between the equipment suppliers and network service providers. This is often termed interoperability between equipments as oppose to interconnectivity between network operators.[10]

The case for standards would seem to be very strong with advantages to both network operators and equipment manufacturers. However, the actual picture is more complex. The economic origins of equipment interoperability are less clear than the origins of network interconnectivity. On the one hand, the split was the product of regulatory and antitrust action in the late 1980s in both the US and in Europe. On the other hand, the broadening scope of the industry was also driven by the technological developments taking place. For example, both Internet Protocol (IP) and Ethernet technologies, on which more of today's telecommunications network depend, emerged from standards bodies which were not at the time associated with the telecommunications industry.

The involvement of many new parties in the value chains associated with convergence in the telecommunications sector has complicated the standards-setting process. There is a proliferation of bodies producing 'standard' specifications which are frequently overlapping. In extreme cases, some specifications are inconsistent or incompatible between different bodies. Thus some industry fora today focus most of their efforts not on writing standards per se, but on profiling standards created by other bodies into interoperability agreements. Strategic forum shopping by industry players complicates matters.

Network operators, in a competitive marketplace, are interested in innovation as a means of competitive advantage. However, as no single service provider (SP) can aspire to own and operate a complete global infrastructure-based network, interconnection is essential. This means that the exclusive exploitation of innovation is often difficult. For network operators, the details of a standard are often not important differentiators between their offers, and they normally have a strong mutual interest in a single standard in order to achieve interconnection and competitive supply of equipment.[11] Today, most network operators regard technical standards as something they 'consume' as part of buying equipment. Several operators have therefore withdrawn from the standards-setting process with an eye on short-term cost savings. This exacerbates conflicts between equipment manufacturers, where instead operators could play a welfare-enhancing role by promoting unique standards. Instead, frequently standards with 'options' are selected as a means of accommodating irreconcilable commercial interests.

On the other hand, for the equipment manufacturers, the technical details of the standard are likely to affect their time to bring innovations to market as well as the market perception of whether they are a 'technical leader' or a 'technical follower'. Equipment manufacturers are certainly interested in innovation as a mean of competitive advantage. For example, in the extreme, all network operators could achieve full interconnectivity by all buying the same equipment from the same manufacturer. In this case, manufacturers would be able to exploit fully the value of their innovations since there would be no need for interworking with other equipment manufacturers. However, operators share a concern to maintain a level of competitive supply and tend to be cautious about exploiting an innovation which locks them into a supply from a single equipment manufacturer. This means that the equipment manufacturers can have ambiguous motivations for open standards. In its turn, these conflicting motivations may, at least in part, explain the fragmentation in standards and the multiplicity of standards bodies.

Regulators are interested in standards as a prerequisite for regulated interconnection. However, even more than network operators, they currently tend to be concerned about the simple presence of standards rather than the technical details and, as a result, many regulators tend not to attempt to follow the detailed technical developments in standards bodies. As an additional note, the unprecedented rate of architectural change in telecommunications networks also poses a strong challenge to standardization efforts. In the past, technological innovation has been confinable to a reasonably static overall architectural framework. For instance, when digital technologies made electromechanical exchanges and analogue transmission systems obsolete, the overall architectural structure of 'access', 'switching' and 'transmission' was unaffected. Also, the basic services running across the network were not affected by this innovation, thus leaving the pricing and the industry business model unchanged.[12] Conversely, in the present environment, the nature of basic services, the pricing of these services – as well as the industry business model – are all changing under the pressure of technological developments. It is consequently hard for the standards bodies themselves to keep their own organizations and membership in step with the pace of the architectural change in the broader industry.

To conclude, we have shown that the real picture of standards is more complex and more ambiguous than a simple investment coordination device. The standards-setting process is certainly alive and well; however, its effectiveness is more nuanced.

13.3.3   Response by Policy-Makers: Technology-Neutral Regulation in the EU

The principle of technology-neutral regulation reflects a trend in the policy-makers' stance on the role of regulation in the face of fast and hard-to-predict technological change. The concept started being discussed in EU circles around a decade ago. Currently, it is a high-level principle enshrined in the 2002 Electronic Communication Network Services (ECNS) framework.[13] It has been instrumental to the achievement of the objective to strengthen the EU internal market, by aiming to eliminate some barriers that, in the face of convergence, prevented a comprehensively similar treatment of communications operators throughout the EU Member States. Nonetheless, technology neutrality is continuously traded off with other high-level principles associated to both the ECNS framework and other EU policies (for example, the i2010 industrial policy initiative in European Commission, 2005). While this principle has widespread recognition industry-wide,[14] countries like Korea have shown that a policy choice of a different trade-off between regulatory, competition and industrial policy can be fruitful.[15]

Policy-makers, understanding of this principle has kept evolving and gaining nuances while communications industry players gained a more sophisticated understanding of its technology dynamics, taking on board the history lesson that technological progress can take unexpected paths. This disrupts firms' large-scale plans as well as affecting even the best-meaning regulatory efforts. For instance, integrated services digital network (ISDN) technology, much hyped in the late 1980s–1990s, was expected to be a neutral technology since its network would have been multipurpose and carried a variety of services. This would have translated into a reduced need for regulation (see the 1987 EC telecommunications Green Paper, pp. 21, 33).[16] Ten years on, it had become more evident that a single technological development (in the form of ISDN) could not live up to its promise to radically reshape consumers' demand and regulation. This contributed to awareness amongst policy-makers that a technology-neutral regulatory framework was needed, rather than just a neutral technology (see the 1997 EC convergence Green Paper, pp. 19–21). This resulted above all in the harmonization efforts in the post-1998 framework review.

We see technologically neutral regulation as having two distinct fundamental connotations. On the one hand, technology neutrality is a feature of the regulatory toolbox. It allows regulatory tools to adapt to a converging environment and be applicable for general regulatory purposes. It reflects regulators' aim to build and deploy the analytical tools needed for future-proof regulation, given that technology and markets present underlying risks. This is the micro side of the principle. On the other hand, taking a macro view of the economy, the principle is also leveraged to justify a *pilatesque* stance on regulatory-mediated technological choice. This is the case the more regulatory decisions enter the field of industrial policy. The principle can thus be seen as a facet of a non-interventionist industrial policy.

This puzzle reflects the underlying conundrum regulators face between static and dynamic efficiency, which at times need to be traded off, and this tension is also evident in the ECNS framework.[17] The principle of technology-neutral regulation reflects regulators' desire to achieve both static (level of investments) and dynamic (innovation) efficiency. Technology-neutral regulation is relatively unproblematic where the state and outlook of markets are unequivocal. It matters most where risk adds to the complexity and may require less orthodox analytical approaches.[18] Regulatory oversight shapes and is shaped by the technological profile of industries.[19] This is because in regulated industries, investment decisions are informed by the national regulatory authority (NRA) approach to risk. Both the NRA's current orientation and its past legacy of access policies will be relevant, a form of regulatory path-dependence.

The current EU regulatory framework faces the challenge of overcoming its underappreciation of the relevance of new services and markets, driven by technological progress.[20] Technology neutrality is an ameliorative approach to regulation insofar as the acknowledgment of path-dependence makes regulators assess that technological choices do matter and their dynamic effects yield more complexity to policy decisions. On the other hand, if in practice the technology neutrality principle leads regulation to hinder firms' welfare-enhancing attempts to mitigate the interdependent risks of technological change, then the application of this principle should be constrained by the greater industrial and technology development societal imperative.

Across the world, institutions and policy-makers draw several lines between competition and regulatory policy on one side and technology and industrial policy on the other in different ways, which impacts the competitiveness of their economies. Technology-neutral regulatory policy stands on one of those boundaries. Whether regulators focus on neutrality to technological competition (within the market) or technological development (for new markets) will depend on what view they will take of Schumpeterian competition. In practice it will depend as well on what tools will be available for regulators to analyze and address market failure in such a context, as will be discussed in the following sections.

In conclusion, the analysis of the origin and rationale of the concept of technology neutral regulation suggests a series of points. First, the

importance given to technology neutrality as one of the high-level principles of the EU communications regulatory framework shows that
policy-makers are keenly aware that technological development in the
communications space is not a one-player game: if it were so, straightforward regulatory measures could ensure the achievement of policy objectives. Second, because of its open-endedness, the concept of technology
neutrality risks being used to mask the ambiguity related to the conflicting nature of regulatory related remits. This is because regulators can at
times de facto perform industrial policy functions, although the regulatory
rhetoric can underplay these for political convenience. Third, successful
regulation thrives on a level of independence from government, as well as
to be accountable to its stakeholders (Majone, 1996); thus technology and
industrial policy is an uncomfortable territory for regulators to tread in.

These considerations may lead us to discard technology neutrality as
insubstantial insofar as it cannot consistently be reflected in regulatory
practice across the board (while specific regulatory measures can be technology neutral, for example, spectrum allocation). Nonetheless, we argue
that a healthy debate on this principle can have a beneficial effect on societal welfare if it can help reassess the future role of regulation and industrial policy and the division of labor within the public sector in its task to
serve societal interests, which at times are conflicting.

13.4 RISK AND INNOVATION IN NETWORK TECHNOLOGIES

13.4.1 A Simple Model of Risk Bearing in Network Innovations

In section 13.3, we saw that while network operators and equipment
manufacturers undertake standardization efforts and regulators attempt
policies which are technology neutral, neither are perfect instruments.
Innovation normally involves several aspects of risk. This could arise
from any number of things which may be unknown at the time a decision
is made to invest in an innovation, such as technology risk, deployment
risk and market risk,[21] where consumers do not purchase the service in
anticipated volumes. Moreover, a single organization – such as a service
provider (SP) – is seldom able to internalize fully the risks of successfully
developing and deploying a technological innovation, independently from
other parties.

In this section we analyze several historical technological innovations in
telecommunications networks and then apply the analysis to a problematic
recent example of the application of regulation to a telecommunications

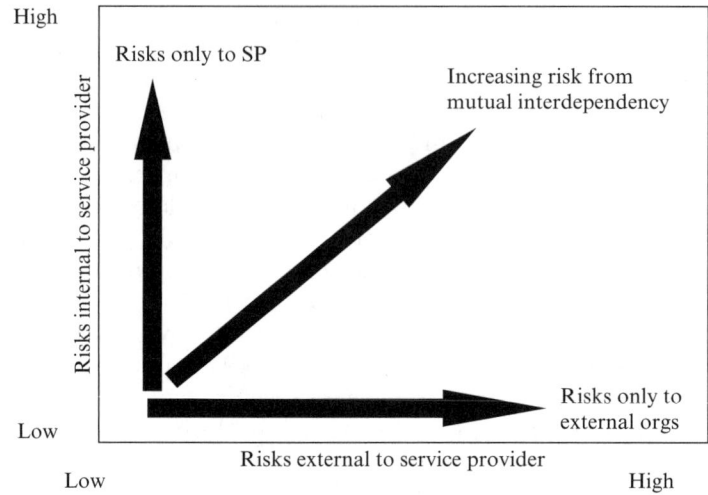

Figure 13.1   Basic model of interdependency risk

service – Ethernet services. Our simple model focuses on the perspective of a single SP, the business of which is influenced by the innovation discussed. Our model separates and measures innovation risk along two components. The first component reflects the degree of risk occurring inside the SP domain, therefore under its control (the vertical axis in Figure 13.1). The second component (horizontal axis) measures risks which fall outside the domain of our reference SP.[22] These outside risks arise from relationships of our reference SP (the innovator) to other organizations which may be at a different level in the value chain or alternatively direct competitors at the same point in the value chain. Where this component is high, our SP requires a great extent of coordination with those external organizations in order to gain benefits from the innovation.

Figure 13.1 represents a set of technological innovations introduced in the past in the communications industry. Technologies that are positioned close to one of the axes are those where an SP could develop or deploy the innovation independently or gain benefit at arm's-length from an externally developed or deployed innovation. In both cases our reference SP is able to internalize fully all risks associated to the change in network technology. Instead, the more the innovation is depicted away from the axes, the higher will the level of mutually dependent risk be. Notice the three scenarios associated to the nature of risk, which we will later argue are relevant to regulatory assessments.

Given the telecommunications industry features of network externalities and path-dependence, our simple model allows us to take into account that an SP which has market power may be able to innovate in a way which is not risky for itself but imposes risks on others. We can then discuss whether such a firm may have a strategic interest to do so, which carries both business and public policy implications (depending on the context). In fact, as the following examples show, there is a huge variety of circumstances in which the effects of innovation can assist third parties, sometimes almost inadvertently.

In the 1980s, the whole telecommunications industry underwent a huge technological change with the introduction of digital technology (digitalization). However, this did not change the basic public switched telephone network (PSTN) service used by most end-users, nor was it necessary for all service providers to change technology in concert. As a result, any one service provider could develop and deploy digital switching within their own control, independently of any changes undertaken by their customers. While this technical innovation carried very large technology risk, there was little mutual dependence risk. As a result we place this innovation high but near to the y axis.

Conversely, the dramatic rise of the Internet and its applications such as the World Wide Web (WWW) and e-mail in the 1990s grew as independent users of telecommunications network outside the domain of the telecommunications service providers. While the Internet has had a dramatic impact on end-users who have bought personal computers (PCs) in order to access it, the telecommunications service providers simply provided their existing services – leased lines and PSTN calls for modem dial access. Again there was little mutual dependence risk and we place this innovation on the right side but close to the x axis.

However, the introduction of mobile telephony in the late 1980s required a completely new network infrastructure as well as a completely new set of end-user equipment. The end-user equipment players and the network operator players had a strong mutual dependency. Digital mobile saw a major standardization initiative in Europe – Global System for Mobile Communications, GSM – which was ultimately highly successful for the players involved. With the high level of mutual dependency, we place this towards the top right-hand corner.

At about the same time, an even bigger standardization project was also being undertaken – Broadband ISDN (B-ISDN[23]). This too required mutual coordinated innovation by both the network operators and the end-user equipment industry. This too we place towards the top right-hand corner. However, this project essentially failed. Instead the end-user equipment (PCs) chose to follow the Internet standards, not the B-ISDN standards, a critical reason being that the end user could access Internet service independently of a network operator's investment in innovation. In essence, the availability of an alternative without mutual dependency risk killed B-ISDN.

Optical transmission developed rapidly in the 1990s and is an interesting case of 'outside' risk, although not to the same extent as for mobile and B-ISDN. Initially, the development and deployment of the technology was entirely 'inside' a service provider's domain and so would have sat firmly on the left-hand axis. However, the nature of optical transmission costs has radically changed the cost structure of providing telecommunications services. This enabled changes in the pricing of telecoms services, which in its turn facilitated market expansion and competition in the provision of services.[24] Moreover, the underlying change in cost structures affected optimal network topology design, which reshaped interconnection costs, thus destabilizing business models and regulatory regimes. Not all industry players reacted fast to such spillover from technology, due to a certain degree of business model and regulatory path-dependence. In many instances, while the cost structures had changed with optical transmission, tariffs and regulation failed to adapt to these changes, leading to suboptimal investments by third parties.[25] In our figure, since the optimal network design for our reference SP depended on demand from interconnected parties (thus on their network design and pricing), the innovation of optical transmission is represented as distanced from the vertical axis.

Furthermore, another major innovation of the 1990s in telecommunications was the Intelligent Network (IN). This allows the flexible creation of services such as the 800 service, premium rate, and Ring Back When Free. While there is a certain amount of mutual dependency between the network operator and its customers for these technology services, they were developed with little overall risk to either the service provider or the end-user. It is therefore located close to the axes of origin.

Finally, the recent example of asymmetric digital subscriber line (ADSL) technology for multimedia broadband in the residential market gives a particular insight. It shows that the staged exploitation of a technological innovation can reduce overall risk (see Figure 13.2). In fact, ADSL-based service requires investment from both the SP and end-users in the technology, leading to some mutual dependence. However, ADSL evolved as a small technology increment from a previous version of the technology, high bit rate DSL (HDSL). This was developed as a means of reducing the costs of delivering leased line products and was deployed wholly 'inside' the SP network without making any change to those products. This deployment absorbed the great majority of the technology risk, and especially of technology development which ADSL was able subsequently to leverage. The kind of dynamic shown by this technology is not unique

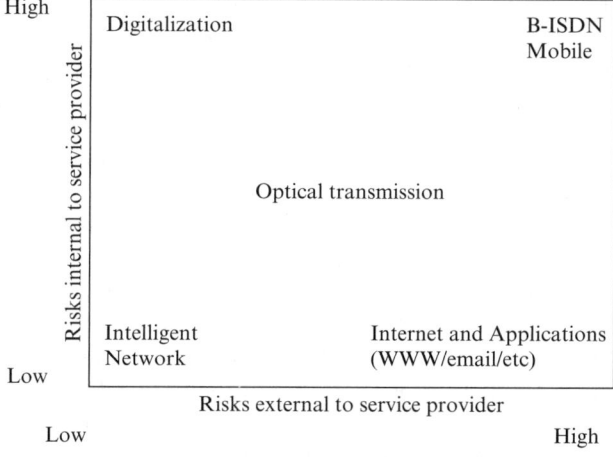

Figure 13.2   Positioning of historic examples

to ADSL and will in fact be part of the development of Ethernet services. In its turn, we highlight that this dynamic bears implications on regulatory practice, which, as the Ethernet case below shows, has at times yielded conflicting outcomes to similar regulatory problems.

13.4.2   The case of Ethernet: the Nature of Innovation and Regulatory Practice

After having discussed several historic telecommunications innovations, we now turn to the analysis of a more recent case. Ethernet services have been the subject of conflicting regulatory analysis and greatly differing regulatory intervention. Some NRAs have deemed Ethernet services to be a leased line and regulated them accordingly, whilst other have found them to be in an 'emerging market' and therefore they have withheld from any regulatory intervention.[26] How could such large discrepancy arise?

We will show how the method just presented can be put to use to test whether markets and services associated to an evolving technology require special regulatory provisions, such as an emerging market connotation. We propose that network innovation can lead to an emerging market only where the risk of technological change is borne interdependently by more than one player.

Ethernet, originally a local area network (LAN) technology, has had a dramatic impact on the way in which data centers are built.[27] It is increasingly the pre-eminent means by which data equipment is interfaced and devices are interconnected (it is now the norm for IT – information technology – equipment to be supplied with an Ethernet socket). We will show that the concept of Ethernet holds different meanings. We will thus use our model to analyze this pervasive technology and then draw policy implications from this instructive case.

We need to keep in mind that Ethernet services (our focus) are a related though distinct concept from Ethernet equipment. The latter follows the Ethernet standards and interfaces and constitutes the switching equipment in Ethernet networks. It is a very attractive technology for service providers as it has many useful technical features and is highly cost-effective. However, somewhat ironically and counter-intuitively, current Ethernet standards imply that Carrier Scale Ethernet services cannot be built upon Ethernet equipment. Ethernet services are supplied by SPs which can connect together the customer's equipment using Ethernet interfaces. As a result, Ethernet equipment and services have been and are still independent and mutually exclusive.

The development and deployment of Ethernet equipment initially had no impact on telecom SPs who continued to supply leased lines and PSTN calls with dial access. However, SPs also spotted a low-risk market opportunity: replace the traditional interfaces on the ends of their leased line services with Ethernet interfaces, thus offering Ethernet Leased Lines as Ethernet Point-to-Point Services (point 1b in Figure 13.2). This change did not result in any major technology shift for the SPs; the leased lines continued to be provided by the same existing underlying network.[28] However, the customer now had a service which was able to directly interface with their own IT and data equipment. The only physical and practical difference between the traditional wholesale leased line service and the Ethernet leased line service comes from deploying Ethernet 'boxes' in place of their equivalents (for example, Synchronous Digital Hierarchy, SDH) at the customer site and at the point of interconnection between the incumbent operator network and the entrant's network.

Completely independently, SPs have also deployed Ethernet equipment within their networks in order to interconnect equipment in a similar way to end-users (SP LAN: point 2 in Figure 13.2). This deployment had no impact 'outside' the sector and has been of limited technology risk to the SPs themselves.[29] However, the outlook for Ethernet technology suggests that there may be room for more innovation. Ethernet equipment is a highly cost-effective technology and there are standards developments under the banner of Carrier Ethernet[30] which can make this technology useable by network operator for Ethernet services. Any decision to deploy Carrier Ethernet is an internal-only decision with any mutual dependency risk (point 3 in Figure 13.2).

As Figure 13.2 depicts, neither present-day Ethernet applications (equipment or services) nor future developments present the condition of interdependent risk which we argue needs to underlie an assessment of emerging market. Hence, while (inadvertent) spillover effects affect how the industry can benefit from such technologies, the nature of the innovation is not such as to require particular regulatory oversight.

Regulation, given its *ex ante* nature, cannot rely on retrospective observation to determine whether innovation constitutes an emerging market (with the connotation of light-or 'lighter'-touch regulation associated to this). The introduction of Ethernet presentation leased line services (in 2005–07 for most European markets) is a clear example of this challenge. In several countries, incumbents started offering Ethernet services to their retail customers, although usually without wholesale equivalents.[31] NRAs across Europe reacted differently.

Above in this chapter, we have tested our simple inside–outside model in order to analyze risk in Ethernet-related technological change. Ethernet has been used as a local area networking (LAN) technology since the 1970s and it is now the predominant technology in this area. In fact, an SP introducing Ethernet services for leased lines is not exposed to interdependent risk.[32]

In another context, network technologists are currently working on extending Ethernet technology – Carrier Ethernet – for more direct use within the network operator's network. Whether this is successful, remains to be seen; however, the risks are internal to the network operator. Our simple model helps us shed light on what could be a more robust approach to regulating markets affected by technologies such as Ethernet. In the first context, Ethernet should be regarded as part of generic technical progress (which ironically emanated from outside the telecoms sector altogether) and which had little or no deployment risk to telecom operators.[33] In this case, the innovation would fall into the category of service enhancement and network upgrade (ERG, 2006). Thus no remedy arising from Significant Market Power (SMP) would have to be materially affected.

This example suggests that some general principles of market analysis need to be developed in order to disentangle what are highly innovative and risky activities due to the interdependent nature of risks (see Scenario C in section 13.5.2 below) from those where development or deployment are exclusively inside or outside the domain of any SP (Scenario B). Our model can thus be used to facilitate a common approach when similar situations arise in the future.

In the case of Ethernet, a superficial analysis would suggest that this technology can simultaneously present elements of risk both internal and external to the action of a service provider wishing to introduce it. This would support a classification within an emerging market, thus hard to grapple with by regulation. Instead, this section demonstrates that the evolutionary path of Ethernet technology shows that such network innovation presents risk which is either exclusively internal or external to the action of SP. As a consequence this dispels the notion that the nature and levels of risk warrant classification of Ethernet Point-to-Point services as an emerging market, while suggesting that these rest well within existing regulated markets (see Figure 13.2).

13.5   CONCLUDING REMARKS: INNOVATION INTERDEPENDENCE AND MARKET REVIEWS

13.5.1   The Current Framework

It is commonly accepted that regulatory frameworks emphasizing the application of the economic tools of competition law are suited to a mature environment of stable outputs, costs and prices and indeed stable industry structures. When, as is frequently the case, technological progress and potentially disruptive technologies lead to economic change and disequilibrium, these tools suffer from the lack of extensive information available (Richards, 2006, 2007). In these circumstances, qualitative assessments, like the simple model we have introduced, can become more important, although they are limited by their judgmental nature.

European regulatory agencies are well aware of this dilemma and while reviewing the application of remedies to established markets – as informed by the European Commission (EC)'s guidance) – they have discussed the issue of emerging markets (European Regulators Group, ERG, 2006). As the ERG points out, a combined reading of the EU Framework Directive and the European Commission's SMP guidelines suggests that regulation in emerging markets should not be 'inappropriate', hinting that emerging markets could be handled under competition law.[34] In particular, the Commission envisages a test to determine the emergent market status as a 'dynamic' application of the Hypothetical Monopolist Test (HMT) on both the demand and supply sides. Finally, the Commission argues that the three-criteria test (TCT),[35] which EU NRAs need to apply to define markets other than those specified by the Commission, is the ultimate test of whether a market is emerging or not. The ERG distances itself from these assertions and policy recommendations, arguing that it is not possible to apply the TCT under circumstances of immature demand and supply. The ERG thus advises NRAs to engage in 'close monitoring' of situations where that is the case and stresses that a simple 'network upgrade' is insufficient to affect the outcome of the TCT (ERG, 2006, p. 20).

While there are indeed good grounds for agreeing with the ERG that the HMT and the TCT are largely superfluous in a situation of dynamic change and uncertainty, by the same token it is not evident what 'close monitoring' defines. That regulation should not according to the Commission be 'inappropriate' is unhelpful to gain any regulatory certainty which may in practice help firms in devising their conduct. From this guidance, NRAs are still largely left in a policy vacuum if emerging markets can neither be defined in principle nor be identified in practice using the established methodologies.

What we wish to emphasize is that similar regulatory discussions are focused on innovation which is implicitly at the level of the single firm, unlike much of network innovation. The latter, as we have highlighted throughout our analysis, relies on multiple parties to upgrade simultaneously or at least to move broadly in compatible steps of an evolutionary path. For instance, barriers to entry may not fall just on one party but involve the activities of several players, each with its incentives to innovate. This contrasts with the legacy of a focus on short-term assessments of SMP on incumbent telecom operators which still informs the EU regulatory regime to a certain extent. We suggest that the key challenge for regulatory policy- and lawmakers is to devise a framework which is able to handle the strategic interplay of many parties over a much longer time period.

13.5.2   Recommendations for Regulatory Policy

A study in a similar vein to our work proposes a taxonomy for emerging markets in the telecommunications industry which centers on the underlying uncertainty about the evolution of future demand (Crocioni, 2008).[36] Our simple model complements this approach by suggesting a way to break down the risk component in technological development which is also focused on the supply side and which could be a first step in regulatory assessments. We argue that there is considerable merit in operating the regulatory framework in such a way that it signals that different scenarios – based on the nature of innovation risk – will indeed be treated differently in regulatory terms. Based on Figure 13.1:

• Scenario A (bottom left quadrant): both the internal and external components of risk are low. This is the case for existing but mostly fully developed service (already largely with sunk investment costs). Traditional quantitative methods can assess the scope for appropriate intervention such as access regulation, quality of service obligations, and so on.

- Scenario B (top left and bottom right): only one of the risk components is high. Here the decision as to whether and how to regulate an 'emerging service' for which the technological characteristics are well understood needs only to take special notice of whether the pace of roll-out may be affected by regulation. Similarly with respect to possible further process innovation (insofar as investment costs are still to be sunk).
- Scenario C (top right): highly interdependent risk. Any regulatory intervention cannot embody any meaningful stance toward investment in speculative and innovative new product or service research and development (R&D) and subsequent deployment. Pre-emptive regulatory forbearance can be a valid policy, subject to regular reassessment informed by strategic analyses of technological change.

In all of the scenarios, the degree of complementarity in innovation will have a major bearing on how risks are shared between different parties and the nature of those risks. The appropriate regulatory responses could be quite different depending upon the particular circumstances. Lacking this, there will be a disincentive to commit to invest in innovatory activities and infrastructures.

Analyses informed by this preliminary step could tackle more consistently the following issues: whether to impose *ex ante* regulation and the assessments of potential foreclosure; the relationship to any policy to promote innovation as part of technology or industrial policies; and the promotion and regulation of standardization and interoperability policies. In general, our simple contribution could be a useful first step to tackle the public policy challenge of ascertaining *ex ante* whether market power is likely to arise and which party or parties may hold it, in a scenario where innovative efforts are highly interdependent across organizations in the communications industry.

NOTES

1. The views expressed in this chapter reflect those of the authors alone and not necessarily those of the organizations to which they are affiliated.
2. In its turn, the discussion on the impact of competition on innovation is still open-ended. For instance, while some cite Arrow (1962) in stressing that perfectly competitive markets augment the incentives to innovate, others are more sympathetic to Schumpeter (1939, 1942), who in his later years highlighted the role of large corporate players on investing in wide-ranging innovation (the so-called 'Mark II' theory).

3. Capron Noll (1970) note: 'Many services normally regarded as natural monopolies have remained natural monopolies only because of the interaction between regulation and technological change.'
4. In the EU, national regulatory authorities (NRAs) refer to guidelines issued by the European Commission's Directorate General for Competition (European Commission, 2002) while performing the task of market analysis.
5. In the UK, Ofcom promoted the transition from the Network Interoperability Consultative Committee (NICC) technical standards committee (focussed on the technical terms of BT compliance) to the Next Generation Networks UK (NGNUK) forum, aiming to facilitate the advent of next generation networks (NGNs). NGNUK is 'a co-ordination forum in which key investors in NGN infrastructure and services will discuss research, consider and, where possible, agree the direction for NGNs in the UK' (www.ngnuk.org.uk).
6. This is due not only to the nature of distribution network (infrastructure) but also to access to content. In general, both direct and indirect network effects affect the provision of communications services. See, for instance, Newberry (1999).
7. A real option can be associated to the managerial ability to respond to new information as it emerges; for example, to defer, to mothball or to exploit an unanticipated opportunity, and so on (Trigeorgis, 1995; Schwartz and Trigeorgis, 2004). Real options are not the only framework to assess risk. An alternative is the analytic approach of the capital asset pricing model (CAPM) which – due to its focus on the overall set of actions of a firm – may be not best placed to give appropriate insight into the risks associated with a specific decision to invest in a specific innovation.
8. The ITU is one of the oldest bodies in the United Nations system, having been founded in 1865.
9. Even in the automotive industry, some of the most radical innovations may require a systemic approach. For instance, hybrid vehicles need a distribution system which supports electricity alongside petrol.
10. Note that standards are closely related to the transactions costs associated with asset specificity. Often an outside supplier is only able to supply inputs to an adequate quality specification if it makes 'buyer-specific' investment. Theoretically, this is not a problem if the investment has zero sunk cost (if there are efficient second-hand markets for the equipment, and there are no significant transaction costs associated with reselling). However, in practice, most investment in telecoms networks involves a significant positive sunk cost, in that future resale involves the firm losing considerable value. With positive sunk cost, suppliers are only likely to be prepared to make the buyer-specific investment if the buyer offers a long-term contract which guarantees that the supplier can get an adequate return on the investment. However, even long-term contracting may not solve the 'problem'. The point is that, once the supplier has sunk the investment, this gives the buyer a stronger bargaining position, and an incentive to try to renegotiate more favourable terms. For a classic industrial organization study on this matter see Joskow (1987).
11. The precise technical specification of some services can however be critical for certain applications and customer groups.
12. This is different from the case of optical transmission innovation, which will be analysed in the following sections.
13. 'The requirement for Member States to ensure that national regulatory authorities take the utmost account of the desirability of making regulation technologically neutral, that is to say that it *neither imposes nor discriminates in favour of the use of a particular type of technology*, does not preclude the taking of proportionate steps to promote certain specific services where this is justified, for example digital television as a means for increasing spectrum efficiency' (European Parliament, 2002; Framework Directive: Recital 18; emphasis added).
14. 'The debate is not so much about whether to embrace technology neutrality as how much, and how soon. "Very few people disagree with the general concept of neutrality,"

says Mr Webb. It is, he says, a "motherhood and apple pie sort of thing" that nobody now wants to admit opposing', *The Economist* (2005). Professor William Webb is Ofcom's Head of R&D.

15. Studies of governmental interventions in Korea provide for several striking comparisons with the EU landscape. See Kim et al. (2008), Jho (2007) and Picot and Wernick (2007).

16. The 1987 Green Paper analysed and assessed technology trends in order to devise the first EC-wide telecommunications regulatory framework (European Commission, 1987).

17. This is evident from the very start of Article 8 of the Framework Directive, which sets out policy objectives and regulatory principles: 'Member States shall ensure that in carrying out the regulatory tasks specified in this Directive and the Specific Directives, in particular those designed to ensure effective competition, national regulatory authorities take the utmost account of the desirability of making regulations technologically neutral . . . The national regulatory authorities shall promote competition in the provision of electronic communications networks, electronic communications services and associated facilities and services by inter alia: . . . (b) ensuring that there is no distortion or restriction of competition in the electronic communications sector; (c) encouraging efficient investment in infrastructure, and promoting innovation' (European Parliament, 2002; Framework Directive: Art. 8.1, 8.2)

18. The European Commission acknowledged this long ago: 'Regulators and market players alike face uncertainty as they look towards the future convergent environment. Regulators will need to have very clear objectives, including those of public interest, and a set of general purpose regulatory "tools"' (European Commission, 1999, p. iv).

19. Capron and Noll (1970) note: 'Many services normally regarded as natural monopolies have remained natural monopolies only because of the interaction between regulation and technological change.'

20. For instance, no direct references to the technology dimension or to innovation can be found among the policy objectives in the pivotal 1999 Communications Review (European Commission, 1999, p. iv).

21. For a detailed discussion on different types of risk (in the context of next generation networks) see Ofcom (2005a, Section 9), where the nature and application of real option theories to regulation are debated.

22. For the sake of simplicity we are aggregating all external risks in the domain of several organizations other than our reference SP. This allows us a two-dimensional representation; the model could be extended so as to include several axes which would lead to a more nuanced outcome, although not altering the key findings.

23. Broadband Integrated Services Digital Network was the major programme of the ITU-T in the late 1980s and early 1990s.

24. The wide-ranging economic and societal consequences of such trends have been widely debated and are usually associated to the concept of 'the death of distance' (Cairncross, 1997).

25. Pre-optical transmission systems required expensive electronics every few kilometers to regenerate the signals. Thus, the overall network costs were mostly a function of distance and bandwidth. However, optical transmission reduced the need for regeneration equipment, previously often a dominant network cost. Many SPs did not change their price structures and still charged for services by distance and bandwidth, focusing on an 'inside only' view of the impact of this innovation. As a result customers still sought to optimize their demand to minimize the distance and bandwidth they needed to purchase and demand was predominantly for short, 'thin' connections. Networks thus configured had widely dispersed nodal electronics. Thus incremental cost analyses (such as those needed for price regulation) of such networks appeared still to show a strong dependence on distance and bandwidth. This was now arising from the nodal electronics demanded rather than the technological need for signal regeneration. This evolutionary path occurred especially in countries (or areas) with weak competition and

cost-oriented regulation based on historic network architectures. Where competition was unfettered and fiercer, a forward-looking approach to pricing resulted in demand for longer, 'fatter' connections, aggregating traffic onto fewer routes and generating a much more consolidated network structure. This evolution required a higher degree of industry awareness and coordination in order to address the element of mutually dependent risk for all parties. By doing so, industry players (firms, regulators) could overcome the prisoner's dilemma nature of the situation and achieve a Pareto-superior outcome.

26. Some regulators have wavered in their policy on whether to include Ethernet amongst the leased lines market (which is subject to regulation). For instance the Spanish NRA, which had initially set Ethernet services apart from the regulated leased lines market, reversed its decision (CMT, 2006). A discussion of variations in regulatory stances towards Ethernet and other telecommunication services is in BT (2007, Annex 1).

27. In the early days of initial deployment, however, Ethernet competed with Token Ring and other LAN technologies (more recently with Fiber Distributed Data Interface – FDDI). For storage area networks (SANs) applications, Ethernet currently competes with Fibre Channel technology: it is yet unclear whether Ethernet will successfully supersede it in this field.

28. This could be for instance synchronous optical networking/synchronous digital hierarchy (SONET/SDH), WDM, asymmetric transfer mode (ATM) or direct fibre. More recently many SPs have exploited existing Multiprotocol Label Switching (MPLS) technology to offer Ethernet services.

29. There is however a commercial risk in that equipment may become redundant or lose its value more quickly than anticipated. In other words, old assets have to compete against new assets if the former are to remain in the market at all.

30. These in continuity fault management (CFM), provider backbone bridging (PBB), PBB – traffic engineering (PBB-TE), and shortest path bridging (SPB).

31. This was not the case in the UK.

32. This is because this technology uses the standardized equipment and interfaces of LANs, but the 'traffic' is actually still carried over the existing telecommunications SDH/ATM networks.

33. There could of course be potential financial risk from devalued assets.

34. This is also stated in the EC Wanadoo Interactive decision (European Commission, 2003).

35. The three criteria are: high and non-transitory entry barriers; tendency towards effective competition; and insufficiency of competition law remedies. These are stated in Article 2 of the Commission recommendation on relevant product and service markets (European Commission, 2007).

36. Crocioni (2008) presents a model where uncertainty is augmented where: innovation and investment incentives are particularly important; or network effects (or switching costs) are relevant; or both of the above interact, adding complexity.

REFERENCES

Afuah, A. (2000). How much do your co-opetitors' capabilities matter in the face of technological change? *Strategic Management Journal, Special Issue: Strategic Networks*, **21**(3), pp. 387–404.

Arrow, K.J. (1962). *Economic Welfare and the Allocation of Resources for Invention. The Rate and Direction of Inventive Activity: Economic and Social Factors*. Princeton, NJ: Princeton University Press.

Baldwin, Carliss Y. and Kim B. Clark (2000). *Design Rules: The Power of Modularity*. Cambridge, MA: MIT Press.

Brandenburger, A.M. and B.J. Nalebuff (1996). *Co-opetition*. New York: Doubleday.

British Telecom (BT) (2007). The application of proportionate regulation in the electronic communications sector. In P. Richards and A. Tarrant (eds), *The economic benefits from providing businesses with competitive electronic communications Services*. Retrieved from http://www.btplc.com/Thegroup/RegulatoryandPublicaffairs/Consultativeresponses/BTdiscussionpapers/index.htm.

Cairncross, F. (1997). *The Death of Distance. How the Communications Revolution will Change our Lives*. Cambridge, MA: Harvard Business School Press.

Capron, W.M. and R.G. Noll (1970). Summary and conclusion. *Technological Change in Regulated Industries*. Washington, DC, Brookings Institution, pp. 197–238.

Commission del Mercado de las Telecomunicaciones (CMT) (2006). Resolucion por la que se aprueba la definición y análisis de los mercados de segmentos de terminación de líneas arrendadas al por mayor y segmentos troncales de líneas arrendadas al por mayor, la designación de los operadores con poder significativo de mercado y la imposición de obligaciones específicas. Barcelona.

Crocioni, P. (2008). Leveraging of market power in emerging markets: a review of cases, literature, and a suggested framework. *Journal of Competition Law and Economics*, **4**(2), 449–534.

Dobbs, I. and P. Richards (2004). Innovation and the new regulatory framework for electronic communications in the EU. *European Competition Law Review*, **25**, 716–30.

The Economist (2005). Spectrum of opinion. **375**(8428), 66.

European Commission (EC) (1987). Towards a dynamic European economy: Green Paper on the development of the common market for telecommunications services and equipment. Brussels.

European Commission (EC) (1997). Towards an information society approach: Green Paper on the convergence of the telecommunications, media and information technology sectors, and the implications for regulation. Brussels.

European Commission (EC) (1999). The 1999 Communications Review. Towards a new framework for Electronic Communications infrastructure and associated services. Communication from the Commission to the European Parliament, the Council, the Economic and Social Committee and the Committee of the Regions. Brussels.

European Commission (EC) (2002). Commission guidelines on market analysis and the assessment of significant market power under the community regulatory framework for electronic communications networks and services. Brussels.

European Commission (EC) (2003). Commission decision of 16 July 2003 relating to a proceeding under Article 82 of the EC Treaty (COMP/38.233 – Wanadoo Interactive). Brussels.

European Commission (EC) (2005). A European information society for growth and employment. Communication from the Commission to the European Parliament, the Council, the Economic and Social Committee and the Committee of the Regions. Brussels.

European Commission (EC) (2007). Commission recommendation on relevant product and service markets within the electronic communications sector susceptible to ex ante regulation in accordance with Directive 2002/21/EC of the European Parliament and of the Council on a common Regulatory Framework for electronic communications networks and services, 2007/879/EC, 28.12.2007. Brussels.

European Parliament (2002). Directive 2002/21/EC of the European Parliament and of the Council of 7 March 2002 on a common regulatory framework for electronic communications networks and services ('Framework Directive'). Brussels.

European Regulators Group (ERG) (2006). Revised ERG Common Position on the approach to Appropriate remedies in the ECNS regulatory framework. Final Version May 2006. ERG 06 33.

Farrell, J. (2003). Integration and independent innovation on a network. *American Economic Review*, **93**(2), 420–24.

Farrell, J. and P.J. Weiser (2003). Modularity, vertical integration, and open access policies: towards a convergence of antitrust and regulation in the Internet age. *Harvard Journal of Law and Technology*, **17**(1), 85–134.

Fransman, M. (2007). *The New ICT Ecosystem. Implications for Europe*. Edinburgh: Kokoro.

Gawer, A. and R. Henderson (2007). Platform owner entry and innovation in complementary markets: evidence from Intel. *Journal of Economics and Management Strategy*, **16**(1), 1–34.

Gunasekaran, V. and F.C. Harmantzis (2008). Towards a Wi-Fi ecosystem: technology integration and emerging service models. *Telecommunications Policy*, **32**(3–4): 163–81.

Henderson, Rebecca M. and Kim B. Clark (1990). Architectural innovation: the reconfiguration of existing product technologies and the failure of established firms. *Administrative Science* Quarterly, **35**(1), 9–30.

Jho, W. (2007). Global political economy of technology standardization: a case of the Korean mobile telecommunications market. *Telecommunications Policy*, **31**(2), 124–38.

Joskow, P.L. (1987). Contract duration and relationship-specific investments: empirical evidence from coal markets. *American Economic Association*, **77**, 168–85.

Kim, Y., H. Jeon and S. Bae (2008). Innovation patterns and policy implications of Adsl penetration in Korea: a case study. *Telecommunications Policy*, **32**, 307–25.

Knight, F.H. (1921). *Risk, Uncertainty and Profit*. New York: Houghton Mifflin.

Langlois, N. (2002). Modularity in technology and organization. *Journal of Economic Behavior and Organization*, **49**(1), 19–37.

Majone, G. (1996). *Regulating Europe*. London: Routledge.

Newberry, D. (1999). *Privatization, Restructuring and Regulation of Network Utilities*. Cambridge, MA: MIT Press.

Office of Communications (Ofcom) (2005a). Ofcom's approach to risk in the assessment of the cost of capital. Final Statement. London.

Picot, A. and C. Wernick (2007). The role of government in broadband access. *Telecommunications Policy*, **31**(10–11), 660–74.

Richards, P. (2006). The limitations of market-based regulation of the electronic communications sector. *Telecommunications Policy*, 30 April, 201–23.

Richards, P. (2007). Technical progress, market evolution and the regulation of the ECS in the EU. *Competition and Regulation in Network Industries*, **8**(2), 165–257.

Robertson, L. and N. Langlois (1995). Innovation, networks, and vertical integration. *Research Policy*, **24**(4), 543–62.

Schumpeter, J.A. (1939). *Business Cycles: A Theoretical, Historical, and Statistical Analysis of the Capitalist Process*. New York: McGraw Hill.

Schumpeter, J.A. (1942). *Capitalism, Socialism and Democracy*. New York: Harper & Brothers.

Schwartz, E.S. and L. Trigeorgis (2004). *Real Options and Investment under Uncertainty: Classical Readings and Recent Contributions*. Cambridge, MA: MIT Press.

Trigeorgis, L. (1995). *Real Options in Capital Investment: Models, Strategies, and Applications*. Westport, CT: Praeger.

14. Next generation mobile networks deployment and regulation in the European Union[1]

Claudio Feijóo, Sergio Ramos and José-Luis Gómez-Barroso

14.1 INTRODUCTION

The ubiquitous broadband revolution will fundamentally change the landscape of European telecommunications. Citing European Commissioner Reding (2008): 'we are currently confronted with a once in a generation opportunity to make sure that Europe promotes and leads the next phase of wireless technology which will be the transition from voice and short text services to the wireless web'. Today mobile telecommunications are a major driver of the European Union (EU) economy, with expected market growth by 2.4 percent in 2009 to reach an overall value of €144.5 billion. Mobile data traffic, including text messages and Internet access, contributes for €33 billion of the total turnover in mobile telecommunications and most of the growth, 8.4 percent (EITO, 2009).

The so-called next generation networks (NGNs) will be the supporting infrastructure of ubiquitous broadband. They are typically defined as multiservice networks, running over Internet Protocol (IP)-based networks, complemented by flexible service platforms and management systems.[2] In particular they will be able to make use of multiple broadband, quality of service (QoS)-enabled transport technologies and in which service-related functions are independent from underlying transport technologies. They will also offer access by users to different service providers and support generalized mobility which will allow consistent and ubiquitous provision of services to users (Kocan et al., 2002; Modarressi and Mohan, 2000). For the purposes of this chapter, an NGN will be simply a single network which delivers multiple applications – voice, data and video – to multiple devices, whether fixed or mobile.

The conditions for the deployment of the access part of NGNs are currently the key topic in the review of the electronic communications regulatory model in the EU (Cave and Picot, 2008). The main issues raised during the debate are the conditions for the return on investments, the type and level of competition and the model required to promote innovation. The subset of NGNs which specifically will support the evolution from today's mobile communications infrastructures are usually called next generation mobile networks (NGMNs). They confront basically the same regulatory issues as NGNs. But they present, in addition, some particularities of interest: the need for adequate spectrum, the evolutionary nature of mobile technologies, and their distinctive regulatory model to date (Ramos et al., 2004).

This chapter's aim is to examine the impact of regulation in the deployment of NGMNs. With that objective the next section introduces briefly the concept of NGMNs and the overall status of the market. From here the examination of the regulatory elements that influence the deployment of NGMNs follows and, finally, the chapter closes with some conclusions.

14.2   THE STATUS OF THE NGMNS

NGMNs are regarded as a future platform for ubiquitous broadband, facilitating the smooth migration from existing mobile infrastructures, and allowing for the commercial launch of new mobile services and applications. Thus, in the mobile carriers' vision, NGMNs will ensure a virtuous cycle of investment, innovation and adoption of services (NGMN Alliance, 2006). From a strict technological perspective NGMNs encompasses fourth generation, 4G-type mobile technologies, such as Long Term Evolution (LTE), Mobile WiMAX (Worldwide Interoperability for Microwave Access), IMT-Advanced[3] (Ring, 2008) and convergent technologies like femtocells, wireless technologies – Near Field Communication, NFC, for instance – completing a 'network of sensors and tags located on surrounding objects' (Schwarz-da-Silva, 2008).

Table 14.1 overviews some of the main milestones and features of the technologies needed for the deployment of NGMN. Furthermore, a number of experts from the industry forecast that mobile broadband connections will overcome fixed ones sometime around 2011–13 (Nerandzic, 2008; Ouvrier, 2008). As an obvious result from these ambitious roadmaps, mobile networks will require access to new spectrum and/or a much more efficient management of it. Without these spectrum improvements it will be impossible to deploy rapidly enough the required technologies that satisfy users' demands or to compete – and complement – satisfactorily with fixed broadband technologies.

In fact the deployment of true NGMNs, albeit slowly, has already

Table 14.1 Summary of technological roadmaps for NGMN deployment

	HDPA (3.5 G)	LTE (4G)	Femtocells	Mobile WiMAX	IMT Advanced	NFC
Theoretical maximum data rates	14 Mbps (downstream) 40 Mbps (downstream in HDPA+) 5.6 Mbps (upstream)	100 Mbps (downstream) 50 Mbps (upstream)	–	50 Mbps	1 Gbps	1 Mbps
Typical data rates	3.6 Mbps (downstream) 2 Mbps (upstream)	–	–	10 Mbps	100 Mbps	400 Kbps
Channel bandwidth	5 MHz	1.25–20 MHz	–	1.25–20 MHz	Up to 100 MHz	2 MHz
Begin of massive deployment	2008–10	2009–12	2009–10 2010–11 (for handsets)	2009–11	2013–17	2011–17
Critical technologies	MIMO	OFDMA	Management	OFDMA	Dynamic spectrum management	Security
Main advantages	Evolutionary from existing 3G	Evolutionary from 3.5 G	Fixed-mobile convergence Increase of coverage	Not a legacy technology	Evolutionary from 4G	Smart environment
Main disadvantages	Transition technology	Time-to-market	Integration in existing networks	Business case for new technology	Still in early stages of standardization process	Business case for deployment

begun. A survey from the Global mobile Suppliers Association (GSA) in September 2007 stated that 138 high-speed downlink packet access (HSDPA) networks had been commercially launched services in 66 countries, compared to 58 commercial launches in 37 countries the previous year. By February 2008 125 HSDPA network operators had commercially launched or were deploying 3.6 Mbps peak or higher downlink data rate capability. At the end of 2008 the number of commercial HSDPA deployments had increased to 237 in 105 countries, and 167 commercial HSDPA networks supported 3.6 Mbps or more, including 81 which supported 7.2 Mbps peak or higher downlink speeds. The latest data (GSA, 2009) show that several operators are launching HSPA+ in 2009 for at least 21 Mbps peak downlink speeds and that there are over 20 mobile operators already committed to deploy LTE.

However, availability of mobile broadband is only one condition needed to reach mass adoption. Affordability of mobile broadband connections, utility and appeal of new mobile applications and services, as well as the availability and the usability of new mobile handsets are also required (Feijóo et al., forthcoming). For these reasons, mobile data services continue to rely on simpler services like messaging which accounted for around 14 percent of total revenues of mobile operators in 2007. While more advanced data services have shown significant growth recently, they still accounted for just around 7 percent of total revenue compared to 5 percent in 2006. The only two major exceptions in 2007 were Japan and Korea, where mobile content average revenue per user (ARPU) was 30.5 percent and 18.9 percent respectively (Netsize, 2008).

No account of the future status of NGMNs can be complete without an examination of the demand side, especially in view of the increasingly important role of users in the so-called mobile 2.0 (Jaokar, 2006). Figure 14.1 shows the forecast for the evolution of mobile social networking users (in millions) compared with those of total mobile subscribers, users of 2G, beyond third generation (B3G) technologies and mobile Internet.[4]

In summary, the mobile sector, particularly in Europe, is at a turning point. As the revenues from voice remain flat or even decline (EC, 2008), there is the need to go further to compete with fixed players on broadband access, to offer new attractive mobile services, applications and content and find fresh business models for them (Ramos et al., 2009).

14.3   REGULATION AND NGMN

From a regulatory perspective, the deployment of NGMN is linked with two main pillars of the electronic communications framework: spectrum

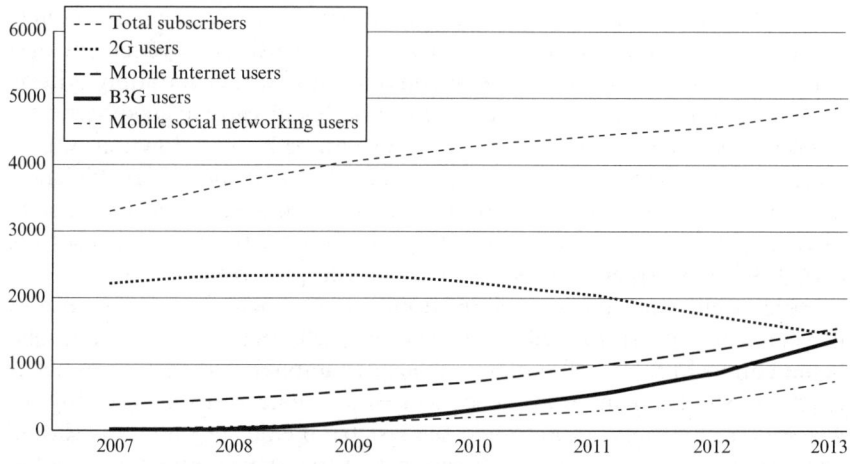

Source: Own calculations from industry data.

Figure 14.1   *Evolution of world mobile subscribers (millions) of different technologies and services: 2G, B3G, mobile Internet and mobile social networking*

management and the triangle 'investment – competition – innovation'. Together, they form what we could name as 'the model' for the future mobile industries from the public perspective. Both of them are currently under review due precisely to their impact on the deployment of NGNs.

14.3.1   Spectrum Management

NGMNs will need new spectrum if a truly ubiquitous broadband connectivity is to be achieved in a competitive environment. The current allocation of spectrum to mobile operators will not be enough to satisfy the roadmaps shown in Table 14.1. Even counting on technology developments to increase the efficiency (bits per Hertz) of radio transmission techniques, the implementation of 4G-type mobile communications will require from 20 to 100 MHz of spectrum per operator (in a design of cellular coverage with high quality of service for high data throughputs). The total budget of spectrum for NGMNs will depend on the availability of spectrum, on the level of competition (that is, the number of mobile network operators in the market[5]) and on the expected return on investments. Higher quality of service will need more spectrums per user.

The first of these factors, radio spectrum availability, is linked with an efficient spectrum usage and management (Feijóo et al., 2009a). In fact, the traditional model for spectrum management appears as excessively complex and lacking flexibility, which has led to the reinforcement of a current opinion favorable to the introduction of market economy criteria in radio electric spectrum management, an idea already argued by Coase (1959). This general principle of efficiency can be broken down into three dimensions: technical efficiency, economic efficiency and social efficiency. Technical efficiency implies that, should there be several technologies to offer a specific service under the same conditions, the one that uses the least resources – frequencies – should be the preferred option. An economically efficient spectrum allocation is one in which there is no way to reallocate spectrum such that everyone is better off, everything else being equal (Morris, 2005). Social efficiency implies that the radioelectric spectrum should be used in those services or applications maximizing the well-being of society. In all cases, the spectrum efficiency analysis should be extended by introducing a prediction of future development or, in other words, optimizing a dynamic definition and not a simply static one.

In fact, many national authorities in charge of spectrum management, the European Commission or the Internet Telecommunication Union (ITU),[6] have released documents supporting the modification of the current spectrum management models and defending the inclusion of market mechanisms. Therefore, that the radioelectric spectrum needs a new management model – or at least a renewed one – is an assertion over which there exists a consensus among public administrations, players with interest in the market and researchers. The new ways of managing the spectrum intend to increase flexibility and transparency, as well as the speed of the response to technological innovations. The actual demand and the value given by the market to each band of frequencies are new governing criteria. Obviously, any change should improve global resource usage efficiency, while maintaining certain inalienable technical requirements.

There are three possible measures[7] leading to change. These are listed in ascending order in the process of assignment of the spectrum usage rights:

- Modifications in the conditions of the license: relaxation of some clauses but, especially, authorization of the transmission of rights (secondary spectrum trade).
- Modifications in the license-assignment mechanisms: usage of auctions.
- Modifications in the definition of the licenses: avoiding binding the license to specific technologies – technological neutrality; or even to specific services – service neutrality.

Although this chapter's aim is not the discussion about the merits of the reform of spectrum management but its relationship with NGMNs, it is worth pointing out that some of these measures are already in use. Secondary spectrum trading has been adopted by the responsible bodies of some member countries (Ofcom, 2008b), although in general only as pilot projects or in frequency bands with moderate economic value and reduced capacity for provoking interferences. The results of spectrum trade so far are very modest (Ofcom, 2008a), but due to existing limitations they are inconclusive about the future behavior of the spectrum market in different circumstances. About spectrum auctions it can be said that they are increasingly used as a more transparent procedure to allocate frequencies to players; however, 'beauty contests'[8] are still a popular mechanism for relevant frequency bands across EU member countries. Again, the objective of the chapter is not to defend auctions as opposed to beauty contests (see Bourdeau de Fontenay, 1999; Cramton, 2002) but to highlight that the allocation and assignment procedures are not harmonized at all. Last, technological neutrality in spectrum has been the subject of several recent 'refarmings' of spectrum, mainly to allow the use of 3G – and beyond – standards for mobile communications in bands originally allocated exclusively for 2G. In general, service neutrality is not yet in place for any relevant spectrum-based service.

In any case the application of these reforms for the deployment of NGMNs leads to a number of short-term opportunities, being the most prominent the so-called 'digital dividend' resulting from the television 'analog switch-off'. Digital dividend, located in the VHF–UHF (very high frequency, ultra high frequency) bands, is the 'beachfront spectrum' (so-called in the USA due to its techno-economic features). The cost of the deployment of a cellular technology is directly proportional to the number of base stations required (Giles et al., 2004). The use of the 700 MHz band, for example, reduces about seven times the number of base stations compared with the 3.5 GHz band for the same coverage conditions (Forge et al., 2007). This reason, together with better indoor coverage and simpler technology, are enough to explain the interest in these classes of frequencies. Nevertheless, if we take into consideration the cost per user at a given quality of service, it is interesting to note that the use of lower frequencies – and, therefore, larger cell sizes – asks for larger spectrum bandwidths to maintain the quality. Alternatively, 'macrocells' are only practically applicable in areas – rural, for instance – where the number of concurrent users does not increase enough to set operations beyond the point of quality degradation (Johansson et al., 2004).

The problem lies, obviously, in the fact that this highly desirable spectrum is already allocated, mostly to the television broadcasters. However, as mentioned, the transition to the digital terrestrial television opens a window of opportunity to a refarm of the spectrum, and to change some of the rules of the game in spectrum, management as mentioned above. In practice this process is not smooth and it is a good example of the difficulties to achieve a solution accepted by all parties, in spite of the European Commission (EC) efforts for the adoption of a common perspective on the use of spectrum available from the digital switchover (EC, 2007).

The following two positions illustrate the problem. Forge et al. (2007) in a study for T-Mobile about the use of digital dividend in the EU conclude that:

> estimates of the accumulated effect indicate as much as an additional 0.6% GDP growth per year for the EU economy by 2020 in the mobile case when compared with broadcast television . . . allocating spectrum to enable wireless broadband could have a dramatic impact on bridging the digital divide by using the new spectrum to provide access for all across the EU's 27 member countries.

It is worth noting that this, sometimes forgotten, important effect on the digital divide is also supported by the other authors, like Cave and Hatta (2008) and Gómez-Barroso and Robles-Rovalo (2008). On the opposite side, a 'counter report' supported by the European Broadcasting Union (EBU) (Oliver and Ohlbaum Associates, 2008) argues that:

> a market for UHF spectrum is likely to fail and not allocate a socially optimal amount of spectrum to terrestrial television. This is because terrestrial TV generates significant public value for society that would not be visible in any hypothetical contest for spectrum with other uses, and cannot easily be replicated through provision of TV using other platforms. The medium-term value that could be created by other uses of UHF spectrum, including rural broadband – appears modest.

Reading these lines, it is easy to realize that the endeavor of a pan-European harmonization of the digital dividend in search for the appropriate economies of scale currently has to confront high barriers. A debate so bitter has, however, several possible solutions: on the one hand, there is spectrum enough, if not to please every player, at least to find a reasonable compromise,[9] and on the other hand, it is possible just to let the evolution of services – addition of interactivity to the broadcasting side or use of audiovisual on the communications side – dissolve the differences between them. As an argument for the first option, it is worth recalling that in the USA, where there is certainly no need to harmonize across member states, the auction for the digital dividend has already taken place in 2008, even including some conditions for 'openness' – still to be seen in its practical implementation – in some of the frequency blocks (FCC, 2008).

Apart from digital dividend there are more examples of immediate influence of spectrum management in NGMN deployment. For instance, although as mentioned earlier the cost of cellular systems is proportional to the number of base stations, it would be possible to arrive at a different efficient solution decreasing the cost of each of them to compensate for the increment in their number.[10] The resulting 'pico' or 'femto' cells will require a fair amount of neutrality in spectrum usage to adapt them continuously to the changing landscape of mobile and wireless communications technologies, business models and users' demands. 'Ad hoc' mobile networks would be another example of much-needed flexibility in spectrum management. Without exhausting the subject, one more example would be sharing spectrum among operators to decrease deployment costs. This could be a result of an 'extended' secondary spectrum trade.

Finally, in addition to making more flexible and decreasing the costs of NGMN deployment, it is worth mentioning that an efficient spectrum framework would increment the number of potential mobile network operators in the market and the overall level of competition in the electronic communications domain (Cave and Picot, 2008). Equivalently, given a certain amount of available frequencies, spectrum management will determine the initial level of competition and the rules for its evolution.

14.3.2   The Mobile Communications Framework: Investment – Competition – Innovation

Investments in NGMNs are arguably the first issue on the agenda of mobile operators from the perspective of a review of the electronic communications regulatory framework. In this regard, since its market started in the early 1990s the EU mobile sector has enjoyed a mostly non-intrusive regulatory model that has led it to be a reference worldwide. As a consequence, the mobile business is characterized by operators' leading positions, controlling as many elements within the value chain as possible (Sabat, 2002). The result from the perspective of mobile operators was the well-known 'walled garden', 'silo' or 'on-portal' model, where content and applications revenues are generated by operators within their own value structure and where users are guided to stay confined as much as possible within this structure. The bottom-end rationale for a walled garden in mobile content is the use of a scarce and costly resource: the mobile networks.

Undoubtedly, in Europe this model had enabled competitive infrastructure development with quite successful results until the migration towards 3G networks started in 2000 (Ramos, 2005). From that point on, the mobile business sector experienced difficulties that forced operators to delay 3G commercial launches by, on average, three to four years (Bohlin et al., 2003). As a consequence during that time mobile players have had to focus more on mobile value-added services than on broadband capacity and, therefore, were not generally considered real broadband players until very recently.

NGMN deployment changes the landscape dramatically. On one side they require considerable investments (De-Antonio et al., 2006; Forge et al., 2005) and therefore mobile operators argue for maintaining the status quo or even for a relaxation of the competition rules – sharing infrastructures or integrating with fixed operators, for instance – or, alternatively, for special conditions – risk premiums – for investors. But on the other side mobile operators also increasingly compete in a convergent scenario with all types of operators. NGNs, and NGMNs as a part of them, will be the first major infrastructure investment for the overall sector to take place in a relatively open competitive scenario. Experience, both for operators and regulators, is scarce.

This fixed–mobile convergence over NGNs will create a stimulated level of competition where now both fixed and mobile operators will contend for the same ubiquitous broadband market.[11] From the regulatory perspective it is not difficult to realize that this type of competition will require a more convergent regulatory framework. Some symptoms of it are already surfacing in the proposals for the next regulatory framework for electronic communications in the EU,[12] in the review of the universal service obligations,[13] in the major examples of the regulation of mobile voice and short message system (SMS) roaming charges, or in the reconsideration of termination charges and interconnection models in the mobile domain. However, it is still unknown whether the approach will resemble the heavy-handed regulation of the incumbent fixed networks or the light touch until now typical of the mobile domain. All in all, the expected consequence of the converged competition would be a more level playing field for operators, independently of the technology they use.

More recently, innovation, the third relevant element in the mobile communications policy framework, has re-emerged with strength. There are two main reasons for it: the opportunities for growth and jobs derived from mobile evolution, and the users' increasing demand for an unrestricted and wide choice of content and applications, including their new role in innovation.

As a consequence the traditional 'walled garden' approach of mobile operators is finally eroding and slowly giving way to the opposite model where the mobile operator is a mere provider of connectivity (Ramos et al., 2002). In between, there are a myriad hybrid models with content and application providers, enablers and brokers sharing revenues with mobile operators (Feijóo et al., 2009b). However, the speed of change seems not to be enough. Citing Holden (2008): 'the level of control exerted by [mobile] operators rankles with, and exasperates, the [application and] content providers, an environment not necessarily conductive for the introduction and mass adoption of innovative mobile services'. Openness, resembling the successful 'open garden' Web 2.0 model of Internet, seems to be the increasingly sought response to incentivize innovation and new business models. What type of openness, how to enforce it and the reaction of players are some of the issues remaining unsolved.

As a summary of the subsection, three major types of uncertainties persist in the mobile communications regulatory framework with regard to NGMN deployment: the conditions for investment in NGMNs (that is, should regulation be the same as before? Should spectrum management be used to ease the deployment?); second, the conditions for competition (that is, should there be a fixed mobile convergence in terms of competition and regulation?); and third, the conditions for innovation (that is, which is the structure of the value-added chain providing more social benefits in the long run?).

14.4 CONCLUSION

Next generation networks will support a renewed electronic communication structure where the market opportunities lie in the provision of ubiquitous broadband connectivity and in the combination of services, applications and content on top of this (Feijóo et al., forthcoming). Mobile operators will deploy the mobile flavor of NGNs to compete among themselves and with fixed operators. In practical terms, NGMNs should provide wide coverage for at least 50 Mbps downstream in the 2015 horizon (Ramos et al., 2009).

However, the deployment of NGMNs by mobile operators confronts hurdles both in the technological domain – uncertainties in the technology roadmaps; and in the market domain – uncertainties about profitable business models and the response of users. Regulation will not stop – or start – the process of deployment of NGMNs, since the rationale behind its deployment resides in the demands and expectations of the users. But regulation, and from a complementary perspective industrial policy, will impact the pace of such a deployment (Cave and Picot, 2008). It can provide a stable and coherent framework for the market to evolve; or the opposite: it could amplify the effect of these uncertainties, delaying the deployment of NGMN.

The most immediate effect of regulation for NGMNs refers to spectrum.

Future – almost present – mobile or wireless access technologies will need more efficient access to broader-spectrum bands to be able to compete in the ubiquitous broadband landscape. Therefore, the reform of spectrum management has a prime role. There are three main conditions for such a reform: (1) more flexibility to accommodate the rapidly changing nature of mobile technologies; (2) a search of techno-economic efficiency to increase the utility of a limited resource like spectrum; and (3) further harmonization in spectrum management mechanisms with the double goal of avoiding new digital divides and increasing the economies of scale to ease the deployment in an already costly scenario. In the authors' view, the main conclusion from the spectrum management discussion is that its potential benefits for NGMN deployment will not be evident until each and every of the measures for reform are included in a comprehensive and coherent EU-wide framework. It would not make much sense to have secondary spectrum trading in place if the conditions to access such spectrum differ across countries or if there is not enough flexibility for the use of it.

Precisely, this is the main difference when comparing EU with other regions (for example, the US): the pressing need for harmonization. With the background of the huge EU success of Global System for Mobile Communications (GSM) standardization (and the initial US delay in mobile communications derived from the many PCS solutions), and the subsequent (and not less huge) failure of the Universal Mobile Telecommunications System (UMTS), the authors are arguing for a different and subtler type of harmonization: it will be about the new conditions for use of spectrum, not the actual technologies using it.

Independently of any of these circumstances, and extrapolating the current pace of network deployment (and user adoption), 2010 and 2011 take shape as critical moments when the spectrum management decisions should be in place. Therefore, the ITU's World Radio Conference (WRC) of 2011 arises as a major milestone in the process leading to 4G and beyond; a date even more significant when the extremely slow process of spectrum management and standardization is acknowledged (Ring, 2008).

Spectrum management, however relevant, is only one part of the equation. It should be framed in a larger perspective: policy-makers and regulators should define a scenario for NGMN deployment as the cornerstone to stimulate innovation that is transferred to markets and users in the form of new services, applications and businesses. There are two minimum conditions for this scenario to be achieved: (1) stability, since the time required for the return on investments for the deployment of NGMNs will be longer than for previous generations of mobile communications; and (2) coherence, in the sense of the definition of a model – comprised policies on spectrum management, investment, innovation and competition in NGN – for mobile operators. Beyond these minimum conditions there could be more ambitious and challenging objectives. These additional objectives are related not to the NGMN deployment as such, but to the still to be fully understood innovation which could originate from a ubiquitous mobile broadband infrastructure. The new mobile innovation would require: (1) a more open ecosystem to decrease the barriers of an heterogeneous and fragmented ecosystem in particular when compared with the Internet (Feijóo et al., 2009b); (2) making the most of the new role of users (Pascu, 2008); and (3) reviewing the institutional framework that supports it (Fransman, 2007), including some additional pieces of the regulatory framework like the intermittently debated convergence with audiovisual. The final challenge consists of meeting these goals while the mobile operators are still in the process of investing in NGMNs.

In conclusion, both elements of reform – spectrum management and regulatory policy framework – have been on the agenda since the stage for the initial liberalization of electronic communications was – approximately – met in about 2002–03 in Europe, though they have never reached actual and complete implementation. In the authors' opinion, the window of opportunity for mobile operators' deployment of NGMNs is and will be open up to the medium term and, to a great extent, it will be independent of regulation as the LTE deployment commitments of operators have proved even during times of economic crisis. However, the regulatory policy framework severely impacts the intensity and speed of the NGMN deployment and, through the ubiquitous broadband this will provide, of a long-awaited new wave of applications and societal empowerment.

NOTES

1. The views expressed are purely those of the authors and may not in any circumstances be regarded as stating an official position of their Institutions
2. The NGN 'Working definition' can be found at: http://www.itu.int/ITU-T/study-groups/com13/ngn2004/working_definition.html
3. IMT-Advanced refers to international Mobile Telecommunications Advanced set of standards for 4G type of mobile communications as defined by ITU-R, Note that although LTE is usually branded as 4G its first release does not fully match the IMT-Advanced specifications.
4. Based on own calculations from market analysts' data. Mobile 2.0 is foreseen to reach the impressive figure of 1000 million mobile social networking users around 2014.
5. In the EU-27, the average – mode – number of mobile operators per member state is four (EC, 2008).
6. As an example, refer to the ITU document 'Market mechanisms for spectrum management', http://www.itu.int/osg/spu/stn/spectrum/workshop_proceedings/STN.MMSM-2007-PDF-E.pdf.

7. Together with these measures, a series of general guidelines has been proposed for the practical introduction of reforms in the spectrum management mechanisms. The interested reader can consult Feijóo et al. (2009a)
8. In a 'beauty contest' licenses are assigned according to the evaluation of a number of technical and economic criteria, typically including anything from type of technology used, deployment of coverage proposed and national added value and employment created to national innovation push.
9. The European Commission already proposes one such possible compromise (EC, 2007).
10. Therefore, in practice, mobile operators will require to be cost-effective in a range of cell sizes (macro, meso, pico, femto) and appropriate allocations of frequency bands and bandwidths (Johansson et al., 2004).
11. Note that integrated operators, that is, operators both with fixed and mobile technologies, are anticipating this trend and they position themselves in this new competition landscape.
12. See http://ec.europa.eu/information_society/policy/ecomm/tomorrow/index_en.htm.
13. Communications from the Commission to the European Parliament, the Council, the European Economic and Social Committee and the Committee of the Regions on the second periodic review of the scope of universal service in electronic communications networks and services in accordance with Article 15 of Directive 2002/22/EC, Brussels, 25.9.2008, COM(2008) 572 final C.F.R. (2008).

REFERENCES

Bohlin, E., J. Björkdahl, S. Lindmark, T. Dunnewijk, N. Hmimda, S. Hultén and P. Tang (2003). Prospects for third generation mobile systems. Report No. EUR20772EN. Seville: Institute of Prospective Technological Studies.

Bourdeau de Fontenay, A. (1999). Auctions vs. beauty contests, is it the question? A new look at access and spectrum allocation in France and in the US. *Communications and Strategies*, **36**, 111–23.

Cave, M. and K. Hatta (2008). Universal service obligations and spectrum policy. *info*, **10**(5–6), 59–69.

Cave, M. and A. Picot (2008). Workshop next ('now') generation access (NGA): How to adapt the electronic communications framework to foster investment and promote competition for the benefit of consumers? Summary, briefing notes and presentations (No. IP/A/ITRE/WS/2008–09 PE 408.554). Brussels: Department Economic and Scientific Policy, European Parliament.

Coase, R. (1959). The Federal Communication Commission. *Journal of Law and Economics*, **2**, 1–40.

Cramton, P. (2002). Spectrum auctions. In M. Cave, S. Majumdar and I. Vogelsang (eds), *Handbook of Telecommunications Economics* (pp. 605–39). Amsterdam: Elsevier Science B.V.

De-Antonio, J., C. Feijóo, J.L. Gómez-Barroso, D. Rojo and A. Marín (2006). A European perspective on the deployment of next generation networks. *Journal of the Communications Network*, **5**(2), 47–55.

EC (2007). Reaping the full benefits of the digital dividend in Europe: a common approach to the use of the spectrum released by the digital switchover. Retrieved from http://eur-lex.europa.eu/LexUriServ/LexUriServ.do?uri=CELEX:52007D C0700:EN:NOT.

EC (2008). Progress report on the single European electronic communications market 2007 (13th report). Retrieved from http://ec.europa.eu/information_society/policy/ecomm/doc/library/annualreports/13th/com_2008_153_en_final.pdf.

EITO (2009). Data services give the wireless market momentum. Retrieved 21 March 2009 from http://www.eito.com/reposi/PressReleases/EITO-PR-20090213.

FCC (2008). Auction of 700 MHz band licenses closes. Winning bidders announced for Auction 73. Retrieved from http://hraunfoss.fcc.gov/edocs_public/attach-match/DA-08-595A1.pdf.

Feijóo, C., J.L. Gómez-Barroso and A. Mochón (2009a). Reforms in spectrum management policy. In I. Lee (ed.), *Handbook of Research in Telecommunications Planning and Management for Business* (pp. 33–47). Hershey, PA: Information Science Reference.

Feijóo, C., I. Maghiros, F. Abadie and J.L. Gómez-Barroso (2009b). Exploring a heterogeneous and fragmented digital ecosystem: mobile content. *Telematics and Informatics* **3**, 282–92.

Feijóo, C., I. Maghiros, M. Bacigalupo, F. Abadie, R. Compañó and C. Pascu (forthcoming). *Content and Applications in the Mobile Platform: On the Verge of an Explosion*. Seville: Institute for Prospective Technological Studies.

Forge, S., C. Blackman and E. Bohlin (2005). The demand for future mobile communications markets and services in Europe. Report No. EUR 21673 EN. Institute for Prospective Technological Studies – JRC – EC.

Forge, S., C. Blackman and E. Bohlin (2007). The mobile provide. Economic impacts of alternative uses of the digital dividend. Summary Report. SCF Associates.

Fransman, M. (2007). *The New ICT Ecosystem. Implications for Europe*. Edinburgh: Kokoro.

Giles, T., J. Markendahl, J. Zander, P. Zetterberg, P. Karlsson, G. Malmgren and J. Nilsson (2004). Cost drivers and deployment scenarios for future broadband wireless networks: key research problems and directions for research. Paper presented at the Vehicular Technology Conference, 2004. VTC 2004-Spring. 2004 IEEE 59th.

Gómez-Barroso, J. and A. Robles-Rovalo (2008). Wireless hopes for universal service in developing countries: an assessment of the Mexican context. *info*, **10**(5–6), 83–91.

GSA (2009). 3G evolution steps confirmed by multiple operators for HSPA+, LTE and EDGE evolution. GSM/3G network update. Available from http://www.gsacom.com/downloads/pdf/GSA_GSM_3G_Network_Update_Feb2009.php4.

Holden, W. (2008). Making music with mobile. Basingstoke: Juniper Research.

Jaokar, A. and T. Fish (2006). *Mobile web 2.0*. London: Futuretext.

Johansson, K., A. Furuskar, P. Karlsson and J. Zander (2004). Relation between base station characteristics and cost structure in cellular systems. Paper presented at the 5th IEEE International Symposium on Personal, Indoor and Mobile Radio Communications, PIMRC 2004.

Kocan, K.F., W.A. Montgomery, S.A. Siegel, R.J. Thornberry Jr. and G.J. Zenner (2002). Service creation for next generation networks. *Bell Labs Technical Journal*, **7**(1), 63–79.

Modarressi, A.R. and S. Mohan (2000). Control and management in next-generation networks: challenges and opportunities. *IEEE Communications Magazine*, **38**(10), 94–102.

Morris, A.C. (2005). Spectrum auctions: distortionary input tax or efficient revenue instrument? *Telecommunications Policy*, **29**(9–10), 687–709.

Nerandzic, D. (2008). Emerging technologies and their implications on regulatory policy. Paper presented at the 17th Biennial Conference of the International Telecommunications Society, The Changing Role of the Telecommunications Industry and the New Role for Regulation.

Netsize (2008). *Netsize Guide. Mobile 2.0, You are in Control.* Paris: Netsize.

NGMN Alliance (2006). Next generation mobile networks beyond HSPA & EVDO. Available from http://www.ngmn.org/uploads/media/White_Paper_ NGMN_Beyond_HSPA_and_EVDO.pdf.

Ofcom (2008a). Progress on key spectrum initiatives. A review and update of the SFR and SFR:IP. Retrieved from http://www.ofcom.org.uk/radiocomms/sfr/ sfrprogress/sfrprogress.pdf.

Ofcom (2008b). Trading guidance notes (December 2008). Retrieved from http:// www.ofcom.org.uk/radiocomms/ifi/trading/tradingguide/tradingguide.pdf.

Oliver and Ohlbaum Associates (2008). The effects of a market-based approach to spectrum management of UHF and the impact on digital terrestrial broadcasting. Executive Summary.

Ouvrier, S. (2008). Driving key technologies for next generation mobile networks. Paper presented at the ICT Mobile Summit 2008.

Pascu, C. (2008). An empirical analysis of the creation, use and adoption of social computing applications. Report No. EUR 23415 EN – 2008. Seville: Institute of Prospective Technological Studies.

Ramos, S. (2005). Contribución al estudio, caracterización y desarrollo del sector europeo de comunicaciones móviles e Internet móvil. Unpublished PhD Dissertation, Universidad Politécnica de Madrid.

Ramos, S., C. Feijóo and J. Gómez-Barroso (2004). Barriers to widespread use of mobile Internet in Europe: an overview of the new regulatory framework market competition analysis. *Journal of the Communications Networks*, **3**(3), 76–83.

Ramos, S., C. Feijóo and J. Gómez-Barroso (2009). Next generation mobile network deployment strategies. *Journal of the Institute of Telecommunications Professionals*, **3**(1), 13–19.

Ramos, S., C. Feijóo, L. Castejón, J. Pérez and I. Segura (2002). Mobile Internet evolution models. Implications on European mobile operators. *Journal of the Institution of British Telecommunications Engineers*, **1**(2), 171–6.

Reding, V. (2008). Digital Europe: the Internet mega-trends that will shape tomorrow's Europe. Retrieved from http://europa.eu/rapid/pressReleasesAction.do? reference=SPEECH/08/616&format=HTML&aged=0&language=EN&guiLan guage=nl.

Ring, S. (2008). IMT-Advanced, a global platform for the next generations of mobile services. Paper presented at the International Converging Mobile Media Conference.

Sabat, H.K. (2002). The evolving mobile wireless value chain and market structure. *Telecommunications Policy*, **26**(9/10), 505–35.

Schwarz-da-Silva, J. (2008). Mobile research at the crossroads. Paper presented at the ICT Mobile Summit 2008.

15. Mobile Internet developments in Europe, East Asia and the US

**Morten Falch, Anders Henten and
Karsten Vandrup**

15.1 INTRODUCTION

The chapter firstly examines how European countries are positioned with respect to the diffusion of new mobile and wireless data technologies and services as compared to the two most advanced East Asian countries, Japan and South Korea, and to the US and Canada. Secondly, the chapter discusses central factors which contribute to the explanation for the differences in the development of the countries and regions in question. For a period of time, Europe dominated the mobile scene with the second generation Global System for Mobile communications (2G GSM) technology, which was developed as a European standard and spread to large parts of the rest of the world. But with regard to the early versions of data services on mobile platforms (for example, i-mode, Wireless Application Protocol (WAP) and general packet radio service (GPRS) – often termed 2.5G), Japan and Korea took the lead, and the same applies to data services on 3G platforms and beyond (3.5G). The US, on the other hand, has trailed behind Japan and Korea and also the leading countries in Europe regarding mobile technologies and services, but is clearly the world leader in networked information technology (IT) on fixed platforms. Lately, however, the US seems to be catching up with Europe in the mobile field. These developments are what the chapter aims at discussing in terms of facts and explanations.

Europe is a diverse market, including Scandinavia, being in the lead with respect to the use of information and communication technology (ICT), as well as countries in Eastern and Southern Europe, where the take-up of ICT is generally less developed. Therefore, the European average is not the best indicator to be used for the analysis of the European development. We have, consequently, chosen to use data from four major European economies instead: France, Germany, Italy and the UK.

Section 15.2 presents the conceptual framework of the chapter. This is followed by section 15.3 comparing the status of mobile data in the three regions. Section 15.4 analyzes the key segments of the mobile ecosystems. Finally, conclusions are drawn in section 15.5.

15.2   BACKGROUND

At the general level, the conceptual framework of the chapter relates to the concepts of systems of innovation (for example, Edquist, 2005) and ecosystems (for example, Fransman, 2007). In fact, systems of innovation approaches and ecosystem approaches are very similar, though systems of innovation – as the concept denotes – gives emphasis to innovation, while ecosystem approaches may have a broader scope. However, the focus of ecosystem analyses is also innovations, as emphasized by Fransman (2007). Moreover, the ecosystem concept connotes the right evolutionary setting for the development of technologies and services in a broader environment. What the chapter aims at analyzing is the development of specific types of 'species' (new mobile and wireless data technologies and services) in different ecosystems or environments.

In addition, diffusion theory (for example, Rogers, 2003) is used. Diffusion processes are seen as a subset of the many processes and elements analyzed in systems of innovation and ecosystem analyses. Diffusion analyses obviously provide a deeper insight into diffusion processes as these are the specific topics of diffusion analyses. While the literature on diffusion traditionally is centered on examining factors determining demand, the literature on systems of innovation and ecosystems often includes supply as well as demand and a wide range of technological, organizational and institutional elements. Furthermore, the systems of innovation and ecosystem approaches quite naturally point at comparative analyses of different systems (in different regions, in our case).

At the more specific level, the purpose of the chapter is to examine the development of mobile and wireless data technologies and services in a selection of European countries (France, Germany, Italy and the UK), Japan and South Korea, and the US and Canada, respectively, and to discuss the possible central explanations for the differences in developments. What immediately springs to mind is the early and rapid development of mobile Internet in Japan and Korea as compared to the European countries as well as the US. Also, the strong position of Short Message Service (SMS) in the European markets and a similar development in the US market somewhat later is remarkable. Explanations for these facts revolve around two phases in the developments: the first phase where the subsequent development is set 'on track'; and the following phase where 'the train is rolling' and it becomes more difficult to change track. There is path-dependency.

The first phase has been debated and analyzed, to a relatively large extent, with respect to comparisons of the developments in East Asia and the developments in Europe and the US respectively (Funk, 2007, 2009; Henten et al., 2004). Funk (2007) provides an explanation for the early and swift development of mobile Internet in Japan, and soon thereafter in Korea. Funk focuses on three main explanatory factors (for the success of the Japanese i-mode system starting in 1999): (1) the implementation of a micro-payments system; (2) a central role of entertainment services; and (3) the deployment of an integrated approach including custom-made end-user devices. Funk discards what he calls cultural explanations and social factors relating to commuting and the relatively low diffusion of fixed Internet in Japan, and he gives emphasis to structural factors (vertical integration) and business model elements regarding the services offered (entertainment) and the financial design (micro-payments).

The present chapter also concentrates on structural explanations. While the first generation of mobile communications and the second generation were based on a modular approach, which was an important reason for the huge success of the European-based GSM standard, mobile Internet requires a more integrated approach, in the initial phases at least (Funk, 2009). This argument is in line with the Henderson and Clark (1990) differentiation between modular and architectural innovations. An architectural (or systemic) innovation requires a more integrated approach in its implementation phase but may later allow for modular approaches once the innovation has gained ground and has settled in the market. While Japan and Korea have followed an integrated approach, Europe and the US have followed a more modular approach – in extension of their 1G and 2G models. Custom-made end-user devices, therefore, did not reach the European and US markets in the early phases of mobile Internet, and the WAP standard, which was implemented in the European and US markets, became subject to several different versions and never became the technology platform envisioned in spite of the massive hype around it.

One of the explanations for these differences in development is related to the relative strengths of the service providers (operators) and equipment and device manufacturers in the Japanese and European markets respectively. In Japan, the operators, and especially NTT, were relatively large companies in comparison with the manufacturers and they could, therefore, set the agenda and require specific systems and terminals. In Europe, large trans-European and transnational manufacturers were one of the results of the liberalization processes, and they opted for a strategy where their equipment and devices could be sold all over Europe and the world at large. The nationally confined telecom operators in the European countries did not have the same say as their counterparts in Japan.

In the second phase, when there are a sufficiently large number of users of a system, other mechanisms become relevant. Funk (2007) deals with these issues in his chapter entitled 'Solving the startup problem in Western mobile Internet markets'. Network effects (Economides, 1996; Varian and Shapiro, 1999) including feedback mechanisms and installed bases, are a central explanatory factor. In Japan and Korea, e-mail systems, which are Internet compatible, became the text-based forms of mobile communication. In Europe, SMS became the dominant mode of text-based communication on the mobile platform, and SMS has later also gained considerable ground in the US market as well. SMS has acquired such a huge following in the European markets that an obvious hypothesis is that it is blocking the development of Internet-based services on the mobile. This could thus be considered as an occurrence of path-dependency (David, 2001) raising barriers to the development of mobile Internet.

There are two aspects of this. One aspect has to do with the core of network economics regarding direct network externalities or effects, that is, that the more users there are of an interactive communication system, the higher the utility for the individual user. If an e-mail-based system is not compatible with the SMS system, users will be reluctant to change to mobile e-mail. Furthermore, the users get habituated with and develop competences in the specific types of systems they use – an argument in line with David's (2007) story concerning the QWERTY typewriter keyboard. The other, and perhaps more important aspect, concerns the fact that operators offering SMS earn a great deal of money on the SMS system. Not only are charges for sending normal SMS messages still high in many cases in Europe, but the turnover and profits from premium services are also high. The incentive to change to text communication based on Internet is, consequently, low.

It should be mentioned that there are many advanced services offered via SMS systems. SMS is not as primitive a system as it is sometimes portrayed. There are many different development possibilities on top of SMS systems. Nevertheless, the SMS-based forms of mobile data communications constitute a side road as compared to Internet-based data communications.

The question is where the US is positioned now with respect to these phases and developments. There is no question that Japan and Korea are still in the lead regarding mobile Internet. But how is the US positioned when compared to Europe? This is what the chapter aims at shedding light on, with available statistical data and explanations. The US never implemented a single mobile standard as the Europeans did with GSM. This seems to have held back the development in the US in the early phases of the 2G developments and it has not, to any large extent, led to integrated approaches between specific operators and specific manufacturers, as in Japan and Korea. Recently, however, US-based companies like Apple (iPhone) and Google have forcefully entered the mobile field.

This indicates an important explanatory factor for the US catching up with Europe – and eventually perhaps the East Asian countries. The US has for many years had a lead in IT and Internet services on fixed platforms, especially on the supply side but also on the demand side. Many of the world's largest IT companies – whether in equipment or services – are US-based. The issue is whether the US is able to leverage this advantage onto the mobile and wireless field.

This discussion is concerned with the overall structural level and the possible explanations for the differences between Europe, East Asia and the US. With respect to the diffusion subset, diffusions of innovations follow an s-curve, according to Rogers (2003), where the early adopters represent about 15 percent of the population. Early adopters will often be characterized by being well educated and having a high income. This was, for instance, the case with the adoption of the Internet in its early phase (Rogers, 2001). Differences in income and education can also be used to explain regional differences in the diffusion of ICT services. The countries with the highest rankings in various e-readiness indexes are in general countries with high levels of income and education. According to Rogers, five attributes of an innovation affect the speed of adoption (Rogers, 2003): relative advantage, compatibility, complexity, observability and trialability.

Relative advantage is by far the most important of these factors and explains about 80 percent of the diffusion. Rogers sees these as technical attributes, that is, as attributes that depend on the technical characteristics of a specific innovation. However, the impact of each of these attributes depends on the market environment in which the innovation is being adopted. This is better reflected in the technology assessment model (TAM) introduced by Davis (1989). This model distinguishes between two different factors: perceived usefulness and perceived ease of use. These relate to the same aspects as relative advantage and complexity but take the user experience rather than the technology as the point of departure. This is important especially in the present context, where we focus on differences in the rates of adoption of the same or similar technologies in different regions.

This model has been applied in a study of user acceptance of advanced mobile services in the Dutch market (López-Nicolás et al., 2008). The study confirms that the TAM model is useful for explaining service adoption. In addition, the study documents the importance of social influence on perceived usefulness as well as perceived ease of use.

The decomposed theory of planned behavior (DTPB) builds on the same two parameters but adds Rogers's compatibility as a third parameter (Hernandez and Mazzon, 2007). The relative advantage or the perceived usefulness of mobile data access depends on what kinds of services they enable, and what the alternatives are. Furthermore, the concept of compatibility relates not only to compatibility of technical standards, but also to compatibility with existing work practices.

The issue of pricing is not really taken into account in these models. But pricing is an important issue when it comes to the adoption of new services on the consumer market. This is addressed in the value-based adoption model (Kim et al., 2007), which is developed especially for analyzing adoption of mobile Internet services. This model operates with four different factors affecting the perceived value and the adoption intention: usefulness, enjoyment, technicality and the perceived fee. It is shown that all four factors have a statistically significant impact.

15.3   STATUS FOR THE DIFFUSION OF MOBILE DATA

The Nordic countries were the first to achieve a wide penetration of mobile services. However, it was not until the introduction of GSM as a common European standard that Europe became a leader in mobile communication. Up to 1998, the penetration of mobile phones was lower within the EU than in the US, but in 1999 the EU penetration increased to 39.6 per 100 inhabitants compared to 31.5 in the US. In 2000, the EU penetration jumped to 63.4. Thereby, the penetration became higher than in Korea (57) and Japan (52.6) – the leading countries in the Asian markets (OECD, 2003).

The successful implementation of one common standard for 2G telephony throughout Europe created a mass market for terminals and enabled roaming in most parts of Europe. Europe dominated the top ten list regarding the penetration of mobile phones, and GSM became the dominant world standard for 2G.

As shown in Table 15.1, the penetration in Europe has remained higher than in the US and in East Asia. However, regarding further mobile generations, the European dominance has dwindled. While NTT DoCoMo was successful in its implementation of i-mode at the turn of the century, European countries were far from successful in their introduction of data

Table 15.1   Mobile subscribers per 100 inhabitants in selected countries

	1996	1998	2001	2003	2005	2008 3Q	3G subscriber (%)
USA	16.6	25.6	45.1	54.5	71.8	87.6	22.8
Canada	11.5	17.7	34.9	41.8	51.64	62.6	9.0
France	4.2	19.2	62.6	67.7	76.71	91.0	18.5
Germany	7.1	17	68.3	78.5	96.04	122.0	19.2
Italy	11.3	35.6	87.1	97.6	122.16	149.3	32.2
UK	11.6	21.9	77.1	89.2	106.26	122.4	25.9
Japan	21.4	37.4	58.8	67.9	75.51	82.3	81.3
South Korea	7.0	30.2	61.4	70.1	79.39	92.6	72.0

Sources: OECD (various years) and Netsize (2009).

services based on the WAP standard at the same time. South Korea has been another early adopter of mobile data services, which now in many ways are more advanced than in Europe. Already in 2005, an overwhelming majority of the subscribers in South Korea had switched to 3G, and in Japan, 50 percent of the subscribers had done the same (OCED, 2007). At that time, most European countries had not yet introduced 3G at all and it was only in Italy and Austria that the penetration of 3G was more than 3 per 100 inhabitants. Although the penetration of 3G is presently increasing rapidly in Europe, the development is still far behind Japan and Korea.

The development in the penetration of 3G in the USA is comparable to that in many European countries. It should, however, be noted that although the percentage of 3G subscribers is comparable or even higher than in the European countries, the total penetration of mobile subscribers is still lower. The development in Canada lacks behind any of the other countries included in Table 15.1.

Although related, 3G and data services are two different things. 3G services include voice as well as data. In fact, voice ARPU (average revenue per user) is larger than data ARPU for 3G subscribers also, and from an operator's point of view an important function of 3G networks is that they can expand the capacity for mobile telephony in high-density areas. Moreover, some data services, for example, SMS, can run on both 2G and 3G networks.

Table 15.2   Penetration of mobile subscribers and average revenue per user (ARPU) (euros)

	Penetration rate, 3Q	ARPU	ARPU Mobile messaging		ARPU Mobile content		ARPU Mobile data	
			absolute	%	absolute	%	absolute	%
USA	87.6	35.47	3.38	9.5	4.33	12.2	7.71	21.7
Canada	62.6	41.76	3.29	7.9	3.23	7.7	6.52	15.6
France	91.0	36.13	3.91	10.8	2.72	7.5	6.63	18.4
Germany	122.0	17.15	3.63	21.2	2.07	12.1	5.7	33.2
Italy	149.3	21.42	5	23.3	2.75	12.8	7.75	36.2
UK	122.4	29.87	8.83	29.6	2.8	9.4	11.63	38.9
Japan	82.3	34.81	6.97	20.0	7.5	21.5	14.47	41.6
Korea	92.6	26.46	2.96	11.2	4.62	17.5	7.58	28.6

Source: Netsize (2009).

Mobile data is a relatively new service area and it is, therefore, difficult to find international comparable data in the official statistics. The most comprehensive source publicly available is the *Netsize Guide 2009* (Netsize, 2009), which includes country information on mobile data services. Table 15.2 is constructed on basis of data from various tables in that report.

According to Netsize (2009), North America is in absolute terms the region that generates the highest income per subscriber on mobile data (Figure 15.1). This somewhat surprising information is supported by other reports, for instance Fierce Mobile Content (2007). However, before the US is nominated as the leader in mobile data, a few reservations have to be expressed. First of all, Asia Pacific includes other countries than Japan and Korea. Both countries have a higher ARPU than the US (Japan has a much higher ARPU). Also in the UK the ARPU is higher than in the US. Secondly, as noted above, penetration is lower in the US and Canada than in Europe. Higher penetration can lead to a lower ARPU as more users with a limited service demand will be included. In addition, high penetration rates reflect that more users have more than one subscription. This is obviously the case in Italy, for example, where the penetration is almost 150 percent. According to a study by Strategy Analytics (2008), 68 percent of the population in North America had at least one mobile phone in 2006, while the same figure for Western Europe was 80 percent. This indicates that double subscriptions are of little importance in North

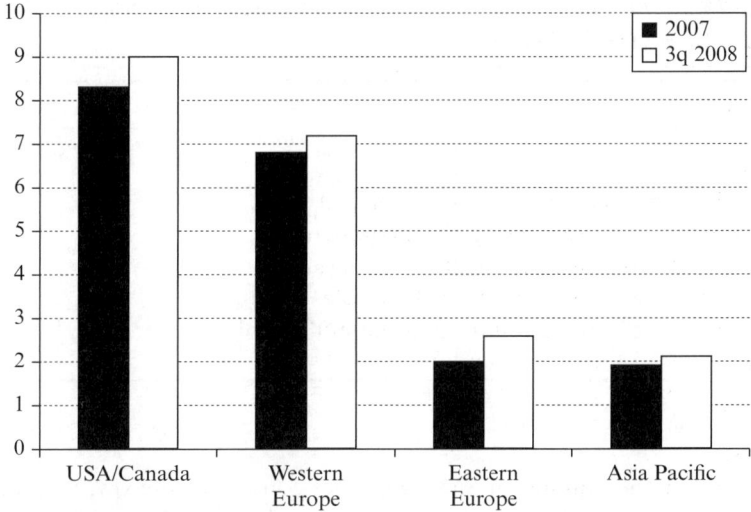

Source: Netsize (2009).

Figure 15.1   Data ARPU(US$ per month)

America, while phone number two or three may account for as much as 30 percent of the subscriptions in Western Europe. Calculations of income per inhabitant may decrease or even annihilate the gap between the US and Canada and Western Europe. Finally, ARPU depends on pricing as well as use.

It is, nevertheless, interesting to note that not only the level but also the growth of ARPU is higher in North America than in the other regions. For instance, the mobile data traffic in T-Mobile's US network more than doubled in one year from 1Q 2007 to 1Q 2008 leading to a 31 percent growth in data revenue (Fitchard, 2009).

Another interesting observation, which might be related to this, is the distribution of revenues generated from messaging and content services. The US is the only country where content services dominate. Mobile messaging includes SMS services and the use of SMS had a slower start in the US than in Europe. According to Strategy Analytics (2008), 50 percent of the mobile users in North America only used voice in 2006. This is compared to 10–20 percent in Western Europe and Asia Pacific. This difference reflects a much more sparse use of SMS in North America at that time. But high growth rates in recent years have compensated for this, so the use per subscriber was at the beginning of 2009 almost double that in Europe (Mobile Europe, 2009).

Table 15.3   Mobile Internet usage across 16 measured countries (Q1 2008) (%)

Country	%	Country	%
US	15.6	China	6.8
UK	12.9	Philippines	3.4
Italy	11.9	Singapore	3.0
Russia	11.2	Brazil	2.6
Spain	10.8	Taiwan	1.9
Thailand	10.0	India	1.8
France	9.6	New Zealand	1.6
Germany	7.4	Indonesia	1.2

Source: Nielsen Mobile (2008).

Although the number of SMS messages continues to grow, its share in non-voice revenues is declining. According to Netsize (2009), SMS used to generate up to 80 percent of the non-voice revenues for a European operator in the early 2000s, but this had by 2008 declined to about 50 percent at least for some of the operators.

When it comes to the number of users of other mobile services than SMS, the US and Europe are far behind the most advanced Asian countries. Capgemini (2008) reports that in 2006 and 2007, 89 percent of the Japanese subscribers used mobile Internet services at least once a month. In the US, only 12 percent were using such services. In Europe, this figure increased from 11 percent in 2006 to 14 percent in 2007.

According to Nielsen Mobile (2008), 37 percent of the mobile subscribers in the US paid a subscription fee for mobile Internet by the first quarter of 2008. However, only 15.6 percent are active users. This compares to slightly lower figures for the UK (12.9 percent) and Italy (11.9 percent), and somewhat lower figures in France (9.6 percent) and in Germany (7.4 percent) (Table 15.3). The lead of the US compared to Europe is due to a growth in the number of users of more than 100 percent from 2007 to 2008. This is compared to a growth in Europe at 50 percent (highest in France with 69 percent) (IT Facts, 2009).

15.4   COMPARISON OF MOBILE ECOSYSTEMS

The central market actors in the mobile ecosystem are the following: equipment and device manufacturers, software developers, network operators, service providers, producers and providers of content, and users.

Table 15.4   Top devices: mobile Internet users

Top devices: % of mobile Internet users (US), Q1 2008

	Device	%
1	Motorola RAZR/RAZR2	10
2	Apple iPhone	4
3	RIM BlackBerry 8100 series (pearl)	2
4	RIM BlackBerry 8800 series (8820, 8830)	2
5	Motorola Q Series (Moto Q, 9h, 9c, 9m, Q Glo)	2

Top devices: % of mobile Internet users (EU), Q1 2008

	Device	%
1	Nokia N95	5
2	Nokia N70	4
3	Motorola RAZR/RAZR2	3
4	Sony Ericsson K800i	3
5	Nokia N73	3

Source: Nielsen Mobile (2008).

In this section, we will limit our analysis to the following segments, as we see these as being the most important in this context: mobile terminals; network infrastructures; services and content; and users.

15.4.1   Mobile Terminals

Ease of use is an important parameter for the adoption of a new service. If the majority of handsets in the market have none or only hard-to-use Internet browsing functionalities, Internet browsing will be low. Until 2004, only a few smartphone handsets with a user-friendly browsing functionality were sold in the US, whilst more models were available in the European markets. Together with a late roll-out of US-wide 3G networks and data services this meant that the US was lagging behind Japan, Korea and Europe.

It follows from Table 15.4 that the European and the US markets are very different with regard to mobile terminal preferences. While Nokia dominates the European market, American manufacturers dominate the US market. In spite of Nokia's global success, it has never been able to penetrate the US market as heavily as it has done in the rest of the world. While Nokia's market share worldwide equaled 40 percent in 2008 (Gartner, 2009), its share in the US was only 10 percent for the same year

(Digital Trends, 2009). Nokia has been engaged in a long-term patent dispute with the rival mobile technology company Qualcomm over licensing of code division multiple access (CDMA) technology, and some industry observers have speculated that the legal wrangling has made Nokia reluctant to commit to the still CDMA-dominated US market (Helsinki Sanomat, 2008).

With the deployment of RIM's Blackberry starting in 2000, and later with Apple's introduction of the iPhone in 2007, the American market got its own drivers for mobile broadband services. The Blackberry was a simple way of getting e-mails on a handset. This was made available first by using standard GSM data rates (max. 13Kbps) and later using 2.5G services like GPRS (70Kbps) and EGPRS (enhanced GPRS) (473 Kbps) data enhancements. The Blackberry fast became the icon of the American business people, allowing them to stay updated on the mailbox while traveling. RIM claimed to have 1 million subscribers in 2003 and 2 million in 2004, increasing to 21 million in 2008 (Impact Lab, 2009). Table 15.4 makes it clear that the top-selling smartphones in the US are the American brands such as Motorola, RIM and Apple, whereas the European market favors European brands. Another reason why the mobile e-mail service was a blockbuster in the US in particular was the lack of SMS use in the US at that point of time. In Europe, text messaging via SMS was and still is by far the most-used non-voice service, and also the service that brings most revenue to the operators.

Blackberry Connect from RIM, Apple's proprietary e-mail platform, or offerings coming with the Symbian and Windows Mobile OS, are the most widespread platforms for mobile e-mail. The most popular devices are those where mobile data functionality has been given high priority in the design. Such phones can be expected to generate far more data traffic than general-purpose devices.

According to Capgemini (2008), only 13 percent of the mobile phone users accessed the mobile Internet in January 2008. However 58 percent and of the mobile users equipped with a smartphone and 85 percent of iPhone users reported that they accessed the mobile Internet within the same period. In particular, the introduction of the iPhone has had a positive impact on the usage of mobile Internet services. Hansberry estimates that as much as two-thirds of the mobile Internet traffic is generated by use of iPhones (Hansberry, 2009).

It should be noted that mobile phones are far from the only terminals to be used for mobile data services. In fact one of the key drivers for the growth in mobile data services within the past year has been use of Universal Serial Bus (USB) modems connecting laptops to the Internet.

15.4.2   Network Infrastructures

Mobile data can be provided by the use of a wide range of different network technologies. Although each standard has its own characteristics, the network technologies are often grouped into generation: 2G (for example, CDMA and GSM), 2.5G (for example, CDMA2000 1RTT – Radio Transmission Technology – and GPRS), 3G (for example, CDMA2000 Evolution-Data Optimized – EVDO; and wideband CDMA WCDMA), and 3.5G (for example, High Speed Packet Access HSPA). Simple data services such as SMS can run on 2G (13 Kbps[1]), but other data services demand at least 2.5G (70–473 Kbps).

Second generation was in the beginning a larger success in Europe compared to the US. Where Europe agreed to deploy the same technology – the 900 MHz GSM system – different standards were used in the US. The outcome was that no single handset could be used everywhere in the US. This stalled the deployment of 2G handsets in the US and also resulted in a later uptake of the following 3G systems.

A likely reason for the slower uptake in the US of 3G technology is that the US has been in the forefront in the deployment of the IEEE 802.11 wireless local area network (LAN) standards (Wi-Fi) and that users in this way have had data services available 'out-of-office' to some extent. Obviously, Wi-Fi is not a mobile technology. However, as it is provided widely in some areas and regions of the US, access to the Internet, for example, from a Wi-Fi based personal digital assistant (PDA) or laptop, is possible in many places, holding back a direct demand for a cellular solution.

Japan was the first to introduce 3G in 2001, and also South Korea was an early adopter (2003). In the major European countries, 3G was introduced in 2003–04, while the US and Canada came a bit later (AT&T as the first US operator introduced 3G in 2004). When it comes to 3.5G, the differences are much smaller. AT&T, for instance, introduced High-Speed Downlink Packet Access (HSDPA) in December 2005, half a year in advance of most European countries (www.3G-americas.org, 2009).

Looking a few years ahead, two technologies are expected to dominate: Long-Term Evolution (LTE) and Worldwide Interoperability for Microwave Access (WiMAX). The expectation has been that WiMAX will dominate in the American and some Asian markets, while LTE will dominate in Europe and in Japan (Vandrup, 2008). However, at least two American operators (Verizon and AT&T) have announced that they will go for LTE. Verizon signed an LTE network deal with Ericsson in February 2009. The American market will thus have two competing technologies, as will a number of other markets.

15.4.3   Services and Content

As the number of mobile users has saturated in the economically developed countries and as the ARPU for voice has decreased, operators have been seeking new revenue streams offering mobile data services. SMS has been a lucrative revenue generator especially in Europe but other non-voice services are also increasingly growing.

The highest data ARPU is found in the US, but the Asia Pacific countries have the highest share of their total mobile turnover deriving from mobile data. The global average is around 20 percent according to *The Netsize Guide* (Netsize, 2009). The Asia Pacific average is almost 25 percent while the Western European and the US and Canada averages are about 21–22 percent.

However, the share of mobile data revenues as compared to total mobile revenues is increasing faster in the US than in the Asia Pacific countries or the European countries. From the first quarter of 2007 to the second quarter of 2008, mobile data revenues increased from 15 percent to 21 percent in the US, while they grew from 18.5 percent to 21.5 percent in Europe and from 23 percent to 24.5 percent in Asia Pacific (Netsize, 2009). This equals an increase of 40 percent in the US – Canada, 16 percent in Western Europe, and 6.5 percent in the Asia Pacific.

As to the categories of mobile data services, Table 15.5 presents the distribution of different kinds of non-voice services for a group of North American, Western European and East Asian countries.

Table 15.5 shows that Internet access is by far the most important mobile data service. Internet access includes many different services but

Table 15.5   Distribution of revenues in mobile content services, 2008 (%)

	Music	Games	TV	Video	Internet	Other	Total
US	5.8	3.4	7.0	2.0	72.4	9.4	100.0
Canada	7.1	5.9	0.7	1.3	75.0	10.0	100.0
France	11.2	5.9	1.2	2.5	66.7	12.3	100.0
Germany	9.9	5.3	1.5	2.0	57.7	23.8	100.0
Italy	9.7	6.9	3.6	2.8	56.1	20.9	100.0
UK	7.9	7.1	0.8	3.1	56.2	24.8	100.0
Japan	4.0	3.4	2.3	2.3	75.7	12.3	100.0
South Korea	17.3	7.6	7.4	4.7	40.6	22.4	100.0

Source: Netsize (2009).

denotes general best effort (quality of service) Internet access. The other service areas are dedicated services managed by the mobile operators. The most surprising observation is possibly the great similarity between all the countries listed with respect to the distribution between the mobile data service categories. Internet access is, however, at a higher level in North America and Japan as compared to Western Europe. Moreover, South Korea presents a slightly different picture because of the great weight of music. Also, the development of mobile TV varies among the countries listed, reflecting the number of mobile TV users in the different countries. The exception is Japan, which is one of the most advanced countries with respect to mobile TV services. The reason may be that the table measures revenues and not usage, and that revenues from free-to-air mobile broadcast services are limited.

The comparison of usage data from Europe with similar data from the US is also illustrative (Table 15.6). From this table it follows that the US is leading in Internet-based services such as mobile e-mail, and that 'info via browser' is more widely used in the US, while received SMS ads and music are more widespread in Europe. This could indicate that Europe, being an

Table 15.6   *Internet activity of European and US mobile subscribers in November 2008 (%)*

	US*	Germany	France	Italy	UK
Sent/received photos or videos	28	23	25	30	30
Received SMS ads	24	27	64	54	36
Accessed news/info via browser	18	9	14	10	19
Used e-mail	17	9	9	12	12
Listened to music	9	20	17	17	22
Accessed social networks	9	3	9	5	9
Played downloaded game	9	8	4	8	10
Purchased ringtones	9	4	3	4	3
Accessed downloaded application	8	4	3	6	6
Watched video	3	5	6	7	4

Note: *September 2008

Sources: Mobile penetration: how many people access a mobile site at least once a week – itfacts.biz, 29 September 2008; and Internet activity of European mobile subscribers in November 2008 – itfacts.biz, 23 February 2009.

early adopter of mobile data services via SMS, has had more emphasis on developing applications which can be provided without Internet access. In the US, where SMS was adopted later, while e-mail and web access were introduced earlier on, the market for such SMS applications has been more limited – especially in the business market.

One of the main discussions concerning the differences in the development and take-up of mobile data services in different regions of the world has dealt with business models. DoCoMo in Japan, for instance, had an early success with the semi-walled i-mode business model. Table 15.5, however, indicates that there are no consistent differences in the openness of business models among the countries listed. In all countries, general Internet access makes up the largest portion and the managed services are generally at a comparable level among countries.

15.4.4   Adoption of Mobile Data Services by Users

The rate of adoption depends on the perceived usefulness of mobile data services compared to the expected costs related to acquiring, learning to use the service, and compatibility with current practice. The importance of these parameters has been confirmed by the empirical studies on the adoption of mobile data services mentioned above (López-Nicolas et al., 2008; Kim et al., 2007).

The perceived usefulness depends on the services available and possible substitutes. With regard to messaging services, SMS is an obvious substitute to e-mail via mobile Internet access. Thus the availability of SMS reduces the usefulness of mobile Internet access.

SMS and to a certain extent mobile e-mail are substitutes to mobile voice telephony. A major driver for the adoption of SMS has been a comparative cost advantage. The cost advantage of e-mail and SMS services depends on the pricing schemes applied. With a flat rate on mobile data services, such services are attractive once a subscription is made.

It is a subject for discussion whether fixed Internet services complement or substitute for mobile data services. A clear analogy is the relationship between fixed and mobile telephony. In its early phases, mobile telephony stimulated use of fixed phones as well. It became possible to call people from fixed phones even if the recipients were on the move. In a later phase, the mobile phone became the primary means of communication, and this has resulted in an ongoing reduction in the number of fixed telephone network subscribers.

When it comes to access to fixed network services, the number of personal computers (PCs) and Internet users is highest in North America and lowest in Europe. With regard to broadband penetration, the picture is

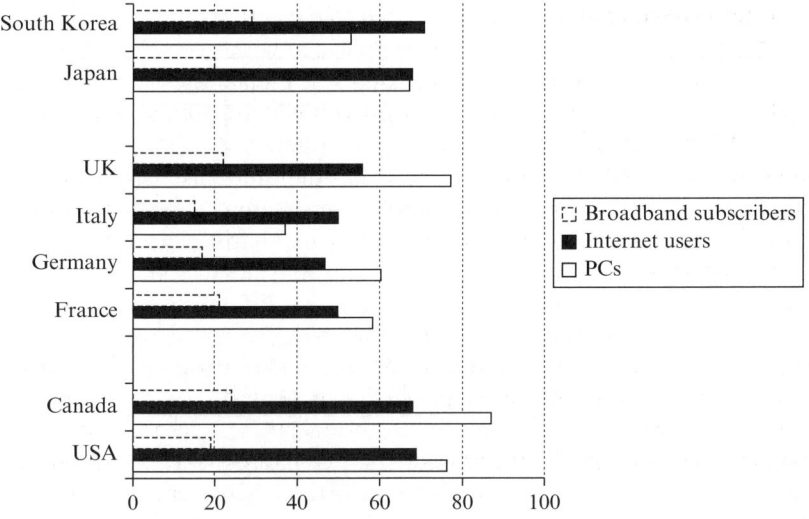

Source: World Economic Forum (2008).

Figure 15.2   PCs, Internet users, and broadband subscribers per 100 inhabitants

more mixed. The highest penetration is to be found in South Korea and the lowest in Italy and Germany (Figure 15.2).

A high penetration in the use of fixed network services can have a positive impact on the perceived usefulness of mobile data services as well. Mobile Internet access enables access to the fixed services, and a developed market of such services adds to the usefulness of mobile access. In addition service providers' experiences can be used to develop services especially designed for mobile terminals.

An often-used method to compare national markets for ICT service is the construction of aggregate indexes for e-readiness. These indexes are composed by a number of different parameters on usage, prices and business environment. The indexes indicate that North America, and particularly the US, is better prepared for the implementation of ICT services than the major European countries. In particular Italy, one of the advanced countries with regard to many kinds of mobile services, ranks low in international comparisons (Table 15.7).

An interesting point to be made in relation to an analysis of the more disaggregated parameters on usage by individuals, business and government is that countries with similar total rankings may have very different profiles (Table 15.8). Germany and Japan rank high in business use,

Table 15.7   E-readiness rankings for selected countries

	E-readiness[1]	Ranking	E-readiness[2]	Ranking
USA	8.95	1	5.49	4
Canada	8.49	12	5.3	9
France	7.92	22	5.11	21
Germany	8.39	14	5.19	16
Italy	7.55	25	4.21	42
UK	8.68	8	5.3	12
Japan	8.08	18	5.14	19
South Korea	8.34	15	5.43	13

Sources: 1. Economist Intelligence Unit (2008). 2. World Economic Forum (2008).

Table 15.8   E-readiness ranking in usage parameters

	Individual	Business	Government	Total
USA	17	8	5	9
Canada	12	16	16	15
France	23	18	10	19
Germany	21	2	38	22
Italy	25	45	47	33
UK	6	11	22	12
Japan	22	3	31	21
South Korea	4	15	7	3

Source: World Economic Forum (2008).

while all European countries except France rank low in government use.

Many mobile data services address mainly the private consumer market and, therefore, the individual usage ranking is the most relevant one. But business and especially government usage are relevant to the evaluation of the supply side, as the development of e-business and e-government services depend on advanced usage in these sectors.

15.5   DISCUSSION AND CONCLUSION

Standardization has been a key to success in early mobile markets. Scandinavia became the first mass market for mobile telephony on the basis of the common Nordic NMT (Nordic Mobile Telephone) standard. Later, the successful promotion of GSM as a common European 2G standard was a major reason for the European lead in mobile communication in the 1990s. Common standards enabled national and international roaming and it facilitated an open and competitive environment leading to low charges and high levels of use.

However, this open, standardized and modularized approach turned out to be a problem for the development of mobile Internet. The first successful mobile Internet system was i-mode, which was implemented in a vertically integrated system by NTT DoCoMo as the dominating actor in Japan. Japan and Korea became the most advanced nations with regard to the diffusion and use of mobile technologies. These two countries have maintained this lead in the 3G–3.5G era, but the analysis provided in sections 15.3 and 15.4 indicates that this lead has become less obvious in certain areas.

The explications for the East Asian lead have been many. Funk (2007) believes that analyses by European and US observers have had an erroneous focus on cultural factors and long commuting times and low fixed Internet penetration. Instead, Funk emphasizes the interplay between technology and business model developments and builds on the theory regarding advantages of integrated business models in the first phases of architectural or systemic innovations (Henderson and Clark, 1990).

The present chapter supports the emphasis on technology and business models as the prime explanatory factors. What the chapter aims at doing is to insert the technology and business model factors into the setting of the differences in innovation systems or ecosystems in the regions in question. As stated by Funk (2007, 2009), the Europeans as well as the Americans have built on the successful 2G approach, but the reason for this (in the European case, at least) lies not only in a mental model building on past experience instead of analyses of the present and future development possibilities, but also on structural factors in the telecommunications sectoral innovation systems and ecosystems in Europe with strong international equipment and device manufacturers and, to a large extent, nationally confined markets for operation. This provides a relation of strength between manufacturers and operators, but it also sets the necessity of a general standardized approach in order to overcome the limited size of the national markets in Europe.

This discussion is concerned with the 'stage-setting' first phase of mobile Internet developments. Once the path has been laid out, factors regarding network effects come into play in a second phase. Instead of the Internet-compatible e-mail development for text-based messaging in the Japanese market, the European markets saw a strong development of SMS. The initial success of SMS bred more success for SMS, based on network effects, and has settled into a system with widespread SMS competences among the users and an interest in maintaining the SMS system from the supply side because of the vast turnover and income from basic and premium SMS. The result is a total dominance of the SMS system for mobile data in Europe. And even though there are, indeed, advanced offerings on top of the SMS system with premium services of many different kinds, and even though history shows that a dominant system may quite quickly be toppled by other somehow superior systems (for instance Word substituting for WordPerfect), the outcome is that Europe, at the moment, is locked into a side road instead of the main Internet road for mobile data developments.

In the US market, the 2G take-off was hampered by competing standards complicating, for instance, interstate roaming. And with respect to 3G, the US still trails behind not only Japan and Korea but also the more advanced countries in Europe. However, when it comes to the use of mobile data services, the situation is less clear. In fact, the US has in absolute terms the highest ARPU for mobile data services. As explained in the chapter, a number of reservations must be expressed before it can firmly be concluded on this background that the US has become the world champion in this 'discipline'. ARPU depends on use as well as pricing and the number of subscribers, and in Europe, ARPU is lowered by the fact that many customers have more than one subscription.

Nevertheless, the various parameters compared in the previous sections indicate an ongoing convergence between conditions on regional markets. While East Asia was well ahead with regard to the introduction of 3G networks, the upgrade to 3.5G has taken place almost simultaneously in the different regions. When it comes to the implementation of WiMAX and LTE, the US may take off ahead of Europe. Thus, a comparison of the various parameters does not provide a clear picture of leaders and laggards with respect to the adoption of mobile data services. However, it is clear that the European lead in the mobile sector has been lost, first to East Asia and now, to a certain extent, to the US as well.

A key issue triggered by this chapter is, consequently, whether this is due to an ongoing convergence based on the globalization in all parts of the value networks, or whether there are some specific factors favoring market development in the US as compared to market developments in Europe.

A striking difference, just a few years ago, between the US market and the markets in Europe was the relatively low use of SMS in the US. This, however, does not apply any more. SMS use has grown extensively in the US market (IT Facts, 2008). But in comparison with the European markets, both 'tracks' are used in the US. E-mail via mobile Internet is taken up primarily by business people, while SMS is used by the general mobile users. The more extensive diffusion of Internet-based messaging may promote other mobile Internet services as well. Once subscribers have access to the mobile Internet, they begin to use some of the other services (for example, content services), which are available.

Another and probably more important factor is the convergence between the mobile and Internet sectoral systems of innovation. The development and use of Internet services are more advanced in North America than in the major European countries. This is reflected in a better ranking in various indexes on e-readiness, for instance the e-readiness index by the Economist Intelligence Unit (EIU, 2009). It is likely that the US market for mobile data leverages on the US lead in fixed Internet services. The US IT giants, Microsoft, Google and Apple, are forcefully entering the mobile field. While mobile Internet was driven by the mobile operators in Japan, it may be the IT companies in the US which will be the drivers of mobile Internet in the US.

In a survey performed by Netsize (2009), respondents could choose between Internet giants, mobile operators, mobile pure players, platform providers and handset vendors as the actors to dominate the 'mobile 2.0' development. Sixty-four percent and 46 percent respectively backed Internet giants and mobile operators as the likely winners, while the other players were at a considerably lower level (Netsize, 2009). Although this only reflects the opinion of industrial observers, it does point at a trend towards the growing importance of Internet players in the field of mobile communications. With the US dominance in this field, it may be an indication of the future strength of the US in mobile data developments.

The above-mentioned development indicates that Europe cannot take a leading future position in mobile communications for granted. With the development of mobile data services, the mobile system of innovation becomes more interlinked with other parts of the ICT sector, and it becomes more difficult to maintain a leading position in mobile communications without a similar position within ICT more broadly speaking. Policies addressing improvement of e-readiness in general might, therefore, be an efficient way to ensure continued development and use of mobile services in Europe. Policies to promote mobile data need to encompass a wide variety of policy areas as the development of mobile data depends on structural factors in the markets on the supply side as well as take-up factors like the general e-readiness of the potential users on the demand side.

NOTE

1. Data rate figures are theoretical maximums, and in practice these figures are lower.

REFERENCES

Capgemini, (2008). Mobile Internet services in Europe and USA: initiatives to drive adoption and usage. *Telecom and Media Insights*, **30**, 1–12.
David, P. (2001). Path dependence, its critics and the quest for 'historical economics'. In P. Garrouste and S. Ioannides (eds), *Evolution and Path Dependence in Economic Ideas: Past And Present* (pp. 15–40). Cheltenham, UK and Northampton, MA, USA: Edward Elgar.
David, P. (2007). Clio and the economics of QWERTY. *American Economic Review*, **75**(2), 332–7.
Davis, F.D. (1989). Perceived usefulness, perceived ease of use, and user acceptance of information technology. *MIS Quarterly*, September, 319–40.
Digital Trends (2008). Nokia wants bigger share of US market. Available at http://www.digitaltrends.com.
Economides, N. (1996). The economics of networks. *International Journal of Industrial Organization*, **14**(6) 673–99.
Economist Intelligence Unit (EIU) (2008). E-readiness rankings 2008 – maintaining momentum. EIU, London.
Economist Intelligence Unit (EIU) (2009). E-readiness rankings 2009 – the usage imperative. EIU, London.
Edquist, C. (2005). Systems of innovation – perspectives and challenges. In J. Fagerberg, D. Mowery and R. Nelson (eds), *Oxford Handbook of Innovation* (pp. 181–208). Oxford: Oxford University Press.
Fierce Mobile Content (2007). North America now leads on data ARPU. http://www.fiercemobilecontent.com. 22 October 2007.
Fitchard, K. (2009). T-Mobile data revenue growth due mostly to SMS, not 3G. *Telephony Online*. http://telephonyonline.com/wireless. 27 February.
Fransman, M. (2007). *The New ICT Ecosystem: Implications for Europe*. Edinburgh: Kokoro.
Funk, J.L. (2007). Solving the startup problem in Western mobile Internet markets. *Telecommunications Policy*, **31**, 14–30.
Funk, J.L. (2009). The co-evolution of technology and methods of standard setting: the case of the mobile phone industry. *Journal of Evolutionary Economics*, **19**, 73–93.
Gartner (2009). Worldwide mobile phone sales. Gartner, Egham. http://www.gartner.com. 2 March 2009.
Hansberry, E. (2009). iPhone claims nearly two-thirds of mobile browser share. http://www.informationweek.com. 3 March.

Helsinki Sanomat (2008). Nokian vaikeudet USA:ssa pahenivat. www.nettisano-mat.com. 11 November.

Henderson, R.M. and K.B. Clark (1990). Architectural innovation: the recon-figuration of existing product technologies and the failure of established firms. *Administrative Science Quarterly*, **35**(1), 9–30.

Henten, A., H. Olesen, D. Saugstrup and S.E. Tan (2004). Mobile communica-tions: Europe, Japan and South Korea in a comparative perspective. *Info – The Journal of Policy, Regulation and Strategy for Telecommunications*, **6**(3), 197–207.

Hernandez, J.M.C. and J.A. Mazzon (2007). Adoption of Internet banking: proposition and implementation of an integrated methodology approach. *International Journal of Bank Marketing*, **25**(2), 72–88.

Impact Lab (2009). The Blackberry story. Available at http://www.impactlab.com/2009/02/08/the-blackberry-story.

IT Facts (2008). Mobile penetration: how many people access a mobile site at least once a week. http://www.itfacts.biz. 29 September 2008.

IT Facts (2009). Internet activity of European mobile subscribers in November 2008. http://www.itfacts.biz. 23 February 2009.

Kim, H., H.C. Chan and S. Gupta (2007). Value-based adoption of mobile inter-net: an empirical investigation. *Decision Support Systems*, **43**, 111–26.

López-Nicolas, C., F.J. Molina-Castillo and H. Bouwman (2008). An assessment of advanced mobile services acceptance: contributions from TAM and diffusion theory models. *Information and Management*, **45**, 359–64.

Mobile Europe (2009). Mobile messaging in 2009 – SMS is still the driver. 5 February 2009. http://www.mobileeurope.co.uk/.

Netsize (2009). *The Netsize Guide 2009*. http://www.netsize.com.

Nielsen Mobile (2008). *Critical Mass – The Worldwide State of the Mobile Web*. http://mmaglobal.com.

OECD (various years). *Communications Outlook*. Paris: OECD.

Rogers, E.M. (2001). The digital divide. *Convergence: The International Journal of Research into New Media Technologies*, **7**(4), 96–111.

Rogers, E.M. (2003). *Diffusion of Innovations*. New York: Free Press.

Strategy Analytics (2008). Understanding the mobile ecosystem. http://www.adobe.com/devnet/devices/articles/mobile_ecosystem.pdf.

Vandrup, K. (2008). WiMAX vs. LTE – the role of the financial crisis. *Nordic and Baltic Journal of Information and Communications Technologies*, **2**(1), 19–21.

Varian, H. and C. Shapiro (1999). *Information Rules*. Boston, MA: Harvard Business School Press.

World Economic Forum (2008). *The Global Information Technology Report 2007–2008*. Geneva: World Economic Forum.

Index

Titles of publications are shown in *italics*.